MODERN INDUSTRIAL CERAMICS

Eugene C. Stafford, D.Ed.
State University College
Buffalo, New York

Bobbs-Merrill Educational Publishing

Indianapolis

To Joann, David, and Amy

Copyright © 1980 by The Bobbs-Merrill Company, Inc.
Printed in the United States of America

All rights reserved. No part of this book shall be reproduced or transmitted in any form or by any means, electronic or mechanical, including photocopying, recording, or by any information or retrieval system, without written permission from the publisher:

The Bobbs-Merrill Company, Inc.
4300 W. 62nd Street
Indianapolis, Indiana 46268

While information, recommendations, and suggestions contained herein are accurate to the publisher's best knowledge, no warranty, expressed or implied, is made with respect to the same since the publisher exercises no control over use or application, which are recognized as variable.

First Edition
First Printing 1980
Cover and interior design by Ted Snelson

Library of Congress Cataloging in Publication Data

Stafford, Eugene C.
Modern industrial ceramics.

Includes index.
1. Ceramics. I. Title.
TP807.S77 666 79-14207
ISBN 0-672-97129-1

CONTENTS

Preface	vii
Acknowledgments	ix
Introduction to Industrial Ceramics/Section	**I**
Unit 1: Basic Ceramic Terms	1
Unit 2: Major Ceramic Industries	3
Unit 3: Major Ceramic Products and Applications	5
Unit 4: Careers in Ceramics	11
Ceramic Materials/Section	**II**
Unit 5: The Earth	15
Unit 6: Ceramic Chemistry	21
Unit 7: Using Ceramic Raw Materials	24
Ceramic Raw Materials Industry/Section	**III**
Unit 8: Procurement	27
Unit 9: Transformation	37
Unit 10: Utilization	42
Unit 11: Disposition	43
Stone Industries/Section	**IV**
Unit 12: Introduction to the Stone Industries	45
Unit 13: Crushed and Broken Stone Industry	47
Unit 14: Dimension Stone Industry	53
Unit 15: Natural Gems	60
Refractory Industry/Section	**V**
Unit 16: Material Science	65
Unit 17: Procurement	68
Unit 18: Transformation	69
Unit 19: Utilization	73
Unit 20: Disposition	75
Kiln Industry/Section	**VI**
Unit 21: Introduction	77
Unit 22: Periodic Kilns	78
Unit 23: Continuous Kilns	82
Unit 24: Kiln Operation	86

Contents

Gypsum Industry/Section   VII
Unit 25: Material Science   93
Unit 26: Procurement   94
Unit 27: Transformation   96
Unit 28: Utilization   99
Unit 29: Disposition   101

Plaster Mold Industry/Section   VIII
Unit 30: Introduction   103
Unit 31: Batching of Plaster   106
Unit 32: Model Making   108
Unit 33: Mold Making   112

Lime Industry/Section   IX
Unit 34: Material Science   121
Unit 35: Procurement   123
Unit 36: Transformation   125
Unit 37: Utilization   129
Unit 38: Disposition   131

Portland Cement Industry/Section   X
Unit 39: Material Science   133
Unit 40: Procurement   138
Unit 41: Transformation   141
Unit 42: Utilization   147
Unit 43: Disposition   150

Concrete Industry/Section   XI
Unit 44: Material Science   153
Unit 45: Batching Concrete   155
Unit 46: Manufacturing Concrete Products   158
Unit 47: Concrete in Construction   161

Clay Industries/Section   XII
Unit 48: Introduction   165
Unit 49: Material Science of Clay   167

Structural Clay Products Industry/Section   XIII
Unit 50: Material Science   171
Unit 51: Procurement   173

Contents

Unit 52: Transformation 175
Unit 53: Utilization 184
Unit 54: Disposition 190

Whiteware Industry/Section XIV
Unit 55: Material Science 191
Unit 56: Procurement 194
Unit 57: Transformation 196
Unit 58: Utilization 213
Unit 59: Disposition 216

Glaze Industry/Section XV
Unit 60: Material Science 217
Unit 61: Procurement 222
Unit 62: Transformation 224
Unit 63: Utilization 231
Unit 64: Disposition 234

Procelain Enamel Industry/Section XVI
Unit 65: Material Science 235
Unit 66: Procurement 237
Unit 67: Transformation 239
Unit 68: Utilization 244
Unit 69: Disposition 246

Glass Industry/Section XVII
Unit 70: Material Science 247
Unit 71: Procurement 250
Unit 72: Transformation 252
Unit 73: Utilization 262
Unit 74: Disposition 265

Abrasives Industry/Section XVIII
Unit 75: Material Science 267
Unit 76: Procurement 269
Unit 77: Transformation 271
Unit 78: Utilization 276
Unit 79: Disposition 278

Index 279

PREFACE

Modern Industrial Ceramics is the first comprehensive book of its type. It is designed and written to provide meaningful complete information on the many (and growing) commercial ceramic product industries in the United States. The subject matter is distinctly *not* approached from an art and craft standpoint but, rather, from an industrial or production observation post. Art and craft considerations are included as secondary and supplemental information and do not constitute a majority of the text.

The book's introductory section defines the basic ceramic terms and discusses the major ceramic industries and their products and applications. Ceramics career information is included as a separate unit in Section I—*Introduction to Industrial Ceramics*.

Separate sections of the book detail such topics as: Ceramic Materials, Ceramic Raw Material Industry, Stone Industries, Refractory Industry, Kiln Industry, Gypsum Industry, Plaster Mold Industry, Lime Industry, Portland Cement Industry, Concrete Industry, Clay Industries, Structural Clay Products Industry, Whiteware Industry, Glaze Industry, Porcelain Enamel Industry, Glass Industry, and Abrasives Industry.

Each section discusses the *material science* of the industry, how the raw materials of the industry are *procured*, how they are *transformed* and *utilized* and, when we have finished using them, how the products are *disposed of*.

This book reflects the latest needs of a new text. *Ecological considerations* are thoroughly discussed and integrated in the text matter throughout the book. *Career awareness* information is first treated in its own unit and then discussed, as applicable, in subsequent units. Finally, *metric measure* equivalents are included with each measurement throughout the text.

ACKNOWLEDGMENTS

Many firms and individuals graciously donated both time and materials to help make this unique text the most up-to-date on the market. The author expresses special thanks to his many friends and contacts in the contemporary ceramics industry for having provided the constant support and constructive criticism necessary to make this text as complete and current as possible. Special thanks are given to the following firms and sources of materials, all of whom supplied important materials for this text.

American Art Clay Co., Inc.
American Olean Tile Company
A. P. Green Refractories Co.
ASARCO Incorporated
Ball Corporation, Gary D. Demaree of
The Barre Granite Association
Bickley Furnaces, Inc.
Brick Institute of America
Buffalo China, Inc.
Cedar Heights Clay Co.
Corning Glass Works
Duncan Ceramic Products, A Division of Duncan Enterprises,
Fresno, California
The Exolon Company
Ferro Corporation
General Refractories Company
Glass Packaging Institute
Gold Bond Building Products,
Division of National Gypsum Company
Great Lakes Dredge & Dock Company
Harbison-Walker Refractories, Dresser Industries, Inc.
Lead Industries Association, Inc.
Longview Lime, Division of SI Lime Company
National Clay Pipe Institute
National Crushed Stone Association
The National Lime and Stone Company
National Limestone Institute Inc.
National Ready Mixed Concrete Association
New York State College of Ceramics at Alfred University
Niagara Mohawk Power Corporation
The Edward Orton Jr. Ceramic Foundation
Owens-Corning Fiberglas Corporation
Portland Cement Association
The Refractories Institute
Shenango China Products
J.C. Steele & Sons, Inc.
Syracuse China Corp.
United States Army Corps of Engineers, Buffalo District
United States Gypsum Company
R.T. Vanderbilt Company, Inc.

SECTION I

Introduction to Industrial Ceramics

Unit 1: Basic Ceramic Terms

Ceramics are earthy, inorganic, nonmetallic materials. They are usually heat-treated at high temperatures during the manufacturing process. There may be two major similarities when defining ceramics. These are as follows: (1) the type of raw material is specified—earthy, inorganic, nonmetallic; and (2) the type of heat treatment used in manufacture—usually high temperatures.

Earthy identifies the major source of ceramic raw materials. About 90 percent of the 103 known elements may be used to make ceramic products. All these elements have the same origin—the Earth!

In reality, probably all of the 103 known elements are used directly or indirectly in manufacturing ceramic products. Ceramic materials are *inorganic* and *nonmetallic*. This means they are not living and are not metal.

Heat treatment is simply heating a material. The amount of heat used for heat treatment varies. Usually, it involves high temperatures. A *high temperature* will produce *incandescence* (visible red heat). Visible red heat occurs at temperatures of 540°–590°C (1004°–1094°F). The major exception to high temperatures in ceramic heat treating is in the manufacture of plaster from gypsum. This requires a temperature of only 170°C (338°F).

Ceramic Products

Our definition of ceramics includes all the ceramic products (Fig. 1-1), whether made by hand or by machine; and it includes all quantities, whether a piece is produced singly or mass-produced by the thousands. Therefore, ceramics includes artware, such as pottery and freehand blown glass (Fig. 1-2) and mass-produced products.

Industrial ceramics involves only those manufactured products which are usually mass-produced by machines. Industrial ceramic products are made from earthy, inorganic, nonmetallic materials. These products are usually subjected to high-temperature heat treatment. *Manufacturing* means to produce a product in a plant or at a place other than its final position of use.

Hobby Ceramics and Art Pottery

Since hobby ceramics and art pottery are quite popular, many people regard them as the *only* ceramics industry. Actually, these two groups are involved with creative handwork, rather than with mass production.

Hobby ceramics usually involves cleaning up slip-cast greenware (Fig. 1-3) and then decorating and glazing the piece. Instruction, materials, and firing are provided by a hobby ceramics shop. Most hobby ceramists are involved for the fun and enjoyment of an avocation.

Art pottery is usually made of plastic clay which has been hand formed or *thrown* by a potter in a studio. A potter may also be a creative artist, as discussed in a later unit. Potters usually do all their own work—from preparing the plastic clay to its final firing. This may include hand forming, throwing, drying, decorating, glazing, and one or more firings. Art pottery may be either functional ware or decorative sculpture. This is creative work; technical knowledge is necessary to produce successful pottery. Potters usually produce their ware for public sale.

2 / *Modern Industrial Ceramics*

Fig. 1-1. Note the variety of ceramic products, ranging from the bathroom sink to a grinding wheel. (Courtesy R. T. Vanderbilt Company, Inc.)

Fig. 1-2. A freehand blown crystal cocktail set. Crystal is a lead glass.

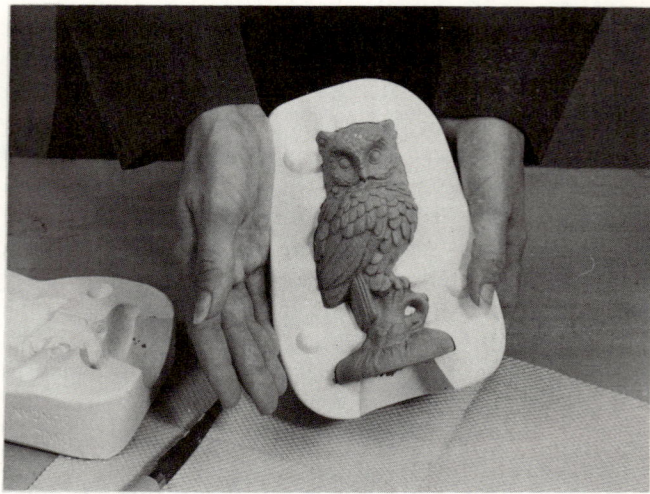

Fig. 1-3. A decorative piece made in a slip casting mold. The object and forming method are typical of hobby ceramics.

Unit 2: Major Ceramic Industries

Eleven major ceramic industries are discussed in this unit. These industries are summarized in Table 2-1. The large number of different ceramic industries emphasizes their *breadth* and why the plural of industry (industries) is used.

Ceramic Raw Material Industry

The ceramic raw material industry involves removing the inorganic, nonmetallic materials such as clay, feldspar, and flint from the earth's crust. Transforming these raw materials into usable form (a powder) is also done by the raw material industry.

Stone Industries

There are three major divisions in the *stone industries*. These include crushed and broken stone, dimension stone, and natural gems.

Refractory and Kiln Industries

High-temperature heat treatment processes are made possible by the *refractory and kiln industries*. Refractories are materials that can resist high temperatures. Kilns are lined with refractory materials.

Hydrosetting Material Industries

The next three industries are known as hydrosetting material industries. Each of these industries first uses *heat* to produce a primary product—a *powder*. There are three basic powdered hydrosetting materials—plaster, lime, and Portland cement. Each of these powders will chemically "set up" when mixed with water. This type of chemical reaction is called *hydrosetting*.

Gypsum Industry

The gypsum industry manufactures plaster from rock gypsum. This process uses heat. Plaster is used to make wallboard, models, and molds. Various other products, such as toothpaste, also include plaster.

Lime Industry

The lime industry manufacturers lime from limestone or other raw materials composed of calcium carbonate ($CaCO_3$). The three types of lime are *agricultural* lime, *quick-lime*, and *slaked* lime.

Portland Cement and Concrete Industries

The chief difference between the major products of the *Portland cement* and *concrete industries* involves *heat*. High-temperature heat treating is used to produce only Portland cement. Portland cement is a fine powder. It is either gray or white in color.

Concrete, on the other hand, is made up of Portland cement, fine aggregates, coarse aggregates, and water. When the Portland cement in the concrete hydrosets, it binds the aggregates into a durable solid. This solid is known as *concrete*.

Clay Industries

There are two major divisions in the *clay industries*. The

Table 2-1. A Summary of the Logical Order of the Major Ceramic Industries Included in This Book

Ceramic Industry	Explanation of Logic and Similarities
1. Ceramic raw material	Procures raw material from the earth
2. Stone	Uses natural material that nature may have heated
3. Refractory and kiln	Provide the means for high-temperature heat treatment
4. Gypsum	1. Raw material requires heat to produce primary product, a powder
5. Lime	2. Powder hydrosets
6. Portland cement and concrete	
7. Clay	Clay is formed, dried, and fired to produce glassy bond
8. Glaze	Glassy coating
9. Porcelain enamel	Glassy coating } Glassy Materials
10. Glass	May be used alone
11. Abrasives	Uses many of the preceding

first division includes the *structural clay products* such as bricks and tiles. The second division is known as *whitewares*, which includes dinnerware and electrical insulators. The clay products are fired to a high temperature—after forming and drying. This makes them durable solids.

Glassy Material Industries

There are three major glassy material industries. The first of these, the *glaze industry*, is associated with the clay industries. Glazes are glassy coatings fired onto clay products to protect and/or beautify them.

Another glassy coating is produced by the second of the glass industries. This is known as the *porcelain enamel industry*. Porcelain enamels are glassy coatings fired onto a metal or glass. The porcelain enamels are used like glazes. They protect or beautify, or both protect and beautify the product.

Everyone recognizes the products of the *glass industry*. Although there are hundreds of glass products, there are only six basic types of glass. These are known by their composition. Most of the glass products used each day (windows, drinking glasses, light bulbs, pop bottles, pickle jars, etc.) are made of *soda lime* glass. About 90 percent of all the glass products manufactured each year is made from soda lime glass. The other five types of glass (by their composition) are lead, borosilicate, 96% silica, fused silica, and aluminosilicate.

Abrasives Industry

There are three major product divisions in the *abrasives industry*. These divisions are *loose*, *coated*, and *bonded* abrasive products. Both natural and synthetic abrasive materials are used.

Studying Ceramics

The major ceramic industries are listed and summarized in Table 2-1. The obvious place to begin studying ceramics is with the procurement or *mining* of raw materials. Thus, the ceramic raw material industry may be studied first. The raw material for the stone industry comes directly from the earth's crust without any heat treatment. Therefore, this industry (no-heat processing) is considered second. All of the remaining ceramic industries use heat-treating equipment and processes. Refractory and kiln industries are thus considered third, since they provide the means for heat treating.

The gypsum, lime, and Portland cement industries are listed fourth, fifth, and sixth. These range from simple to complex and from low- to high-temperature heat treatment. All of them use *heat* to produce a primary product in the form of a powder which *hydrosets*. The liquid stage of hydrosetting is usually used to form a secondary product.

Heat treating is used to make a finished product in the clay industries, instead of *hydrosetting*. The *clay industries*, listed seventh, use heat treating to finish a product after forming and drying. This process, called *firing*, creates a glassy bond which makes clay products durable.

The glassy bond in fired clay is similar to the other glassy material industries—glaze, porcelain enamels, and glass. Glaze, eighth, and porcelain enamels, ninth, are both glassy coatings fired onto other materials. Glass, listed tenth, is usually thicker than the glassy coatings and is often used alone. The abrasives industry, eleventh, uses materials and processes similar to the preceding industries. For example, clay, feldspar, and the firing processes are used to produce grinding wheels.

Unit 3: Major Ceramic Products and Applications

The definition of ceramics and the descriptions of the major ceramic industries indicate a large variety of ceramic products. The major ceramic products and their applications are listed in Table 3-1. This will help to identify the various ceramic products.

You may have used some of the products listed here without realizing they were ceramic. Probably, you have taken some of these ceramic products for granted. They are very common and frequently used. Many of the products can be seen in Figs. 3-1 through 3-10.

Fig. 3-1. An electric kiln for firing clay and glazes. The interior is lined with refractory bricks. The kiln shelves and posts are also made of refractory materials.

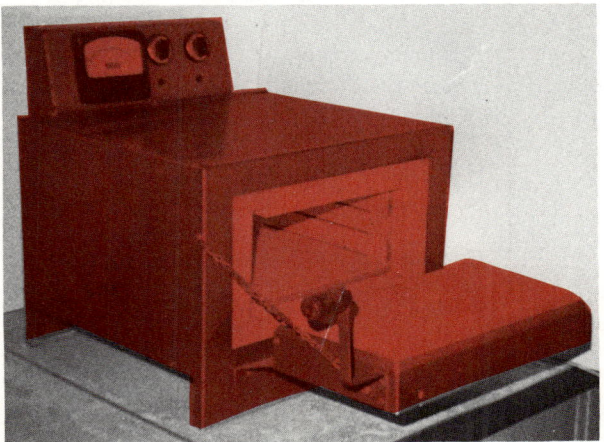

Fig. 3-2. An electric kiln for firing porcelain enamels. The kiln door and interior are lined with insulating firebrick. The kiln floor is protected by a refractory kiln shelf.

Fig. 3-3. A red sandstone wall. The steps are made of slate.

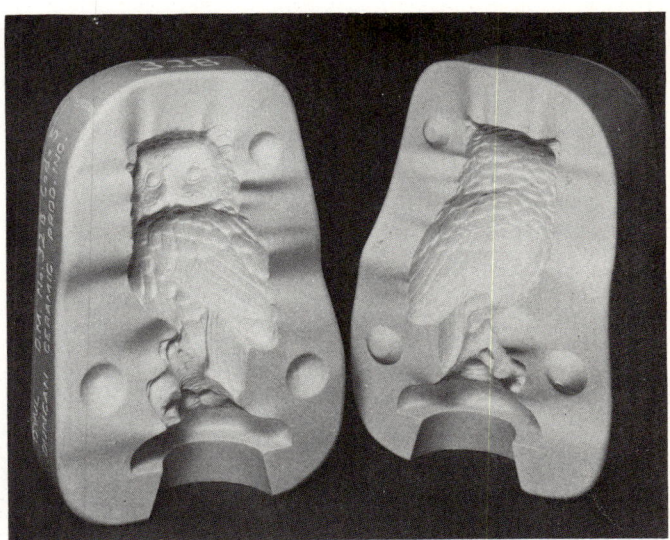

Fig. 3-4. A two-piece slip casting mold made from plaster.

Table 3-1. Major Ceramic Products With Typical Applications

Ceramic Industry	Typical Major Products	Applications
1. Ceramic raw materials	Powdered earthy inorganic nonmetallic materials such as: Clays (dry powdered form) Feldspar Flint Whiting	Refractory products (Fig. 3-1, 3-2) Clay and glaze products (Fig. 3-7) Porcelain enamel products (Fig. 3-8) Glass products (Fig. 3-7, 3-9) Abrasive products (Fig. 3-10)
2. Stone	Crushed and broken	Breakwall, riprap Concrete products
	Dimension	Sills (Fig. 3-5) Walls (Fig. 3-3)
	Natural gems	Jewelry Diamond saws

3. Refractory and kiln	Refractories:
	  Firebrick
	  Refractory shapes
	  Castable refractory
	  Ceramic fibers
		Kilns (Fig. 3-1, 3-2)
		Furnaces

	Kilns: Periodic	Electric kilns (Fig. 3-1) Combustible fuel kilns
	Continuous	Tunnel kiln Rotary kiln
4. Gypsum	Rock gypsum	Portland Cement
	Plaster	Plaster models Plaster molds for forming clay (Fig. 3-4)
	Wall board	Dry wall for interior walls of homes (Fig. 3-5)
5. Lime	Agricultural lime Quicklime Slaked lime	Soil improvement Sewage treatment Mortar (Fig. 3-5, 3-6)
6. Portland cement and concrete	Portland cement	Concrete Mortar cement paint
	Precast concrete	Concrete bricks, blocks, pipe (Fig. 3-6)
	Poured in place concrete	Walls (Fig. 3-5) Highways

Major Ceramic Products and Applications / 7

Ceramic Industry	Typical Major Products	Applications
7. Clay	Structural clay products	Bricks (Fig. 3-5) Tiles Sewer pipe
	Whitewares	Artware (Fig. 4-2) Dinnerware (Fig. 3-7) Electrical insulators
8. Glaze	Glazes	Glassy coating for clay products (Fig. 3-7)
9. Porcelain enamels	Sheet steel products	Appliance housing (Fig. 3-8) Cooking ware Sinks
	Cast iron products	Bathtubs Sinks
	Specialty products	Colored glass blocks Labels on glass products Jewelry
10. Glass	Soda lime glass	Windows Bottles and jars (Fig. 3-9)
	Lead glass	Crystal (Fig. 1-2, 3-7) Optical lenses
	Borosilicate glass	Ovenware Laboratory glassware
	96% silica	Chemical ware Sun lamps
	Fused silica	Electrical insulation Relay line
	Aluminosilicate	Top of stove ware High temperature thermometers
11. Abrasives	Loose abrasives	Grinding and polishing To make coated and bonded abrasive products
	Coated abrasive products	Sheets Disc Belts

	Bonded abrasive products	Nonvitreous:
		  Saws
		  Wheels
		Vitreous:
		  Grinding wheels (Fig. 3-10)
		  Slipstones

8 / *Modern Industrial Ceramics*

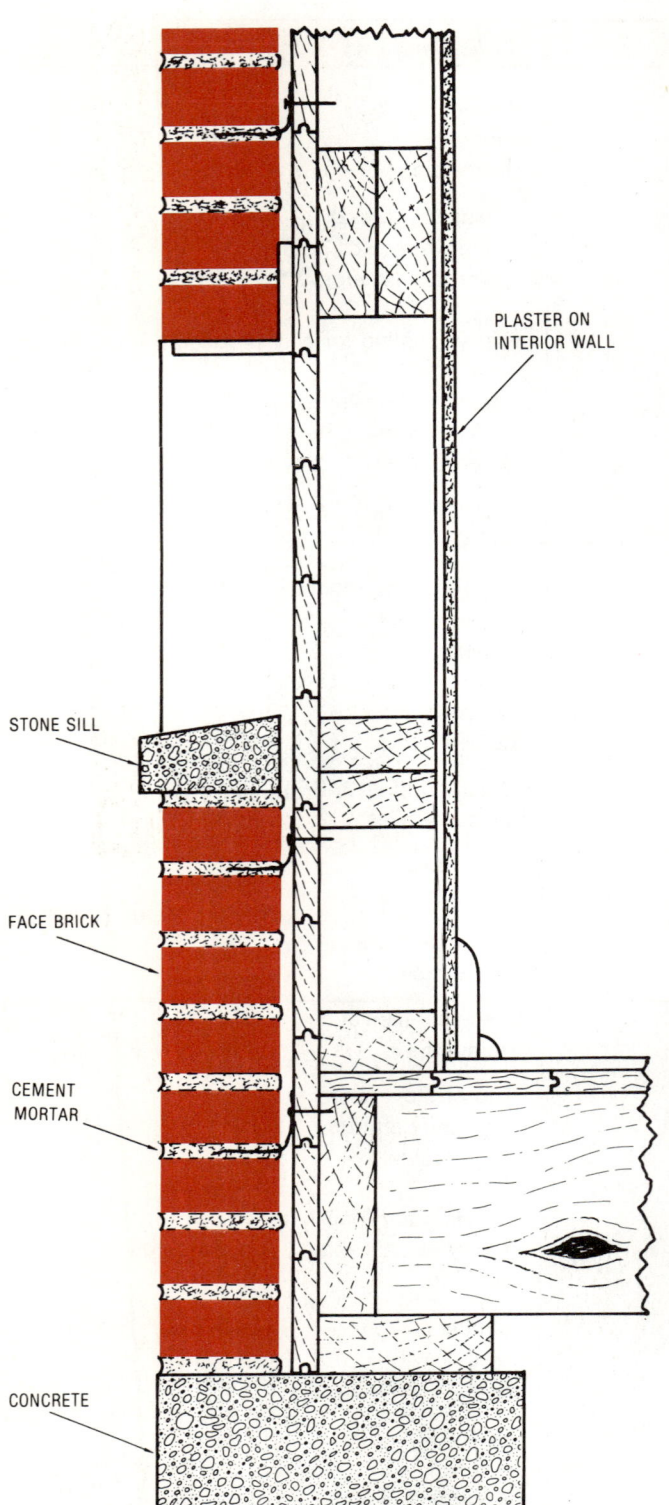

Fig. 3-5. A cross-section drawing of a brick veneer house showing the location and use of five ceramic products—stone, brick, mortar, concrete, and plaster wallboard.

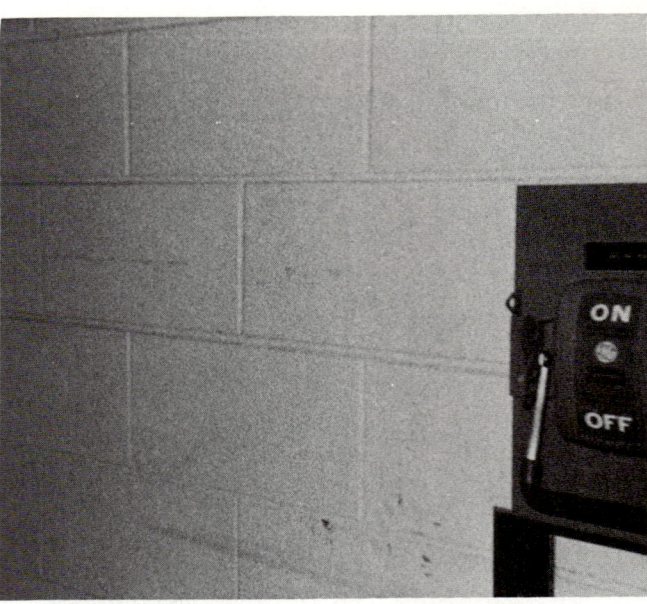

Fig. 3-6. A wall made of precast concrete blocks. Mortar was used to bond the blocks together.

Fig. 3-7. A table with formal place settings using crystal and fine china. Light is provided by the glass chandelier. (Courtesy Niagara Mohawk Power Corporation)

Major Ceramic Products and Applications / 9

Fig. 3-8. Porcelain enamels are used on both the exterior and interior of home appliances like this washer and dryer. (Courtesy Niagara Mohawk Power Corporation)

Fig. 3-9. Glass containers are made in a variety of shapes and sizes. Food, beverages, cosmetics, and drugs are some of the things packaged in glass containers. (Courtesy Glass Packaging Institute)

10 / *Modern Industrial Ceramics*

Fig. 3-10. A grinder with two 6-in. diameter vitreous bonded grinding wheels.

Unit 4: Careers in Ceramics

The variety of job or occupational titles in any industry is almost endless. However, the *United States Department of Labor, Bureau of Labor Statistics* has organized the numerous titles into nine major occupational groups. These groups and their projected growth are shown in Fig. 4-1.

People in most of these occupational groups are directly involved with ceramic materials. The clerical and service workers are not directly involved with ceramics. However, these jobs are necessary for an industry to function. Clerical workers include secretaries, clerks, and bookkeepers. Service workers include custodians, security people, maintenance, and similar positions. People in *all* of the occupational groups are consumers of ceramic products.

Ceramists

A ceramist or ceramicist is anyone who works with earthy, inorganic, nonmetallic materials. They usually subject the material(s) to heat treatment at high temperatures. In effect, the ceramists take the cheapest, most abundant material, the earth itself, and subject it to heat which transforms it into products of value to people. These products help make possible all our industries—both manufacturing and construction. Ceramic products apply to all facets of our life. In fact, the ceramic products make our rising standard of living possible.

Ceramists with specific qualifications are usually employed in three of the major occupational groups:

1. Professional and Technical Workers
2. Managers, Officials, and Proprietors
3. Sales Workers

Specifically, these ceramists are the:

1. Ceramic artist or designer (Fig. 4-2)
2. Ceramic engineer
3. Ceramic scientist (Fig. 4-3)
4. Glass scientist
5. Professional ceramic engineer

Industrial Positions

Typically, ceramists are responsible for one of the following industrial positions:

1. Research and development or R&D (see Figs. 4-2 and 4-3)
2. Production
3. Quality control (Figs. 4-4 and 4-5)
4. Sales
5. Management

A person working in *research and development* is involved in scientific and/or practical research. This person will develop new products or improve existing ceramic products. The ceramist in *production* manufactures ceramic products or constructs with ceramic products. *Quality*

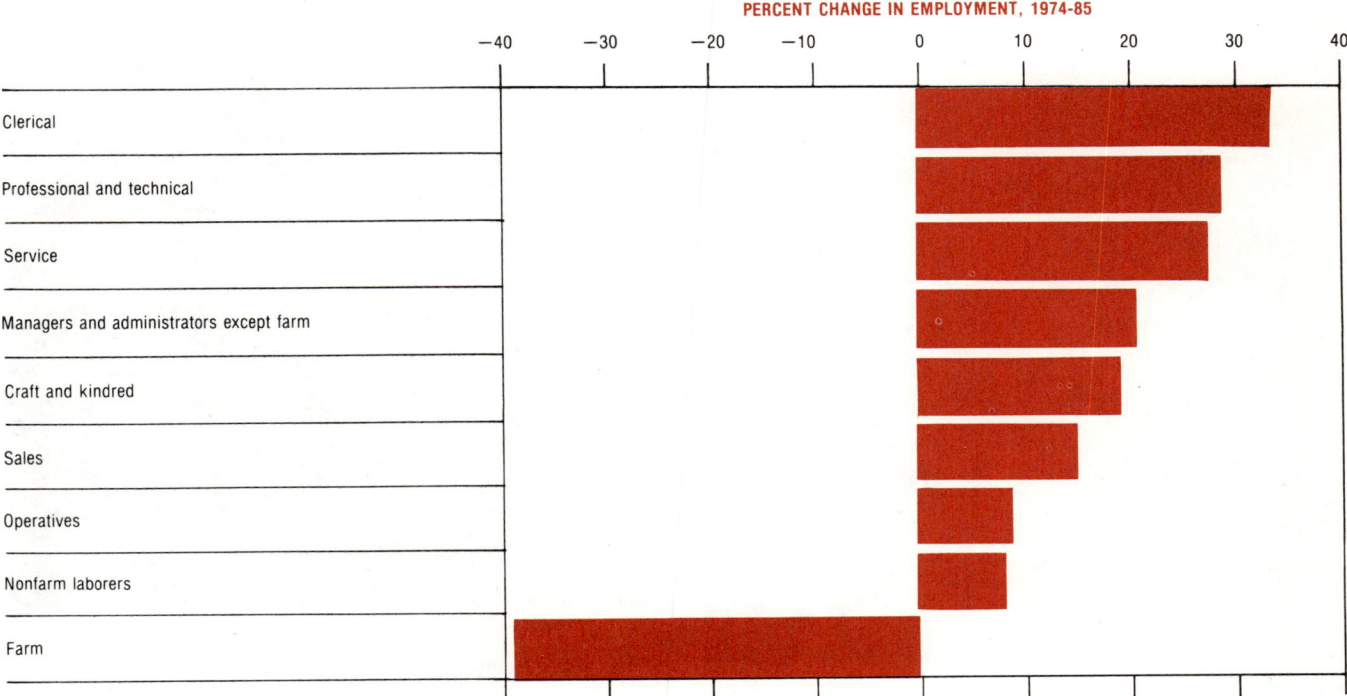

Fig. 4-1. Projected growth among the nine major occupational groups through the mid-1980's. (Courtesy Bureau of Labor Statistics)

12 / *Modern Industrial Ceramics*

Fig. 4-2. Ceramic artists using kick wheels to make thrown pottery. Bisque and glaze fired ware are on the shelves. (Courtesy New York College of Ceramics at Alfred University)

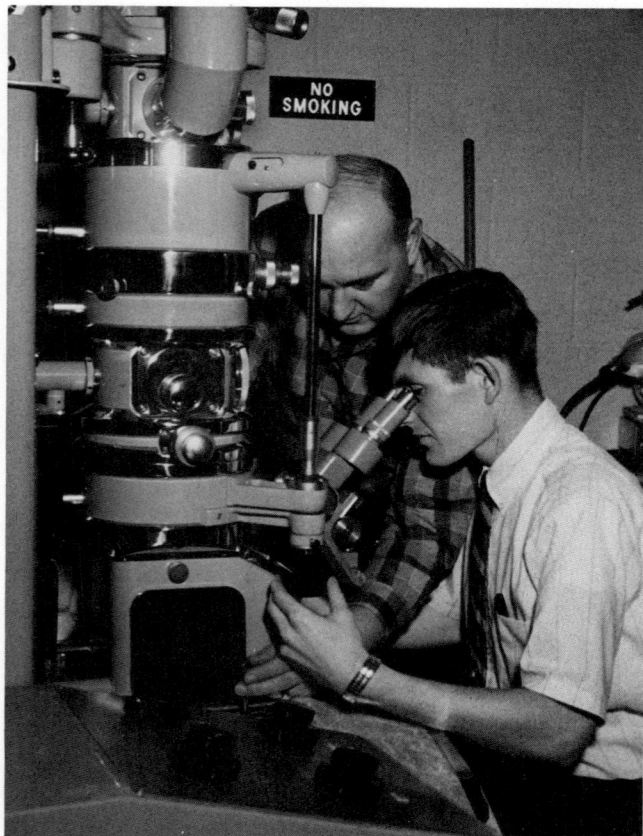

Fig. 4-3. A future ceramic scientist receives instruction on the use of the transmission electron microscope which is used in industry for research and development. (Courtesy New York State College of Ceramics at Alfred University).

Fig. 4-4. Quality control and research and development use thermogravimetric analysis as part of their work. This equipment automatically records the temperature versus loss in weight as a sample is heated. (Courtesy New York State College of Ceramics at Alfred University)

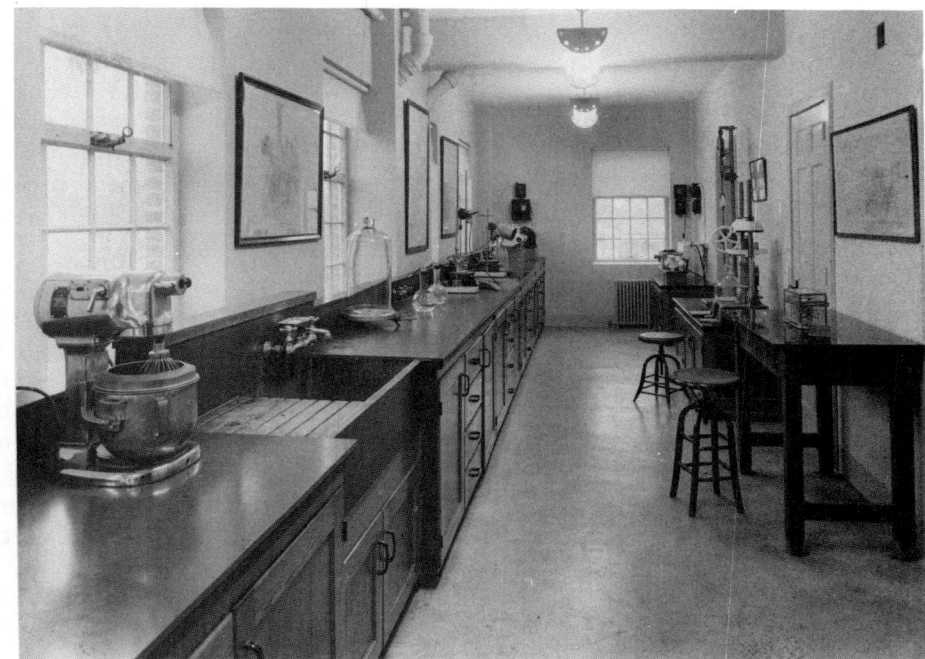

Fig. 4-5. A quality control laboratory. (Courtesy R. T. Vanderbilt Co., Inc.)

control checks, inspects, and maintains standards for the ceramic products. *Sales* people advertise, promote, and market ceramic products. Sales workers include the sales persons and sales engineers. The sales engineers have the education and experience to provide technical assistance with the products they sell. Management is responsible for cooperation and continuity within a given department (for example, production). *Management* has the same responsibility for the overall operation of a firm. Thus, those ceramists in managerial positions may be involved in one, two, or even all of the previously described industrial positions.

Ceramic Artist or Designer

A ceramic artist or designer usually has a degree in fine arts or design. This person makes ceramic art objects (see Fig. 4-2) or designs ceramic products (see Fig. 1-1). This person has earned a Bachelor of Fine Arts (B.F.A.) or Bachelor of Science (B.S.) in ceramics or glass. The ceramic artist or designer would study the broad field of art, and also would take extensive studio work in a concentrated area, such as ceramics or glass.

The terminal degree for ceramic art is the Master of Fine Arts (M.F.A.). This degree usually requires 60 credit hours. Master degree programs usually require only 30 hours, if a doctorate can be earned in that field.

The ceramic artist/designer is usually employed by the gypsum, clay, glaze, porcelain enamel, and glass industries. Also, this person might be self-employed as a potter or glass blower, or might work for a university.

Ceramic artists and designers usually work in the development phase of research and development, if a separate design department does not exist. They design and/or make ceramic products. They may do some technical or engineering work as they create. Thus, they may occasionally work closely with ceramic engineers to start a new product in production.

A self-employed ceramic artist may prefer to be called a *potter* or *creative glass blower*. Potters and creative glass blowers may or may not hold college degrees, such as the B.F.A. They may have learned their craft by apprenticeship or by on-the-job training without formal education. Some may have both a formal education and apprenticeship of some nature.

Potter

Potters are primarily concerned with designing, hand building and throwing with clay, decorating, glazing, and firing art objects. Many of them become quite technical in their work. Others stress their creativeness.

Creative Glass Blower

Creative glass blowers specialize in blowing art glass objects. Like the potters, some glass blowers are quite technical in their work. Others stress creativeness.

Ceramic Engineer

Ceramic engineers generally have a degree in ceramic engineering. They are employed to work in some way with ceramics. People with this job title have earned a Bachelor of Science (B.S.) degree in ceramic engineering as their minimum education. The humanities, mathematics, science, and technology are studied in undergraduate school. An unique part of the bachelor's degree is the study of scientific theory and its practical application to technology. This is done in laboratory courses.

Today, ceramic engineers frequently continue their education to earn master and/or doctoral degrees. The degrees earned are the Master of Science and/or Doctor of Philosophy in ceramic engineering. Occasionally, master degrees are earned in related fields, such as ceramic science, glass science, material science, mathematics, physics, chemistry, and even business administration. The engineers with advanced degrees are usually the ones who receive promotions and challenging assignments. Ceramic engineers with advanced degrees are more likely to work in management or research and development.

Ceramic engineers are employed by all of the major ceramic industries (see Table 2-1), except the stone industries. Thus, they may work with only a single ceramic material or with several materials. They can be employed by the allied industries. This is due to the universal use of ceramic products. Also, employment is available in the armed forces, government, service industries, utilities, and universities.

The ceramic engineers typically work in one of the following industrial positions:

1. Research and development (R&D)
2. Production
3. Quality control (see Fig. 4-4)
4. Sales
5. Management

Ceramic Scientist

Ceramic scientists generally have a degree in ceramic science. They work in research and development. People with this job title have earned a bachelor of science degree in ceramic science. This degree requires more extensive study of mathematics, physics, and chemistry than is required by the degree of the ceramic engineer. While the bachelor's degree is the educational minimum, ceramic science majors frequently continue their education in graduate school to earn master and/or doctoral degrees.

Ceramic scientists are employed by all the major ceramic industries, except the stone industries (see Table 2-1). Allied industries, armed forces, government, and universities frequently employ ceramic scientists.

Typically, the ceramic scientists work in research and development (see Fig. 4-3). They may also work in management after achieving experience and/or advanced degrees.

Glass Scientist

Glass scientists usually have a degree in glass science. They are employed by the glass industry. These people have earned a bachelor of science degree in glass science as their minimum education. Their studies were concentrated on glass science, rather than on the total ceramic technologies. Glass scientists can also continue their education to earn master and/or doctoral degrees.

The glass industry is the major employer of glass scientists. Usually, they work in research and development, production, or sales. Some of them may work in quality control (see Fig. 4-5) or in management. Within the ceramics field, they may also be employed by the refractory, glaze, or porcelain enamel industries. The allied industries that are involved with electronics, duplicating machines, cameras, and similar products which utilize glass may employ glass scientists.

Professional Ceramic Engineer

Professional ceramic engineers generally have a bachelor of science degree in ceramic engineering. They may have earned advanced degrees. These people have taken one or more examinations administered by the state government. Also, these people have had industrial work experience before receiving the Professional Engineer's License. This allows them to use the initials *P.E.* after their name. Their license distinguishes them from other engineers. Most of the licensed ceramic engineers work independently or as consultants. Otherwise, their employer(s) and involvement with ceramics are similar to that of the ceramic engineers.

SECTION II

Ceramic Materials

Unit 5: The Earth

Our earlier definition of ceramics stated that ceramics are earthy, inorganic, nonmetallic materials. *Earthy* identifies the major source of the ceramic raw materials. Therefore, we should begin our study of ceramics with a study of how the planet Earth was formed.

At first, the planet Earth was a gaseous and molten mass, according to scientists. During this early phase, the heavier materials like iron and nickel tended to sink into the center of the mass. This formed the *core* of the earth. The gradual cooling of the exterior caused a *crust* to form. Igneous rocks were formed during this cooling phase.

Weathering

An atmosphere was created as the slow cooling process released gases and water vapor. As the atmosphere of the Earth was developed, it led to the process of *weathering*. The Earth's crust has never stopped weathering. Weathering decomposes and/or wears away the exposed crust. Chemical and/or physical changes occur during this process.

A prime example of a chemical change is the salt water in the oceans. Water dissolved *halite* or native salt, (Fig. 5-1) out of the earth's crust. Eventually, the salt was carried into the oceans.

Physical changes are more obvious, since you can see the results. Debris such as dirt, clay, sand, and silt results from rocks wearing away. Further evidence is found in the smaller particles that make up the rocks known as *sandstones*.

Water has undoubtedly caused many of these changes, since water can exist in all the three states of matter—solid, liquid, and gas. While the earth was cooling, the water was probably very hot (or even *steam*). This caused the chemical changes. Rain, snow, and ice result in moving water such as a stream. The moving water carries rocks and smaller particles. These particles often cause further abrading or wearing away. During winter, water seeps into any cracks in a rock and freezes and expands. The rock is then fractured (cracked). Glaciers are another form of ice. The glaciers caused many of the physical changes to occur, especially as the glaciers receded.

Changes in the earth's crust also are due to various other forces. These forces are sandstorms, landslides, ava-

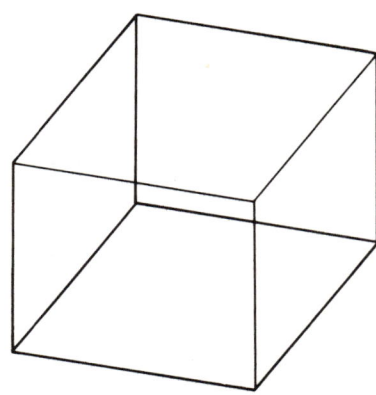

Fig. 5-1. The mineral *halite* has a cubic crystal. Halite is the native salt which chemically is sodium chloride (NaC1).

15

lanches, earthquakes, and volcanoes. Water and the other forces have all contributed to the breakdown of solid rock since the earth was formed.

Minerals and Rocks

Two substances in the Earth's crust are of special interest to ceramics. They are *minerals* and *rocks*.

Minerals

Minerals are defined as being *all* of the following:

1. A natural material in the earth
2. A solid homogeneous crystal (Fig. 5-2) or crystalline (many small crystals) chemical element or compound
3. Inorganic (other than plant or animal)
4. Having a definite chemical composition

A mineral is homogeneous or all one element or compound. Examples are: *halite* (see Fig. 5-1), which is salt or sodium chloride (NaCl), and *quartz* (Figs. 5-3 and 5-4), which is silicon dioxide (SiO_2).

Rocks

Rocks are an aggregate (assembly) of one or more minerals (Fig. 5-5). Examples of rocks composed of only one mineral are limestone [calcite (see Fig. 5-2)], and rock gypsum (Fig. 5-6). Basalt, diorite, and granite are examples of rocks composed of two or more minerals, (see Fig. 5-5).

There are three ways in which rocks were formed. Thus, there are three types of rocks—igneous, sedimentary, and metamorphic.

Igneous Rocks

Igneous rocks are formed when magma cools. *Magma* is the hot molten rock that originates from within the earth. When a volcano erupts, magma flows to the surface. Then it is called lava. When the magma cools, the chemical elements and compounds in it may separate into crystals. Slow cooling results in large well developed crystals, such as quartz (see Figs. 5-3 and 5-4) and *feldspar*, which are any of a group of crystalline minerals that consist of aluminum silicates with either potassium or sodium, and calcium that are an essential constituent of many crystalline rocks. Rapid cooling results in many small crystals or in no crystals. *Obsidian* is natural or volcanic glass that was formed when the magma cooled so rapidly that no crystals were formed.

Two examples of igneous rocks are *basalt* and *granite* (see Fig. 5-3). Basalt is primarily composed of two minerals—pyroxene and plagioclase feldspar. The minerals in basalt are an example of rapid cooling, since the crystals are so small.

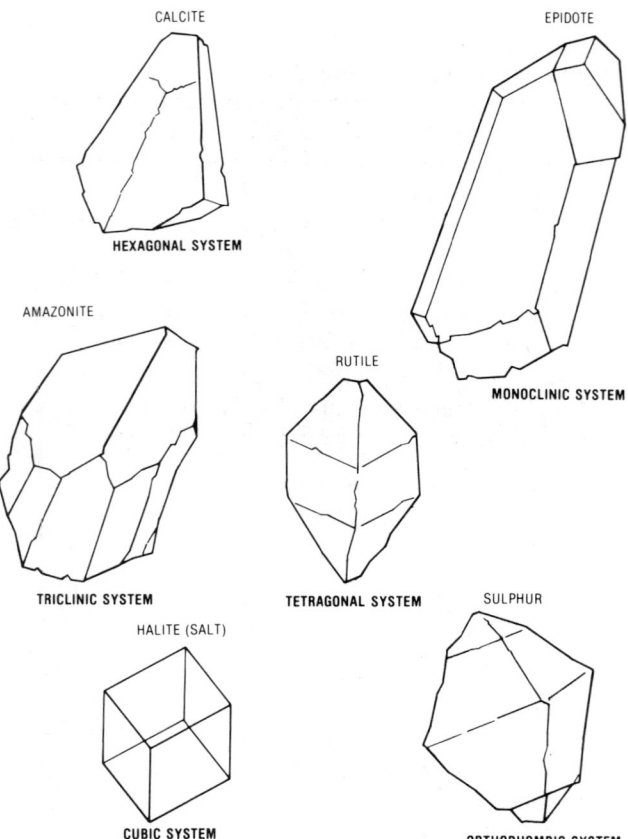

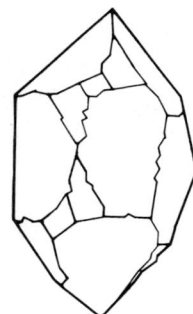

Fig. 5-3. A quartz crystal has many faces. It is clear or transparent and colorless. The quartz crystal is classified as rhombohedra, which is a subdivision of the hexagonal system.

Fig. 5-2. Crystals are grouped into six main systems. The idealized crystal shape and an example of each are shown above. Crystals have well developed faces only when they grow unhampered. Thus, many incomplete crystals exist in the earth.

The crystals of the minerals in granite are usually larger and intergrown. This is typical of slow cooling. You can easily see the minerals in a piece of granite with a hand lens. Granite varies in color from black and white to various shades of red and brown. The minerals present give granite its characteristic color. The black shade is hornblende and biotite mica. The white and red colors are due to the presence of two different feldspars. The clear area that looks like ice when viewed with a hand lens is

The Earth / 17

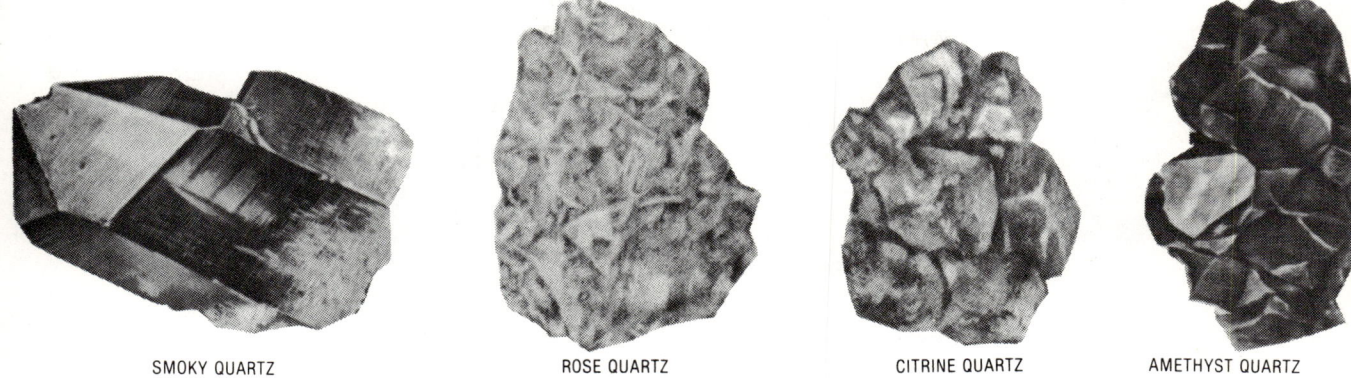

SMOKY QUARTZ   ROSE QUARTZ   CITRINE QUARTZ   AMETHYST QUARTZ

Fig. 5-4. When quartz is colored, it is given a descriptive name like smoky, rose, milky, or citrine quartz. Amethyst is quartz with traces of manganese which provides the pale to dark violet or purple colors.

erosion to the lower elevations where they are deposited in layers. Eventually, these particles become cemented together. Clastic sedimentary rocks like conglomerate, sandstone, and shale (Fig. 5-7) are formed. Clastic particle sizes and terms are defined by the *Wentworth Scale* in Table 5-1.

BASALT

DIORITE

GRANITE

Table 5-1. Wentworth Scale*

Clastic Particle	Size Limit (diameter)
Boulder	Above 256 mm (10.1 in.)
Cobble	64 mm – 256 mm (2.5 in. – 10.1 in.)
Pebble	4 mm – 64 mm (0.2 in. – 2.5 in.)
Granule	2 mm – 4 mm (0.1 in. – 0.2 in.)
Sand	1/16 mm – 2 mm (0.002 in. – 0.1 in.)
Silt	1/256 mm – 1/16 mm (0.0002 in. – 0.002 in.)
Clay	1/256 mm (0.0002 in.) or less

*This scale is based on average particle size.

Fig. 5-5. Basalt, diorite, and granite are examples of igneous rocks.

quartz. If present, the biotite and muscovite micas reflect light. Thus, the minerals usually present in granite include hornblende, feldspars, quartz, and micas.

Sedimentary Rocks

Sedimentary rocks are the result of *clastic* (fragments of preexisting rocks) and chemical deposits. All rocks slowly decompose as a result of chemical and physical weathering. Rock particles (clay and sand) are carried by

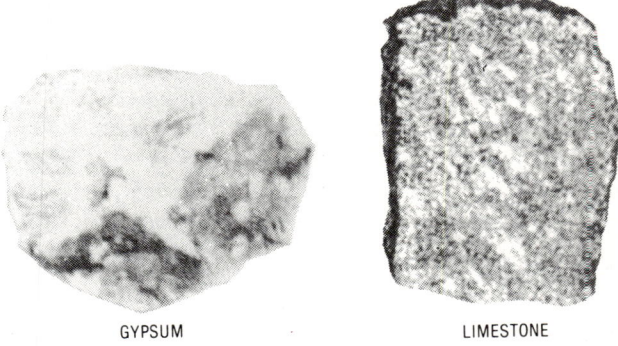

GYPSUM   LIMESTONE

Fig. 5-6. Gypsum and limestone are examples of chemical sedimentary rocks.

18 / *Modern Industrial Ceramics*

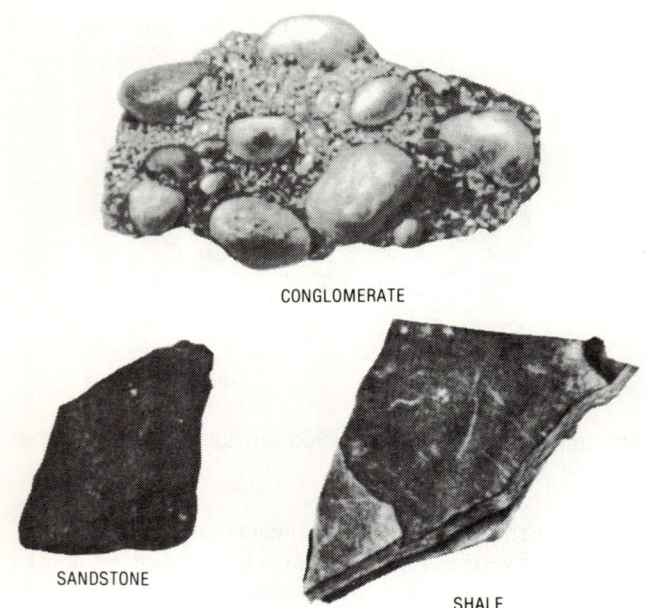

Fig. 5-7. Conglomerate, sandstone, and shale are examples of clastic sedimentary rocks.

Water is collected in the lakes or oceans. Dissolved chemicals begin to precipitate out of the water. Minerals formed from the precipitated chemicals accumulate in layers. Eventually, the layers are cemented together to form chemical sedimentary rocks like limestone and gypsum (see Fig. 5-7).

Metamorphic Rocks

Metamorphic rocks (Fig. 5-8) are formed by altering other rocks. Metamorphic means to change in form. Either high temperatures or high pressures (or both) produce rocks with different textures and/or minerals from the original. Volcanic actions and earthquakes cause the conditions for rock changes. However, the rock changes may occur deep within the earth, since both pressure and heat increase with depth. Some of the metamorphic rocks are listed in Table 5-2 and can be seen in Fig. 5-8.

Summary of Types of Rocks

The three types of rocks are summarized here. These descriptions are helpful when identifying rocks.

Igneous rocks are the rocks that have cooled and solidified from a molten state. They are classified on the basis of their texture and composition. Texture involves the size, shape, and arrangements of the constituent minerals. Composition is the quantity of the different minerals present. Texture is closely related to the cooling history of the rock. Generally, the slower the rate of cooling, the coarser the crystals. Thus, the rocks which cool at or near the surface usually cool more rapidly and are fine-grained; whereas, those rocks that cool deep below the surface, cool slowly and are coarse-grained. The color of igneous rocks is closely related to their composition. Those rocks which contain minerals rich in aluminum, potassium, silica, and sodium are commonly light-colored. Those rocks which contain minerals rich in calcium, iron, and magnesium are usually dark in color.

Fig. 5-8. White marble, slate, and quartzite are examples of metamorphic rocks.

Table 5-2. Metamorphic Rocks

Type of Parent Rock	Parent Rock	Metamorphic Rock
Igneous	Granite	Gneiss (Pronounce nice)
Sedimentary	Limestone Sandstone Shale	Marble Quartzite Slate
Metamorphic	Slate Phyllite	Phyllite Schist

Sedimentary rocks are formed from sediments derived from preexisting rocks. These sediments are deposited in bodies of water (lakes and the oceans). Then, they accumulate and become *lithified* onto rock. They are lithified by compaction (made tighter), dehydration (water is removed), cementation of the sediments (particles bonded together), and some recrystallization. The three main types of sedimentary rock are *clastic* (from rock and mineral fragments), *chemical* (precipitated minerals from solution) and *organic* (derived from plants and animals). *Sediments* are materials that are deposited out of the air or more frequently in water. The three kinds of sediments are clastic, chemical, and organic. The clastic sediments are fragments of rocks or individual minerals. The chemical sediments are materials precipitated in fresh or salt water. The organic sediments are plant or animal fragments and fossils.

Metamorphic rocks are formed by changing the existing rocks, called "parent rocks," to a new form. The process of the change is called metamorphism. It includes increases in temperature, pressure, and the addition of chemically active mixtures of hot gases and solutions. As a result of metamorphism, the metamorphic rock usually has different minerals, a different structure, and a different texture from the parent rock. The metamorphic rocks are classified into "foliate" and "nonfoliate" types, depending on whether the minerals have an alignment allowing the rock to split along a given direction (foliation). Metamorphic rocks form in regions of the crust which contain bodies of cooling magma and where rocks are tightly squeezed and folded.

Rock Cycle

Actually, rock formation is a cyclic process (Figs. 5-9 and 5-10). This cycle can be stopped at any point by natural processes and/or by human intervention. Minerals, rocks, and/or unconsolidated debris result when the cycle is interrupted (see Fig. 5-9). In terms of time, the rock cycle processes (see Fig. 5-10) are usually slow.

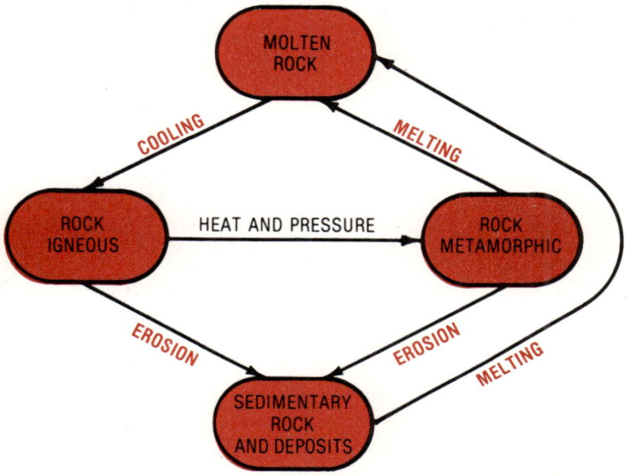

Fig. 5-9. Simplified graphic diagram of the rock cycle.

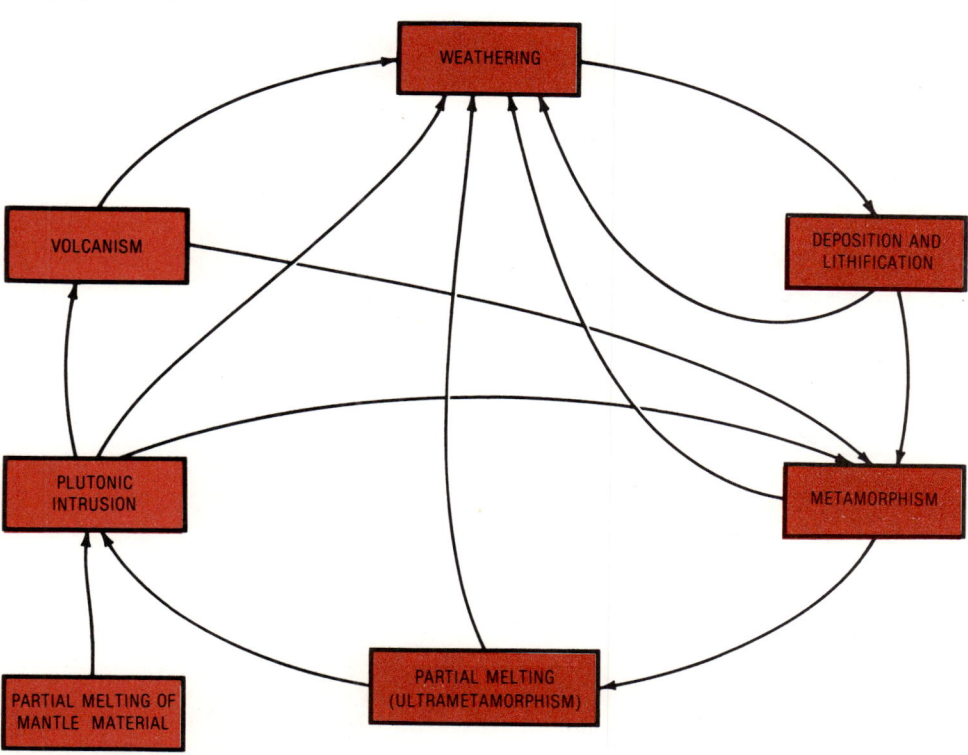

Fig. 5-10. Diagram of the rock cycle with geological terminology.

20 / *Modern Industrial Ceramics*

Composition of the Earth

The composition of the earth is an appropriate summary for this unit. Our planet Earth is nearly a perfect sphere. Basically, the earth is composed of three parts—crust, mantle, and core (Fig. 5-11). Geologists have used data from seismographs to determine these three parts of the earth. Seismographs record earthquakes and similar waves.

The *crust* of the earth averages 25 miles (40.2 kilometers) in thickness. The thickness is least, possibly 10 miles (16.1 kilometers), in the deep trenches beneath the ocean and is greatest in the mountains. This relatively thin layer supports all life on earth, including human life. The crust is the source of the ceramic raw materials. Geologists have and may now be attempting to drill through the earth's crust and into the *mantle*. The mantle is about 1,800 miles or (2,896.2 kilometers) in thickness. Geologists have concluded that the mantle is composed primarily of igneous rocks like basalt.

Seismograph tests show the earth's *core* to consist of a liquid outer core 1,300 miles (2,091.7 kilometers) in thickness and an inner solid core of 800 miles (1,287.2 kilometers). Similar tests have led to the conclusion that the core is composed of iron and nickel. A sectional view of the composition of the Earth is diagrammed in Fig. 5-12.

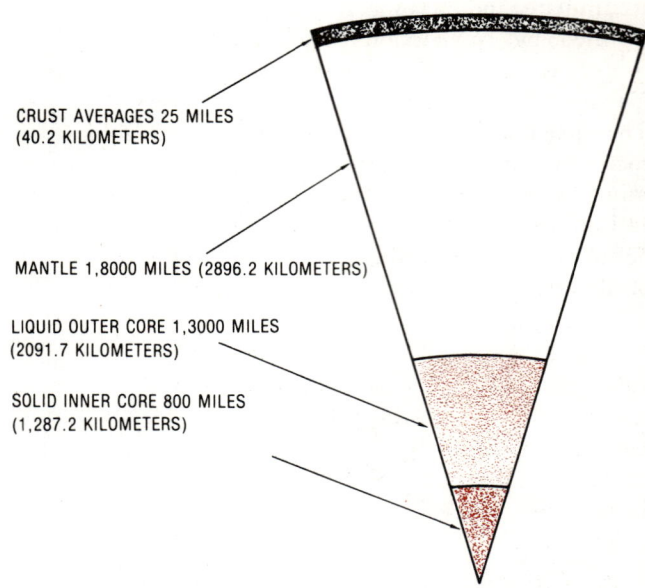

Fig. 5-12. Sectional view showing the composition of the earth.

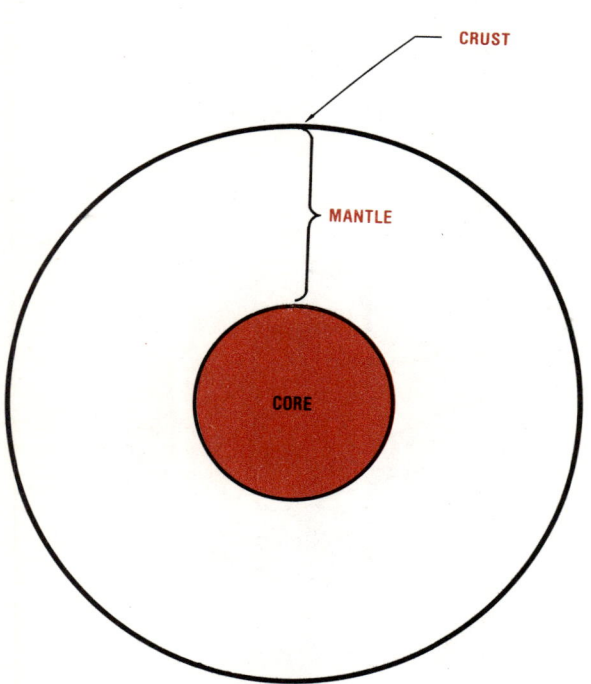

Fig. 5-11. Diagram showing the composition of the earth.

Unit 6: Ceramic Chemistry

The three terms *element*, *compound*, and *mixture* are often used in chemistry. Ceramic chemistry involves all these terms.

Elements

An element is composed of only one kind of atoms, and it cannot be broken down into simpler substances by ordinary chemical change (Table 6-1). Oxygen (O) is the most abundant element in our world (Fig. 6-1).

Compounds

A compound is composed of two or more elements chemically combined in definite proportion. The chemically combined elements cannot be identified by their individual or original properties. For example, the elements hydrogen (H) and oxygen (O) are chemically combined in the ratio of two parts hydrogen to one part oxygen to form water (H_2O).

Mixtures

A mixture is a physical combination of two or more elements and/or compounds. However, the elements and/or compounds retain their individual properties after mixing. For example, salt mixed with sand can be removed by adding water to dissolve the salt. Thus, the salt is not chemically combined with the sand. The combination of sand and salt is a mixture.

Ceramic raw materials are usually in the form of compounds called *oxides*. When oxygen (O) chemically combines with another element (see Tables 6-1 and 6-2), the compound is referred to as an oxide (Fig. 6-2). Rust or iron oxide (FeO) is a very common oxide. The percentages of elements and oxides available in the earth's crust are shown in Figs. 6-1 and 6-2.

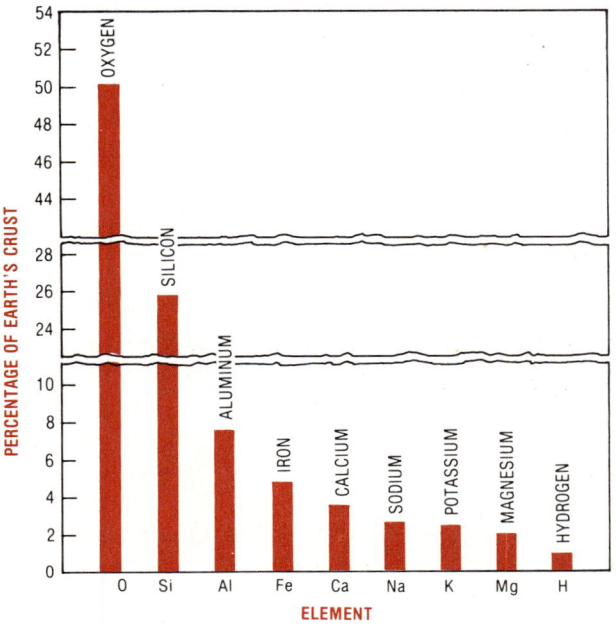

Fig. 6-1. Percentage of elements found in the earth's crust.

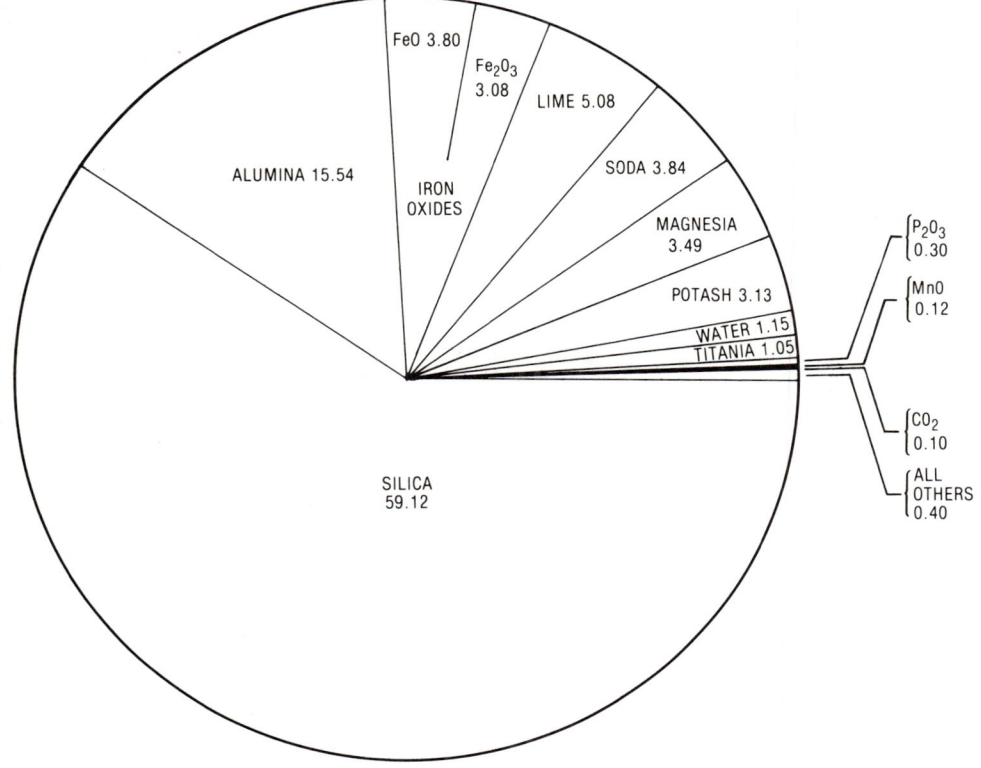

Fig. 6-2. Percentage of oxides available in the earth's crust.

Table 6-1. Elements and Their Symbols

Name of Element	Symbol	Atomic Number	Name	Symbol	Atomic Number	Name	Symbol	Atomic Number
Actinium	Ac	89	Hafnium	Hf	72	Promethium	Pm	61
Aluminum	Al	13	Helium	He	2	Protoactinium	Pa	91
Americium	Am	95	Holmium	Ho	67	Radium	Ra	88
Antimony	Sb	51	Hydrogen	H	1	Radon	Rn	86
Argon	A	18	Indium	In	49	Rhenium	Re	75
Arsenic	As	33	Iodine	I	53	Rhodium	Rh	45
Astatine	At	85	Iridium	Ir	77	Rubidium	Rb	37
Barium	Ba	56	Iron	Fe	26	Ruthenium	Ru	44
Berkelium	Bk	97	Krypton	Kr	36	Samarium	Sm	62
Beryllium	Be	4	Lanthanum	La	57	Scandium	Sc	21
Bismuth	Bi	83	Lawrencium	Lw	103	Selenium	Se	34
Boron	B	5	Lead	Pb	82	Silicon	Si	14
Bromine	Br	35	Lithium	Li	3	Silver	Ag	47
Cadmium	Cd	48	Lutecium	Lu	71	Sodium	Na	11
Caesium	Cs	55	Magnesium	Mg	12	Strontium	Sr	38
Calcium	Ca	20	Manganese	Mn	25	Sulfur	S	16
Californium	Cf	98	Mendelevium	Md	101	Tantalum	Ta	73
Carbon	C	6	Mercury	Hg	80	Technecium	Tc	43
Cerium	Ce	58	Molybdenum	Mo	42	Tellurium	Te	52
Chlorine	Cl	17	Neodymium	Nd	60	Terbium	Tb	65
Chromium	Cr	24	Neon	Ne	10	Thallium	Te	81
Cobalt	Co	27	Neptunium	Np	93	Thorium	Th	90
Copper	Cu	29	Nickel	Ni	28	Thulium	Tm	69
Curium	Cm	96	Niobium	Nb	41	Tin	Sn	50
Dysprosium	Dy	66	Nitrogen	N	7	Titanium	Ti	22
Einsteinium	Es	99	Nobelium	No	102	Tungsten	W	74
Erbium	Er	68	Osmium	Os	76	Uranium	U	92
Europium	Eu	63	Oxygen	O	8	Vanadium	V	23
Fermium	Fm	100	Palladium	Pd	46	Xenon	X	54
Fluorine	F	9	Phosphorus	P	15	Ytterbium	Yb	70
Francium	Fr	87	Platinum	Pt	78	Yttrium	Y	39
Gadolinium	Gd	64	Plutonium	Pu	94	Zinc	Zn	30
Gallium	Ga	31	Polonium	Po	84	Zirconium	Zr	40
Germanium	Ge	32	Potassium	K	19			
Gold	Au	79	Praseodymium	Pr	59			

Symbols and Formulas

A chemical *symbol* for each element is used in chemistry as a type of shorthand. These symbols can be found in Tables 6-1 and 6-2. The symbol (O) represents one atom of oxygen. We exhale carbon dioxide (CO_2), a compound that contains two atoms of oxygen. The chemical *formula* (CO_2) indicates the chemical composition of the compound. Formulas of compounds may be simple—like CO_2, H_2O, and $NaCl$; or they may be complex—like $K_2O \cdot Al_2O_3 \cdot 6SiO_2$—a potash feldspar.

Oxides

Ceramics chemistry is concerned with oxides and their reactions when combined, heated, or melted. Therefore, the characteristic of the oxide is emphasized by writing the oxides separately with *high periods* (·) between them to indicate they are chemically combined. A potash feldspar is $K_2O \cdot Al_2O_3 \cdot 6SiO_2$, rather than $2(KAlSi_3O_8)$ which is the typical chemistry formula. The three oxides which compose this feldspar are obvious, since they are indicated in the ceramic chemical formula.

These are the advantages of ceramic chemical formulas:

1. *High periods* separate the oxides in the formula.
2. Each oxide in the formula is indicated.
3. Chemical composition of the formula is easily read.

Formulas in Equations

Formulas are used in chemical *equations* to indicate what happens in a chemical reaction. Chemical equations are the shorthand for writing equations. For example:

$$\text{limestone} \xrightarrow{\Delta} \text{calcium oxide} + \text{carbon dioxide}$$
$$CaCO_3 \xrightarrow{\Delta} CaO + CO_2$$

The Greek symbol *delta* (Δ), is used to indicate heat in a chemical equation.

Oxide formulas and equations will be used to show chemical reactions throughout our study of ceramics.

Table 6-2. Periodic Table

The elements in this table are arranged in order of increasing atomic number. Elements of similar properties are placed one under the other, yielding eight (8) groups of elements. Within each group there is a gradation of chemical and physical properties, but in general a similiarity in chemical behavior. From group to group there is a progressive shift of chemical behavior from one end of the table to the other.

Group / Period	I	II										III	IV	V	VI	VII	VIII
1	1 H Hydrogen											\-\-\-\-\-METALLOIDS AND NONMETALS\-\-\-\-\-					2 He Helium
2	3 Li Lithium	4 Be Beryllium										5 B Boron	6 C Carbon	7 N Nitrogen	8 O Oxygen	9 F Fluorine	10 Ne Neon
3	11 Na Sodium	12 Mg Magnesium	\-\-\-\-\-\-\-\-\-\-\-\-\-\-\-TRANSITION METALS\-\-\-\-\-\-\-\-\-\-\-\-\-\-\-									13 Al Aluminum	14 Si Silicon	15 P Phosphorus	16 S Sulfur	17 Cl Chlorine	18 Ar Argon
4	19 K Potassium	20 Ca Calcium	21 Sc Scandium	22 Ti Titanium	23 V Vanadium	24 Cr Chromium	25 Mn Manganese	26 Fe Iron	27 Co Cobalt	28 Ni Nickel	29 Cu Copper	30 Zn Zinc	31 Ga Gallium	32 Ge Germanium	33 As Arsenic	34 Se Selenium	35 Br Bromine	36 Kr Krypton
5	37 Rb Rubidium	38 Sr Strontium	39 Y Yttrium	40 Zr Zirconium	41 Nb Niobium	42 Mo Molybdenum	43 Tc Technetium	44 Ru Ruthenium	45 Rh Rhodium	46 Pd Palladium	47 Ag Silver	48 Cd Cadmium	49 In Indium	50 Sn Tin	51 Sb Antimony	52 Te Tellurium	53 I Iodine	54 Xe Xenon
6	55 Cs Cesium	56 Ba Barium	57 La Lanthanum	72 Hf Hafnium	73 Ta Tantalum	74 W Tungsten	75 Re Rhenium	76 Os Osmium	77 Ir Iridium	78 Pt Platinum	79 Au Gold	80 Hg Mercury	81 Tl Thallium	82 Pb Lead	83 Bi Bismuth	84 Po Polonium	85 At Astatine	86 Rn Radon
7	87 Fr Francium	88 Ra Radium	89 Ac Actinium															

LANTHANIDES (RARE EARTH METALS)	58 Ce Cerium	59 Pr Praseodymium	60 Nd Neodymium	61 Pm Promethium	62 Sm Samarium	63 Eu Europium	64 Gd Gadolinium	65 Tb Terbium	66 Dy Dysprosium	67 Ho Holmium	68 Er Erbium	69 Tm Thulium	70 Yb Ytterbium	71 Lu Lutetium
ACTINIDES	90 Th Thorium	91 Pa Protoactinium	92 U Uranium	93 Np Neptunium	94 Pu Plutonium	95 Am Americium	96 Cm Curium	97 Bk Berkelium	98 Cf Californium	99 Es Einsteinium	100 Fm Fermium	101 Md Mendelevium	102 No Nobelium	103 Lw Lawrencium

Unit 7: Using Ceramic Raw Materials

You will find many ceramic materials are involved as you study ceramics. However, most of these materials are used in very small quantities.

Natural Raw Materials

In Table 7-1, the eleven natural ceramic raw materials are listed. They are presented in the order they will be presented in the book.

Raw materials may be natural or manufactured. Therefore, Table 7-1 is limited to only the *natural* ceramic raw materials. Also, these materials are those used in larger quantities. Natural raw materials that are used in large quantities include sedimentary clay (Fig. 7-1) and talc (Fig. 7-2).

Technology

Technology is an important relationship between humans and materials. This is the human relationship to manufactured things, and it deals with *how* and *what* we do to materials. Specifically, humans *procure*, *transform*, *use*, and *dispose of* materials. The technical terms are:

1. Procurement (gathering raw materials)
2. Transformation (changing raw materials to usable products)
3. Utilization (using the products made from the raw materials)
4. Disposition (discarding used products)

A production continuum using these terms is diagrammed in Fig. 7-3. The terms are defined in Section III—*Ceramic Raw Material Industry*.

We begin the study of the ceramic industries with the *ceramic raw material industry*. This is the basic industry that provides the raw materials used to make ceramic products. Other companies depend on this industry for

Table 7-1. Natural Ceramic Raw Materials

Raw Material	Ceramic Industry Using this Material	Composition Mineral or Chemical	Type	Source
1. Granite (Fig. 5-5)	Stone	Hornblende Feldspar(s) Quartz Mica(s): biotite, muscovite	Igneous rock	Vermont
2. Limestone (Fig. 5-7)	Stone Refractories Lime Portland cement Concrete	Calcite: Calcium carbonate ($CaCO_3$) Dolomite: Dolomitic limestone ($CaCO_3 \cdot MgCO_3$)	Sedimentary rock	Universal
3. Marble	Stone	$CaCO_3$ or $CaCO_3 \cdot MgCO_3$	Metamorphic rock	Universal
4. Sandstone (Fig. 5-6)	Stone	Sand size particles usually some silica (SiO_2)	Sedimentary rock	Northeast U.S.
5. Clay (Fig. 7-1)	Refractories Portland cement Clay Glazes	Kaolinite (Hydrous Alumina silicate ($Al_2O_3 \cdot 2SiO_2 \cdot 2H_2O$)	Clay: unconsolidated debris Shale: sedimentary rock (Fig. 5-6)	Universal
6. Gypsum (Fig. 5-6)	Gypsum Portland cement	Hydrous calcium sulfate ($CaSO_4 \cdot 2H_2O$)	Sedimentary rock	Central U.S., Michigan, New York, California, Nevada, Utah, Montana

Raw Material	Ceramic Industry Using this Material	Composition Mineral or Chemical	Type	Source
7. Feldspars	Clay Glaze Porcelain enamel Glass Abrasive	A few examples: Potash spar ($K_2O \cdot Al_2O_3 \cdot 6SiO_2$) Soda spar ($Na_2O \cdot Al_2O_3 \cdot 6SiO_2$) Calcium spar ($CaO \cdot Al_2O_3 \cdot 2SiO_2$)	Mineral	Universal
8. Nepheline syenite	Clay Glaze Porcelain enamel	$K_2O \cdot 3Na_2O \cdot 4Al_2O_3 \cdot 9SiO_2$ Note similarity to feldspar	Mineral	Lakefield, Ontario, Canada
9. Talc (Fig. 7-2)	Clay Glaze	Magnesium silicate ($3MgO \cdot 4SiO_2 \cdot H_2O$)	Mineral	California, Nevada, New York
10. Borax	Glaze Porcelain enamels Glass	Hydrous sodium borate ($Na_2O \cdot 2B_2O_3 \cdot 10H_2O$)	Mineral	California, Nevada
11. Silica	Refractories Portland cement Clay* Glaze* Porcelain enamels* Glass Abrasives* *Silica is called flint when used in these industries. Flint is cryptocrystalline quartz	Quartz: Silicon dioxide (SiO_2)	Mineral if quartz (Fig. 5-3)	Universal

Fig. 7-1. A sedimentary clay that fires red.

their raw materials for many reasons. Some companies may be small, too diversified, or use many different types of materials. These factors make procurement and/or transformation infeasible for many of them.

The raw materials industry first procures the inorganic nonmetallic materials such as clay, feldspar, and flint from the earth's crust. Then these materials are transformed into a usable form. This is usually a *powder*.

26 / *Modern Industrial Ceramics*

Fig. 7-2. Ceramic grade talc from the state of New York. *NYTAL* is a trade name for New York talc. (Courtesy R. T. Vanderbilt Company, Inc.)

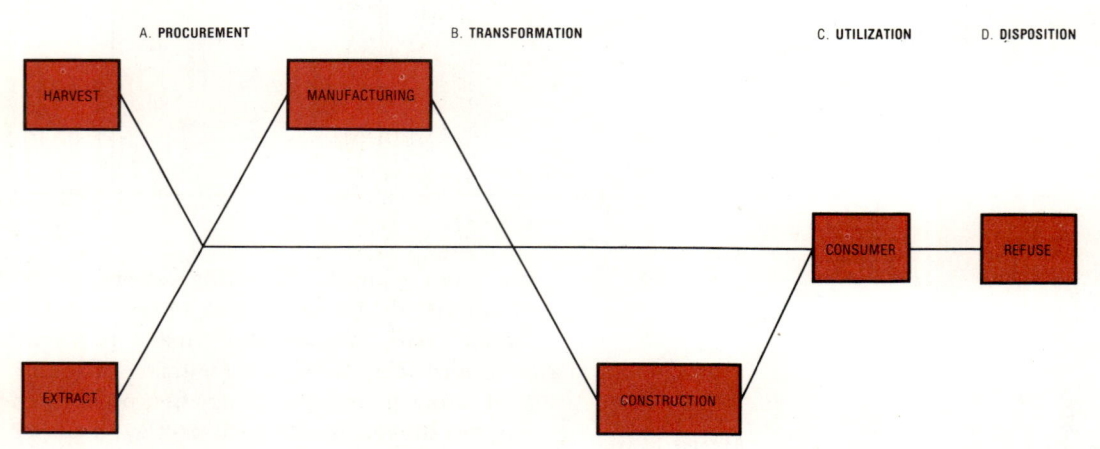

Fig. 7-3. Diagram of a typical production continuum. This diagram shows how the materials move through the four phases. For example, gravel is extracted and is transported directly to a consumer who paves his driveway with the gravel. The same gravel could be sent to manufacturing where it is made into ready-mix concrete. Then, the concrete is used to construct a sidewalk for a home.

SECTION III

Ceramic Raw Materials Industry

Unit 8: Procurement

The means by which natural resources are obtained from nature is called *procurement*. *Harvesting* and *extracting* are the two types of procurement.

Harvesting

The methods by which the materials are taken from the biosphere are *harvesting* processes. The *biosphere* refers to living or organic materials. Fishing, logging, and milking dairy cows are examples of methods of *harvesting*. Harvesting processes are not applicable to the ceramic industries, since those processes involve organic materials. Harvesting processes include:

1. Cutting
2. Digging
3. Netting
4. Peeling
5. Pickling
6. Slaughtering
7. Topping
8. Trapping

Extraction

Extracting processes, on the other hand, are used by the ceramic industries to obtain raw materials from the earth's crust. Materials are removed from the atmosphere, hydrosphere, and lithosphere by the extracting processes. The earth's crust is composed of the following:

1. Hydrosphere
2. Lithosphere
3. Unconsolidated debris
   a. Inorganic debris
   b. Organic debris
4. Biosphere

Hydrosphere

The hydrosphere is the water layer, including both fresh and salt water, on the crust of the earth. Approximately 75 percent of the crust is covered by water, thus it is called the hydrosphere.

Lithosphere

The lithosphere is the rock layer exposed to the atmosphere and directly below the hydrosphere. It is the solid part of the earth.

Unconsolidated Debris

Unconsolidated debris results from weathering processes. This debris can be described as various types of loose particles. Of course, the size of the particles varies extensively, and they are classified by the Wentworth Scale (see Table 5-1).

Gravel, sand, silt, and clay are the common forms of unconsolidated inorganic debris. Most of these forms of debris are used by ceramists. Dirt or soil is a mixture of inorganic and organic debris in varying ratios. Dirt is of little interest to ceramists; however, it is quite necessary to human survival.

Biosphere

The biosphere is composed of the living—flora and fauna—and organic materials. *Flora* includes all plant life, while *fauna* includes all animal life. Ceramics is not directly involved with the biosphere. However, the extraction processes may affect the biosphere, especially its *ecological balance*.

Ecology is the relationship of the flora and/or fauna to their environment. The ecological balance is often delicate. It requires the careful attention of everyone. Restoring the land after the raw materials have been extracted has recently become an important task.

Extraction Methods

Typical extraction methods used by the different material-based industries are:

1. Adsorption
2. Desorption
3. Distillation
4. Drilling and pumping
5. Electrolysis
6. Evaporation
7. Excavating
8. Fractionating condensation
9. Liquefying and rectifying
10. Precipitation
11. Washing

The ceramic industries usually are concerned only with excavating. Specifically, the ceramic raw material industry is concerned with the extraction processes of:

1. Prospecting
2. Developing
3. Winning, or mining

Prospecting

Prospecting involves locating and identifying the raw materials in the earth's crust. Those people who locate and identify the various raw materials include:

1. Prospectors
2. Geologists
3. Geoscientists or geochemists and geophysicists
4. Historical geologists
5. Physical geologists
6. Mineralogists
7. Petrologists

Prospecting frequently involves working outdoors. It is an interesting and challenging type of work with unusual rewards.

Purpose

The *purpose* of prospecting is to discover a deposit of raw material and to:

1. Observe the character, type, and amount of *overburden* (see below)
2. Observe the thickness and extent (breadth and depth) of the deposit
3. Observe the drainage conditions
4. Observe transportation facilities available and/or needed
5. Obtain reliable samples representative of the deposit

Overburden is the material above a deposit. It may include vegetation (flora), dirt, other unconsolidated debris, and solid rock. The character, type, and amount of overburden are extremely important factors. These factors determine the type of winning or mining process that will be used.

Methods of Prospecting

The methods of prospecting include: (1) *geological studies*, and (2) *exploratory prospecting programs*. Both these methods are done systematically, and are used frequently and quite extensively. These methods of prospecting may be done in the order listed or at the same time. Usually, it is geology, historical and/or physical, that helps the prospectors to make intelligent decisions before they do their field work. Decisions as to *where to look* and *what to look for* can be made by studying the geology of an area. Geological maps, topographic maps, and aerial photographs are studied before most of the field work is attempted. Of course, accidental discovery, old-fashioned prospecting by walking and visual survey, and part-time prospectors will hopefully always continue. However, most of the major deposits will be discovered by systematic geologic study. Exploratory prospecting will follow the study.

Exploratory Prospecting

Exploratory prospecting is the actual field work that involves discovering the external signs of a deposit. External signs that provide clues to the possible existence of a deposit include:

1. Outcroppings of an actual deposit
2. Vegetation and/or lack of vegetation
3. Springs
4. Ponds

Vegetation provides a clue as to the type of material beneath the surface. For example, vegetation does not thrive on clay. Clay particles retain water after a rainfall. When clay finally dries, it will become hard and shrink. Thus, poor plant growth could indicate the presence of a large deposit of clay.

A horizontal line of *springs* can indicate a deposit of clay. The water seeps downward through the soil until it reaches a stratum or layer that is impervious to it. Then the water collects and forms a spring. Clay is a material that causes the water to collect, resulting in a spring.

A pond is an area where water does not readily drain away. An impervious material, like clay, causes the accumulation of surface water and, therefore, provides clues to the prospector.

External Signs in Prospecting

External signs are usually discovered by aerial photographs and/or visual observation. Exploratory prospecting methods include:

1. Aerial photographs or remote sensing
2. Visual observation
3. Boring or drilling
4. Test pit

Aerial photographs should be studied before conducting visual observations to establish where to look and to save time. These photographs are made from aircraft with special cameras. The photographs can be of several types. For example, black-and-white photographs, color photographs, infrared (IR) film, radar, computer analysis of film, and space photographs are some of the aids used today. Aerial photographs are used with topographical and geological maps for complete study of an area.

Visual observations can be made while walking or while riding in a surface vehicle or aircraft at low altitude. However, *walking* is the preferred method, since it is done slowly and methodically close to the earth's crust. Also, a person can climb obstructions and cross bodies of water that could stop a surface vehicle. Aircraft are usually too far from the surface of the earth for accurate observations.

Drilling

Boring or drilling is done both manually and mechanically. Hand drilling may be done while conducting visual observations. The purpose of drilling is to confirm external signs and/or obtain samples for laboratory study. Manual drilling methods include augers, core drills, and well drills. Manual well drills are seldom used today. Mechanical methods permit deeper penetration and larger samples. This method includes the rotary auger, well or churn drills, and core drill (Fig. 8-1).

Rotary Auger Drills

Rotary auger drills are capable of drilling to a depth of 100 feet (30.5 meters) with a 4-in. (10.2 cm) diameter auger. These are used primarily for wildcat exploration in soft nonstratified formations such as clay. Subsurface contamination prevents completely accurate samples by this method. Therefore, auger drills serve primarily to locate and prove the extent of a deposit. Once a deposit is located, other drilling methods may be used for a more complete picture of the material.

Fig. 8-1. A small core-drilling, trailer-mounted unit being used for a limestone survey. Note the hose attached to the top of the hollow drilling shaft. The hose provides water for cooling the drill. (Courtesy New York State College of Ceramics at Alfred University)

Well Drill

The well or churn drill is used to a limited extent after rotary auger exploration. This drill is used to test the quality of a deposit.

Core Drill

The core drill (Fig. 8-2) provides an accurate sample. The sample can be visually examined and evaluated in the field and/or laboratory. Also, a core provides a permanent record of the area being explored.

Test Pits

Test pits may also be dug to obtain a vertical profile or sample of a deposit. Small test pits 30-40 in. (76.2–101.6 cm) in diameter can be excavated with a pick and shovel. Excavating by machines, as with a small backhoe or similar mechanical shovel, provides larger test pits.

Development

The first extraction process discussed was prospecting. The second process is the development of a newly discovered deposit. *Development* is the process of preparing to mine the deposit that has been located.

The people doing development work are listed here in the order in which they become involved in a project:

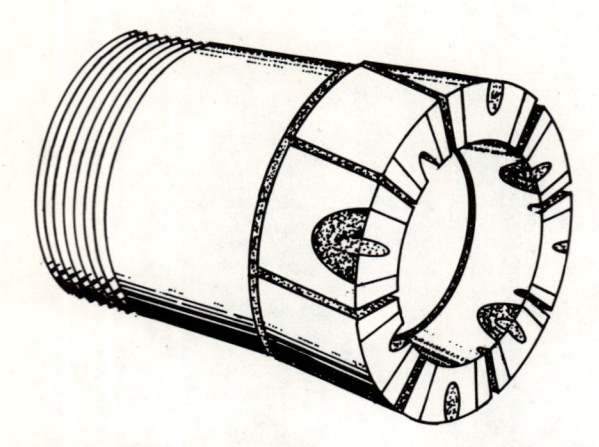

Fig. 8-2. A diamond core-drill tip does the actual cutting. Since it is hollow, water can flow to cool the drill tip and it can accommodate the core sample.

1. Geologists
2. Mining engineers
3. Surveyors
4. Mineral economists
5. Attorneys
6. Mineralogists
7. Petrologists

Usually, these people work as a team. Thus, several of them may be working on their tasks or on a project at the same time. Typical development work that is done by these people includes the following:

1. Conducting a market survey
2. Surveying property
3. Buying the property or leasing the mineral rights
4. Building access roads to the deposit
5. Moving machines to the site
6. Removing the overburden if necessary
7. Sinking a shaft and digging tunnels if necessary
8. Constructing buildings and other facilities.

Winning

The first and second extraction processes discussed were prospecting and developing. The third process is called *winning* or *mining*. Winning is the most visible of the three extraction processes. You may be familiar with some of its aspects. *Winning is the process of removing raw materials from the earth's crust. Mining is the common name for winning.*

Careers

People who are involved with *winning* or mining include:

1. Mining engineers
2. Ceramic engineers
3. Mechanical engineers
4. Industrial engineers
5. Miners
6. Heavy machine operators
7. Operatives (semiskilled workers)
8. Laborers (unskilled workers)

Job Titles

Job titles 1 through 4 usually require a bachelor degree (B.S.) as the minimum education. People in those jobs may have also earned master and/or doctorate degrees. However, most of the workers in this group have bachelor degrees.

Job titles 5 and 6 usually require high school, vocational or technical school, apprenticeship, and/or a two-year or associate degree as the minimum education. Titles 7 and 8 usually require a high school education and on-the-job training as a minimum requirement.

Winning Processes

There are three major *winning* or mining processes. They have the following technical names:

1. Superficial
2. Subterranean
3. Subaqueous

Possibly, you know a location where one of these processes is being conducted. If you visit the site, always ask permission to look around; *use* the required safety equipment.

The type of *winning* operation used depends on the *depth* of the deposit and whether there is any *water* near the deposit. Each of the three methods has its advantages and disadvantages. These are considered by engineers before starting the actual work.

Superficial

Superficial is the most common of the winning or mining processes. The winning processes are done on the lithosphere. They are known by a variety of names, depending on the material being *won*—or mined as:

1. Open-pit (Fig. 8-3)
2. Quarry (Fig. 8-4)
3. Strip mine
4. Hydraulic mining or hydrolicking

Superficial winning is generally the least expensive of the winning operations. All the advantages and disadvantages of superficial winning have to be considered to determine whether it is the best method for removing a deposit. The advantages of superficial winning listed by Professor George A. Kirkendale are:

1. The entire deposit can be removed. Pillars of the

Procurement / 31

Fig. 8-3. Open-pit winning of clay in Ohio. A dipper-stick type of power shovel is used to excavate the deposit. (Courtesy Cedar Heights Clay Co.)

Fig. 8-4. A quarry after blasting. Note the irregular shapes due to the use of explosives. (Courtesy Gold Bond Building Products, Division of National Gypsum Company)

deposit must be left standing to support the roof when mining under ground.
2. No timbering is required. The cost of timbering a mine adds considerably to the cost of mining.
3. There are no tunnels and shafts to keep open, drained, and ventilated.
4. There is no danger to personnel from cave-in of tunnels and shafts.
5. There is no danger of fire or explosion of fumes.
6. There is no artificial lighting to maintain.
7. The open pit provides a larger working face.
8. A larger area can be blasted at one time.

Kirkendale also listed some disadvantages of superficial winning. They are:

1. Removing the overburden costs time and money.
2. The deposit is exposed to the weather. Rain may soak the deposit, such as clay, and make excavating impossible. Also, the pits become filled with water and have to be drained.
3. The water content of a deposit varies due to weather conditions. This makes consistent beneficiation difficult.

Ecology, the ecological cycle, the hydrologic cycle, and land reclamation are other factors which can be considered disadvantages. Each factor is an important consideration. Usually, they are interrelated.

Overburden Removal

Overburden removal and its disposition is an important factor in superficial operations. The character, type, and amount (breadth and depth) are determined while prospecting. Disposition of the overburden is usually decided during the development phase. It may become a by-product of the operations. For example, the topsoil may be sold to housing developments and the subsoil may be sold for fill dirt. Today, these materials, both topsoil and subsoil, may be required by law to be used for land reclamation. This is a consideration of disposition (see Unit 11). However, it does help us to know how each process may be dependent on the other processes that are utilized.

Three methods of removing the overburden and the material in a superficial deposit include:

1. Manual
2. Animal power
3. Mechanical power

Manual

Undeveloped nations and very small operations may still use picks, shovels, and wheelbarrows. Art potters, hobbyists, and similar small firms may use this method, since only small quantities of material are needed. Also, it is relatively inexpensive.

Animal Power

Animal power, for example, the horse-drawn scraper, could also be utilized by the very small operator for the reasons previously cited. However, the type of material definitely affects how it is won or mined.

Mechanical Power

Mechanical power has been and will continue to be the major approach to superficial winning. The *size* of mechanical powered operations will vary extensively—from a mine with a single farm tractor to a mine with

32 / *Modern Industrial Ceramics*

several huge excavators. Mechanical equipment may include:

1. Standard and heavy-duty construction machines
   a. Farm tractors with accessories such as scrapers, scoops, and forklifts.
   b. Industrial construction tractors (Fig. 8-5). These may have a:
      (1) Blade which is pushed (usually a *bulldozer*)
      (2) Scoop (front mounted)
      (3) Scraper (pulled)
      (4) Ripper (pulled)
   c. Front loader or high lift
   d. Power shovel—dipper-stick type with ability to rotate (see Fig. 8-3)
      (1) Power provided by
         (a) Steam boiler and engine
         (b) Gasoline engine
         (c) Diesel engine
         (d) Electric motor
      (2) Locomotion
         (a) Pneumatic tires
         (b) Railroad tracks
         (c) Caterpillar treads
         (d) Barge-mounted (a simple form of dredge)
   e. Backhoe—many different sizes ranging from mounting on a small tractor to larger units
   f. Crane equipped with:
      (1) Clamshell
      (2) Dragline
   g. Earthmover or excavator

2. Specialized machines:
   a. Shale planer
   b. Hydrolicking
3. Transport or movement equipment (methods used to move material within pit, from pit to stockpile, to plant, and/or to consumer):
   a. Surface
      (1) Trucks
         (a) Dump
         (b) Flatbed
      (2) Tractor with trailer
         (a) Dump
         (b) Flatbed
         (c) Low boy
      (3) Earthmovers or excavators
      (4) Railroad
         (a) Flatbed
         (b) Hopper
         (c) Gondola
   b. Above, below or on the surface
      (1) Continuous-belt conveyors
      (2) Screw or auger conveyors
   c. Water
      (1) Barge
      (2) Laker
      (3) Ship

Specialized mechanical equipment that is used in quarrying dimension stone, natural gems, and in some other unique situations has not been listed above. These machines will be considered when the procurement processes for a specific ceramic industry are discussed.

Fig. 8-5. Mining clay in South Carolina with a front-loaded scoop on a crawler-type tractor. Note the safety bars and roof to protect the operator. (Courtesy R. T. Vanderbilt Company, Inc.)

Subterranean Process

Subterranean mining is a winning process that is done beneath the surface of the earth. Actually, this work is done inside the lithosphere of the earth. Working within the earth makes this process more expensive. It does have several advantages:

1. No overburden to remove.
2. Deposit is not exposed to the weather. Therefore, year-round work can be conducted.
3. Deposit does not become wet or frozen, since it is not exposed.

The disadvantages of subterranean winning are:

1. Pillars of the deposit remain to support the roof and are thus wasted for use as a raw material.
2. Timbering and/or similar reinforcement is often necessary, which increases mining costs.
3. Ventilation and draining of water are necessary. Fumes and dust are serious problems.
4. Danger from cave-in, fire, or explosion is possible.
5. Artificial lighting is necessary.

Common names for this type of extraction are:

1. Underground mining
2. Subsurface mining
3. Shaft-and-tunnel mining
4. Deep-pit mining

Several factors make the subterranean processes the preferred method of winning. They are preferable when the:

1. Overburden is several feet to hundreds of feet in thickness.
2. Deposit is a scarce material.
3. Deposit lies beneath homes or other buildings.
4. Weather is severe.

Equipment

Subterranean equipment is similar to that developed for coal and metallic ore mining. Standard equipment, similar to that used for surface work, is used if the deposit is 15 feet (4.572 meters) or more in thickness. The major difference is that these machines are powered by electric motors. Thus, there are no harmful fumes.

The electric motors may operate on alternating current (ac) or direct current (dc). If they are operating on direct current (dc), *batteries* can be used to power the machines. This allows the machines greater flexibility. For machines with ac motors, long, heavy-duty extension cords on retractable drums or overhead wires are used to provide the alternating current.

Subterranean equipment machines include:

1. Winning operation machines
   a. Front loader
   b. Caterpillar-type tractors
   c. Railroads
2. Transport or materials handling machines
   a. Trucks
   b. Conveyors
      (1) Continuous-belt
      (2) Screw or auger
      (3) Bucket elevator

Pneumatic tools are also used in these operations (Fig. 8-6). Compressed air is a safer type of power, since it does not produce fumes.

Fig. 8-6. Pneumatic tools are powered by compressed air. These tools are noisy, but do not produce fumes. (Courtesy Gold Bond Building Products, Division of National Gypsum Company)

Specialized Equipment

Specialized equipment is often necessary for subsurface work. These machines are usually electrical and do the same tasks as those listed above. They differ from those machines in height and capacity (Fig. 8-7). For example, a truck 30 in. (76.2 cm) in height hauls rock gypsum from the mine face to a bucket elevator which transports it to the surface. The operator of this special truck sits sidewise in a position similar to the driver of a small sports car.

34 / *Modern Industrial Ceramics*

Fig. 8-7. A low-profile lead-acid battery powered truck for underground mining. This truck can carry 8 tons. The driver sits sidewise. (Courtesy Lead Industries Association, Inc.)

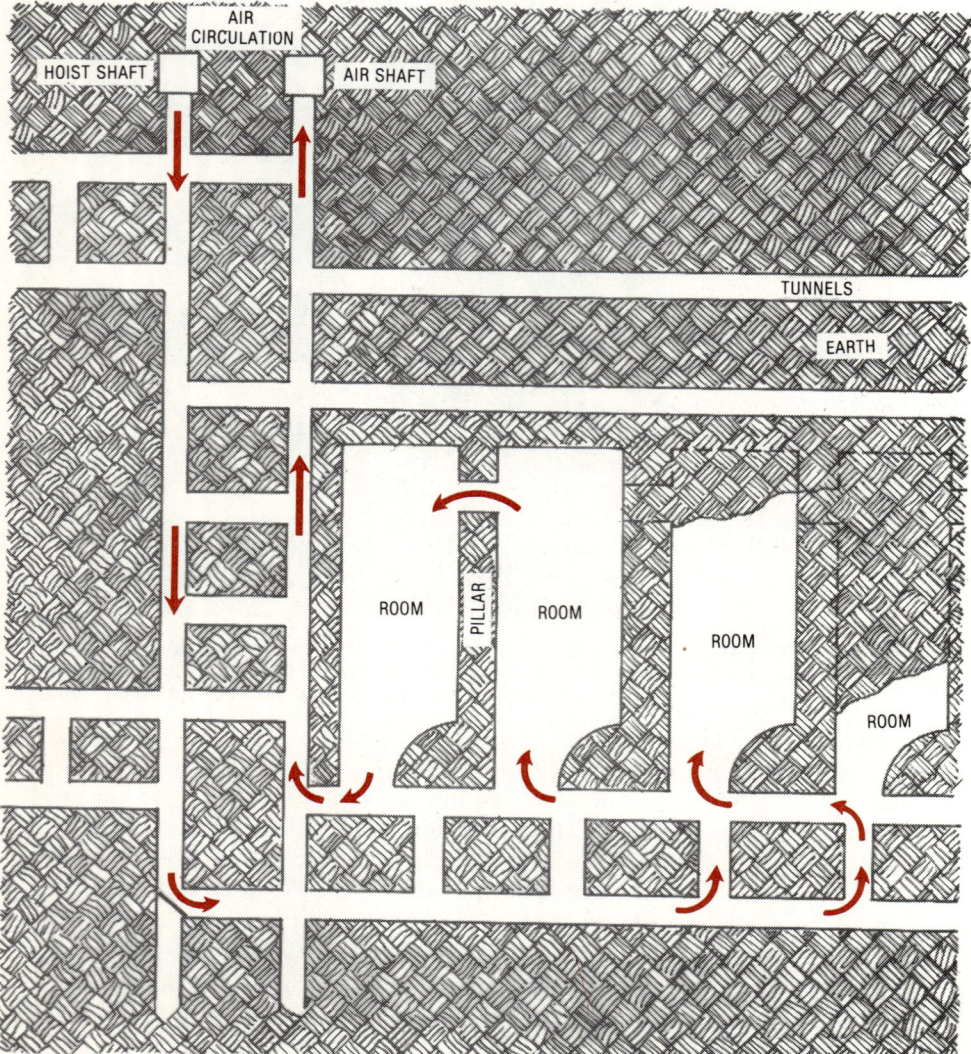

Fig. 8-8. A typical floor plan of room-and-pillar underground mining. A series of rooms are created as the deposit is removed. The pillars support the roof.

Other special machines include:

1. Personnel carriers
2. Drilling units
3. Timbering or reinforcement units (for example, a roof bolt drill unit)
4. Blasting units
5. Scaled-down railroads

Subterranean operations require shafts, tunnels, timbering or other reinforcement, rooms, and pillars. Room-and-pillar mining is shown in Fig. 8-8.

Subaqueous

Subaqueous winning processes are done under water. The equipment floats on the surface of the water and excavates the deposit on (or underneath) the bottom. This method is quite expensive. It is used only when the other winning processes cannot be used.

The subaqueous winning process becomes necessary when an open pit fills with water. This opening and filling may be done to create a lake from which a barge can operate. *Dredging* is a common name for subaqueous winning.

Dredges may be classified as:

1. Barge-mounted
   a. Power shovel (dipper dredge) or backhoe
   b. Bucket ladder
   c. Crane (Fig. 8-9)
      (1) Clamshell
      (2) Dragline
   d. Cutter head with suction (Fig. 8-10)
   e. Pump
2. Ship
   a. Crane
   b. Hopper dredge (Fig. 8-11)

The dredged material may be placed in a barge, stored aboard ship, or pumped into a holding area. Holding areas include ponds and tanks.

The advantages of dredging include:

1. Flexibility
2. Standard equipment utilized
3. Usually no overburden to remove
4. Free from dust problems

The disadvantages of dredging are:

1. A tug is usually required to move a barge-mounted dredge and/or barges for raw material.
2. Barge transportation of materials is limited to waterways.
3. Ice and frozen equipment result from freezing weather.
4. Few ceramic raw materials are available under water—except clay, shells, and coral.

Fig. 8-9. A barge-mounted crane equipped with a clamshell. A dragline could be used in place of the clamshell. This dredge can lift up to 18 cubic yards of material. (Courtesy Great Lakes Dredge & Dock Company)

36 / *Modern Industrial Ceramics*

Fig. 8-10. A cutterhead suction dredge. The cutterhead has been raised above water (left foreground). Dredged material is pumped to a holding area via the floating pipeline (right). (Courtesy ASARCO incorporated)

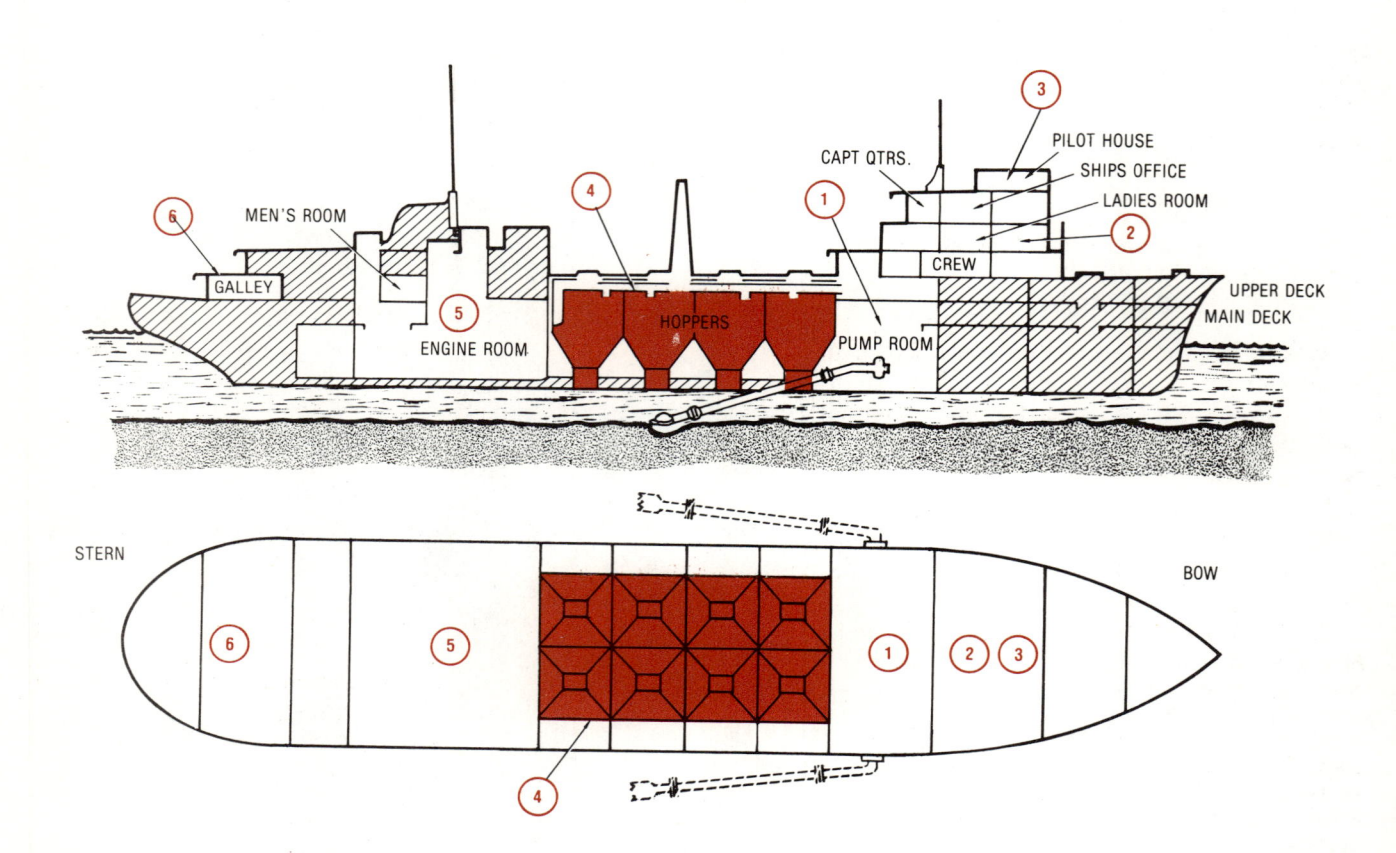

Fig. 8-11. A hopper dredge works like a vacuum cleaner. It uses water instead of air to transport materials to the hoppers. When the hoppers are full, the dredge moves to a holding area; then pumps move the material into storage. (Courtesy United States Army Corps of Engineers, Buffalo District)

Unit 9: Transformation

The processes by which natural resources or synthetic materials are converted into usable products or power are called *transformation*. Transformation can also be defined as a systematic series of actions taken to achieve desired changes in a material. The ten principle processes of transformation are:

1. Agglomeration
2. Assembling
3. Classifying
4. Comminution
5. Conditioning
6. Cutting
7. Finishing
8. Forming
9. Mixing
10. Separation

Definition of Terms

Professor Charles B. Scofield defined each of the above processes in 1972. The following definitions are consistent with his work.

Agglomeration includes those methods used to form larger pieces from smaller particles:

1. Compressing
2. Granulating
3. Heating and sintering
4. Solution evaporation

Assembling involves those methods by which workpieces or sections are attached to one another:

1. Bonding
2. Mechanical fasteners
3. Mechanical-physical joints
4. Combination of bonding, mechanical fasteners, and mechanical-physical joints

Classifying is the method by which the grade or quality of materials is determined:

1. Chemical properties
2. Physical properties
3. Combination of chemical and physical properties

Comminution is the method used for physically reducing larger pieces of materials into smaller pieces. Some common methods are:

1. Bubbling
2. Crushing
3. Cutting
4. Dispersing
5. Exploding—percussion
6. Pulverizing

Conditioning involves the methods used to bring about a desirable change in the physical properties of a material, such as:

1. Cooling
2. Drying
3. Heating
4. Magnetizing
5. Stressing
6. Steaming
7. Soaking

Cutting includes the methods used to make workpieces according to specifications by removing part of the original material. These methods are either:

1. Chip producing, or
2. Nonchip producing

Finishing involves the methods used to beautify and/or protect materials:

1. Impregnating
2. Surface alterations
3. Surface coating

Forming involves the methods used to shape materials to specifications, without removal or addition of material:

1. Material redistribution
2. Plastic deformation

Mixing includes the methods used to combine or recombine materials to form compounds or mixtures:

1. Chemical
2. Mechanical

Separation includes the methods used to isolate an element, compound, or mixture:

1. Chemical
2. Electrical
3. Magnetic
4. Mechanical
5. Thermal

Transformation Systems

Transformation systems are a series of processes that operate in an organized manner. Their purpose is to convert natural resources and/or materials into manufactured products. There are two major systems:

38 / Modern Industrial Ceramics

1. Production
2. Power

Production is the making of goods either in a plant or on a site at the final position of use. The two types of production are:

1. Manufacturing
2. Construction

Manufacturing is the making of products in a plant or place other than its final position of use. It involves the in-plant systems developed to yield movable products. Concrete blocks or clay bricks are manufactured ceramic products. When these blocks and bricks are assembled into a structure, such as a house, we call this activity *construction*. Construction is building on site in its final position of use. It involves the systems required to produce a manufactured object to remain on site.

Those systems developed for the purpose of harnessing energy to perform work are together known as *power*. A hydroelectric plant is an example of this definition.

Ceramic Raw Material Industry

All ten of the major transformation processes are used to manufacture ceramic products. However, the sequence and number actually used by a specific plant will vary. They rarely, if ever, occur in the alphabetical order that was used in the preceding *Definition of Terms*.

The ceramic raw material industry usually uses only three of the ten processes discussed. Only the three processes are necessary to produce their product, a powdered material. The three processes are listed here in the order in which they are usually used:

1. Comminution
2. Classifying
3. Separating

Another term for the three preceding processes is *beneficiation*. The refining of raw material into usable form is beneficiation. Comminution of large clay lumps into powder by crushing and grinding is an example of this process.

Comminution

Comminution is the method used to physically reduce larger pieces of material into smaller pieces. Comminution is accomplished by machines that apply force by one of these methods:

1. Reciprocating pressure
2. Continuous pressure
3. Impact

Crushing reduces the larger pieces of material as they are mined into a variety of smaller pieces. Some of the smaller pieces may be as small as powder or dust. A material is called *crushed* when it is reduced to about 14-mesh size and has sharp, irregular edges. The *hardness* of the material being crushed affects the amount and size of the smaller pieces. For example, shale will break down into many smaller pieces when it is crushed, while granite will produce fewer smaller pieces.

The crushers vary in size from the small laboratory-type machines to the units capable of handling a feed as large as a truckload (Fig. 9-1). The following list includes those crushers most frequently used (Fig. 9-2):

1. Reciprocating pressure
   a. Jaw crusher
   b. Gyratory crusher
2. Continuous pressure
   a. Single-roll crusher
   b. Roll crusher
3. Impact
   a. Hammermill

The terms *primary* crushing and *secondary* crushing are used when this process is repeated. Repeated crushing usually depends upon the hardness of the material.

Fig. 9-1. This is the clay after winning. It is being fed into a crusher by the truckload. The load may weigh as much as 10 tons. Industrial-size crushers are capable of reducing 100 tons per hour. (Courtesy R. T. Vanderbilt Company, Inc.)

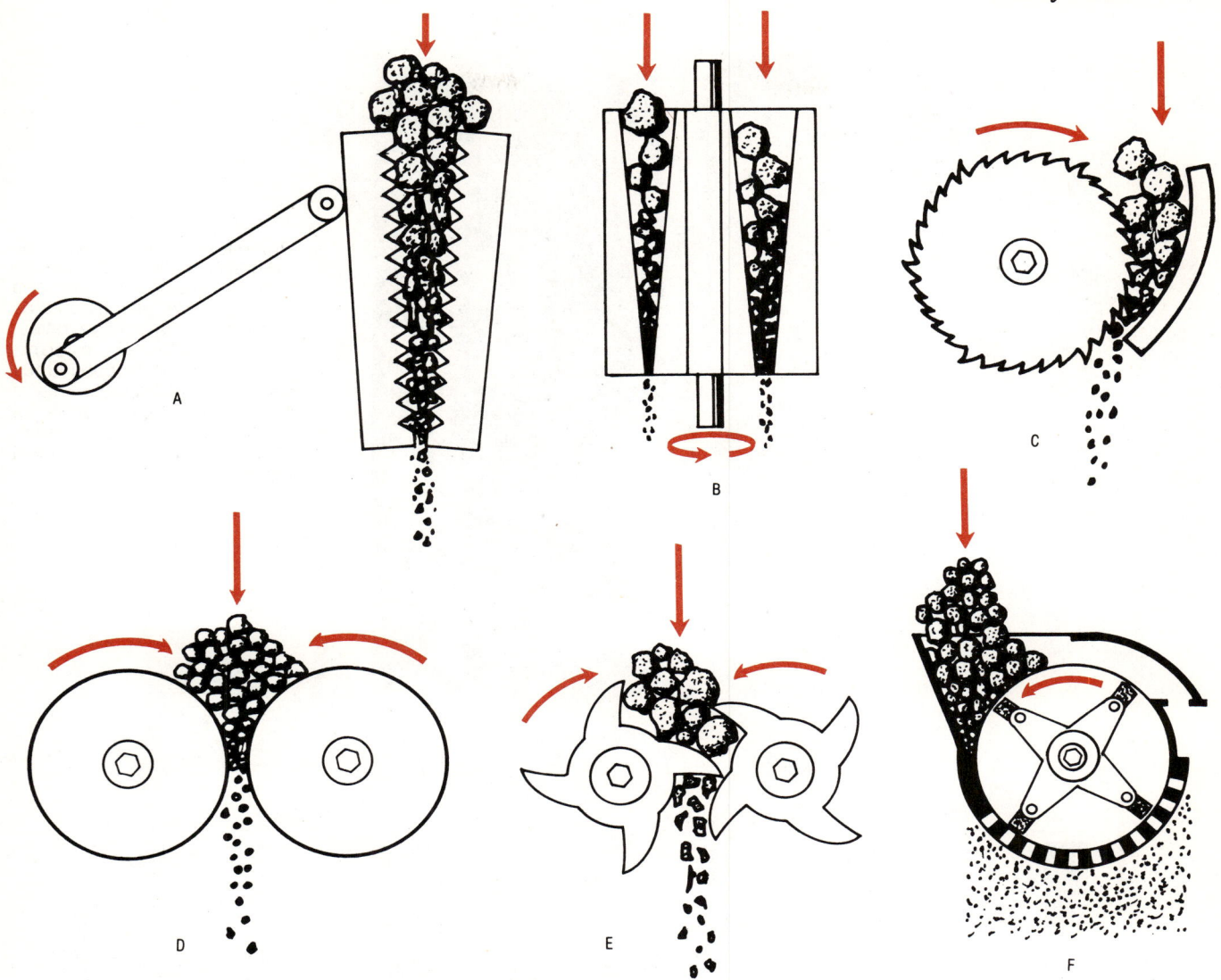

Fig. 9-2. Typical crushers. A jaw crusher applies pressure like a nut cracker (A). A gyratory crusher has an inner cone that oscillates but does not revolve (B). A single-roll crusher has teeth which help force material through the machine (C). A smooth roll crusher (D); toothed roll crusher (E); and a hammermill (F). A hammermill has several hammers which strike the material, causing size reduction.

Further reducing the particle size to 20 mesh or finer is called *grinding*. These grinders vary in size from laboratory units to those with a capacity of several tons (Fig. 9-3). The following list includes those grinders most frequently used (Fig. 9-4):

1. Reciprocating pressure
2. Continuous pressure
   a. Dry pan or muller (Fig. 9-5)
   b. Ball mill (Figs. 9-3 and 9-6)
   c. Rod mill (see Fig. 9-4)
3. Impact
   a. Ball mill
   b. Rod mill

Classifying

Classifying is the method by which the grade or quality of materials is determined. The ceramic raw material industry classifies materials by the size of the particles which pass through various sizes of sieves. Sieves are screens with a mesh number. The mesh number indicates the size or number of openings per linear inch. The fine screens, (2 to 400 mesh), are usually made of woven bronze wire. The coarse screens,—¼ in. (6.4 mm) and larger, may be made from perforated steel plates.

Classifying may be done wet or dry. Several screens may be stacked together with the coarser screens at the top to separate the feed into several sizes. Machines shake or vibrate the screens. This allows the material to feed through, and prevents clogging.

40 / *Modern Industrial Ceramics*

Fig. 9-3. An industrial-size ball mill that is 10 feet (3.05 meters) in diameter. A ball mill is a hollow cylinder which contains grinding media and revolves. Size reduction is accomplished by both continuous pressure and impact. This is due to the combined action of revolving and tumbling media. (Courtesy R. T. Vanderbilt Company, Inc.)

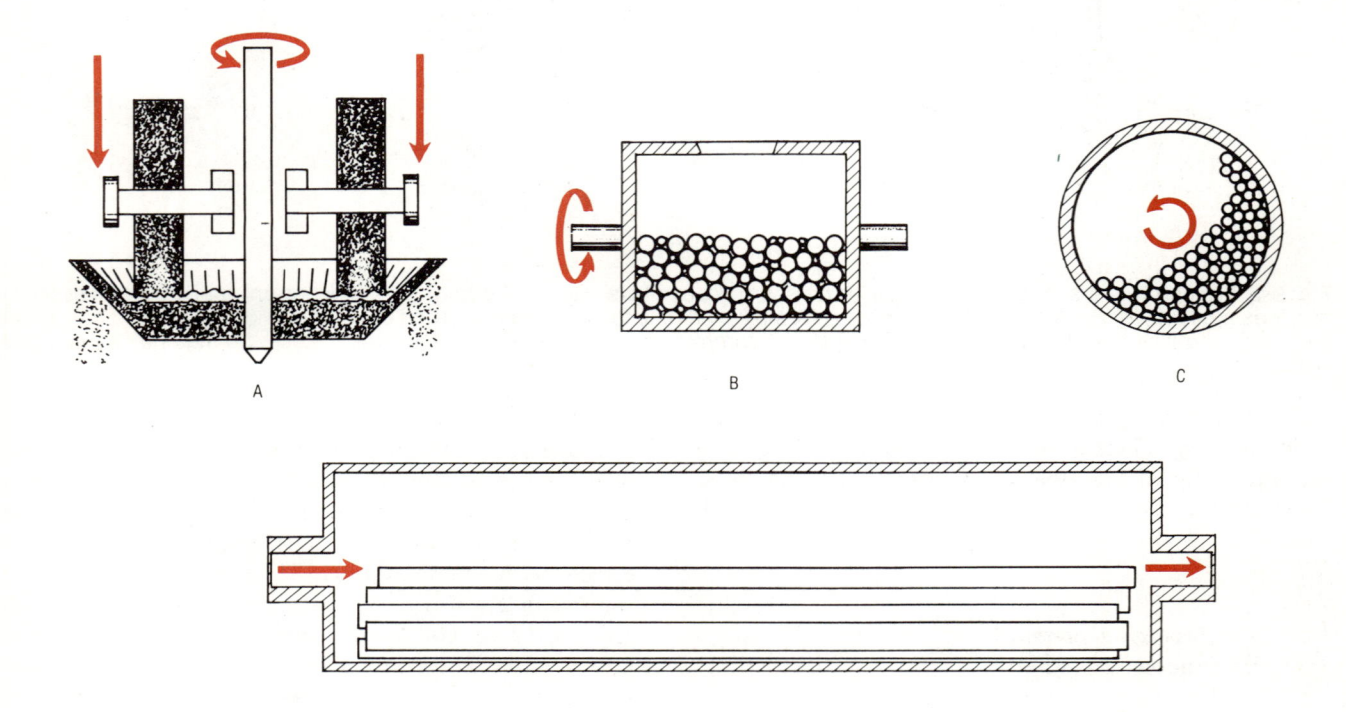

Fig. 9-4. Typical grinders. A dry pan or muller has two heavy wheels which rotate around the circular pan (A); ball mill is a hollow cylinder which revolves and tumbles hard media (B); rod mill is similar to the ball mill, but it is longer and uses rods rather than balls for media (C); and end view of a ball or rod mill which shows how the media revolves and tumbles (D).

Separation

Separation includes the methods used to isolate an element, compound, or mixture. The ceramic raw material industry uses some separation processes such as drying and/or magnetic separation.

Drying is used to remove surface moisture. This is necessary, since damp material tends to clog the machines. The drying is done in open-air sheds, or the material may be force-dried by hot air. A rotary drier is often used for force drying. This involves a large revolving drum. The material is fed through the drum as it revolves.

Magnetic separation involves the use of one or more magnets to remove the iron compounds. This is done while the material is either a powder or a liquid. Usually, the magnet is stationary. A conveyor moves the material either underneath or above the magnet.

Fig. 9-5. A laboratory-size dry pan or muller. The two wheels apply continuous pressure as they move in a circular path.

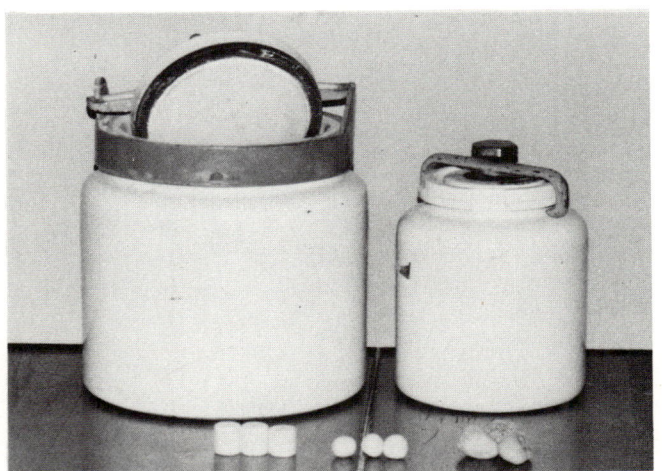

Fig. 9-6. Laboratory-size ball mills. Ball mill jars in operating position (top) and two sizes of ball mill jars—1 gallon (lower left) and 1 quart (lower right), with typical grinding media shapes (lower foreground). Note the use of a rubber gasket and clamp to seal the jars.

Unit 10: Utilization

The ways we use natural resources and human-made products to fulfill our needs is called *utilization*. We have three types of needs:

1. Physiological
2. Sociological
3. Psychological

The needs of the body, such as food and shelter, are *physiological* needs. The needs created by society, such as highways and schools, are *sociological* needs. Our personal needs, such as hobbies and worship services, are *psychological* needs.

Ceramic Raw Material Industry

The ceramic raw material industry produces a variety of refined materials. These materials are made into ceramic products to help meet our needs. The ceramic industries discussed earlier are the major consumers of refined ceramic raw materials. Therefore, the ceramic raw material industry usually provides materials to industries which then manufacture the products.

Manufactured Products

The transformation of material, as it is won or mined from the earth, into a usable form or product is truly manufacturing. Beneficiation by comminution, classifying, and/or separating produces a powdered material or manufactured product.

Primary Products

The products of the ceramic raw material industry are called *primary* products. They are purchased to be used in another or *secondary* product. For an example—powdered clay, a primary product, is purchased from a clay company and made into flower pots. The clay (in the final form of a flower pot, a secondary product), is sold again.

Secondary Products

The average person is usually not aware of the primary products made by the ceramic raw material industry. These products are not used directly to meet our daily needs. However, many of the secondary ceramic products are quite obvious in our daily lives. Dinnerware and glass containers are typical examples. However, we are not always aware of all the secondary ceramic products. For example, the refractories and plaster lath used inside furnaces and walls are hidden from our view.

Awareness of Products

Helping you to develop an awareness of *where* ceramic materials are used is the reason for including the topic of *utilization* in this book. The focus will be on *where* ceramic materials are used in manufactured products and/or construction.

Unit 11: Disposition

The processes we use to remove residues, waste, and products no longer desired are *disposition* processes. Disposition is accomplished either by reclaiming or by discarding.

Reclaiming is done by recycling and reusing. Both of these reclaiming methods recover the materials. Then the recovered materials may be transformed into new products. Glass bottles are a typical example of both recycling and reusing.

Discarding involves storing or discharging the product into the earth's crust, water, or atmosphere. Also, discarding is often unpopular and has caused problems in the past. Reclaiming reduces the need for discarding and the amount of waste remaining for disposal.

It is important that you know how ceramic products are disposed of after they have been used. Thus, disposition has been included here. Since ceramic products are physical objects, we will be concerned with disposing of solids.

Products of the ceramic raw material industry are not a disposition problem, since they have been physically refined. If it is necessary to dispose of, (for example, a powdered clay), it could be discarded by returning it to the place of origin—the earth's crust.

Disposition Problems

Superficial and subterranean winning can provide disposition problems. The major problem is disposing of the residues. These include disposing of the overburden and similar unwanted material.

Today, good ecology practices usually require that the overburden be stockpiled. Then, when the open-pit deposit has been removed, the overburden is used to help in returning the site to productive land. Trees or other crops are planted to complete the process of returning a site to its natural state.

Another problem exists with underground mines. The problem is disposing of the material that has to be removed before the deposit is reached. Sometimes, this material can be used to refill the underground areas remaining in the mine after winning. Refilling is a desirable practice, since it prevents future cave-ins. Worked-out underground mines may also be caved-in by using explosives.

Land Reclamation

Land reclamation is practiced by many of the ceramic industries. The land reclamation methods used are common to all those industries that remove materials from the earth's crust. The methods used include re-covering an area with fertile soil, planting grass or other crops, and/or planting trees. Reforestation is frequently used when the land is in a remote area.

The objective is to return the land to a productive state. Land reclamation benefits human life, plants, and animals. In fact, it helps to maintain the balance of nature. This is necessary to insure the continuation of life on this earth, especially as we know it today.

Summary

Procurement, transformation, utilization, and disposition are summarized in Fig. 11-1. The vertical flow sheet shows how this industry relates to the way we use materials. Also, the flow sheet emphasizes the fact that this is a basic industry—the *beginning* of the cycle. This industry is responsible for two major facets of the cycle—procurement and transformation. Everyone becomes directly or indirectly involved with the remainder of the cycle as we use and dispose of the goods.

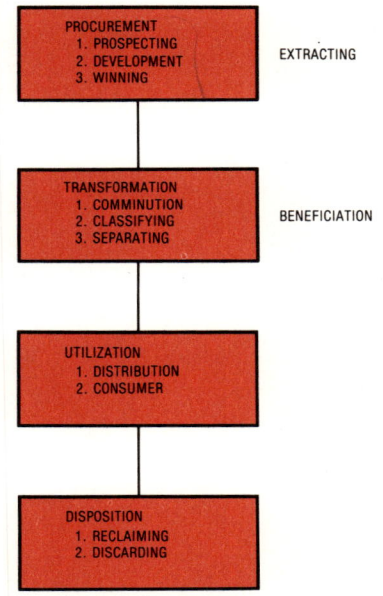

Fig. 11-1. A typical vertical flow sheet for the ceramic raw material industry. This industry is involved with the procurement and transformation phases. They usually deliver material in the form of a powder to the consumer.

SECTION IV

Stone Industries

Unit 12: An Introduction to the Stone Industries

The stone industries have three major divisions. The name of each division largely describes the products of that division. These divisions include:

1. Crushed and Broken Stone
2. Dimension Stone
3. Natural Gems

Crushed and Broken Stone

The two major products of the crushed and broken stone industry are shown in Fig. 12-1. The stockpiles behind the truck contain *crushed stone*. The truck is hauling *broken stone*. Note the difference in particle size. Crushed stone is merely broken stone that has been reduced *(crushed)* by machines. Thus, it receives its name and the reason why its particle size is smaller.

Dimension Stone

The dimension stone industry manufactures stone in definite shapes and sizes, such as *veneer* for walls. This typical product is the darker colored material below each set of windows in Fig. 12-2. Note how the color and texture of the dimension stone contrast with the brick, concrete, glass, and metal.

Natural Gems

Natural gems are those that are made into products such as jewelry. Their shapes are one of the three shapes in Fig. 12-3. The three shapes are:

1. Faceted gem—the enlarged side view of a brilliant cut, a diamond, for example
2. Cabochon—the actual size of a star gem in a high round "cab," a sapphire, for example
3. Baroque—actual size, a tiger eye, for example

All these shapes have three dimensions that make them more interesting and beautiful than could be done on a two-dimension (printed) illustration.

Organization

The three major stone industry divisions will be

Fig. 12-1. Products of the Crushed and Broken Stone Industry. (Courtesy National Crushed Stone Association)

46 / Modern Industrial Ceramics

introduced in this section in the order above. Basically, this sequence also develops from the simple to the complex products. The content will be organized as was done for the ceramic raw material industry to include:

1. Material science
2. Procurement
3. Transforming
4. Utilizing
5. Disposing

This organization of the content will be used for all the material-based industries discussed here. The terms were defined in Section III—*Ceramic Raw Material Industry*.

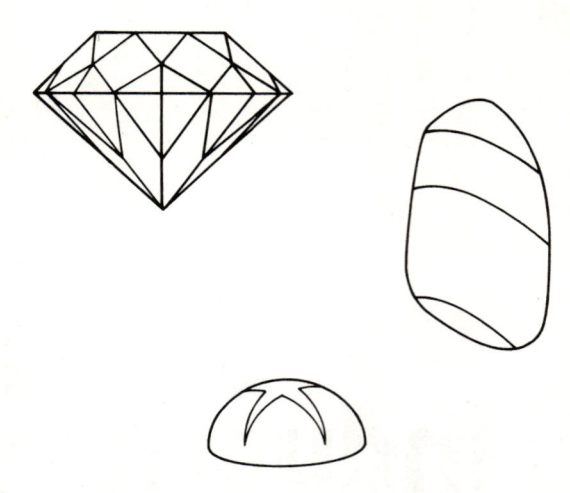

Fig. 12-3. The *natural gems* industry markets three basic types of products.

Fig. 12-2. A common use for dimension stone.

Unit 13: Crushed and Broken Stone Industry

The major products for this industry are crushed stone and broken stone. Both these products are used to benefit us in numerous ways. The layout of a plant which produces both products is shown in Fig. 13-1.

Material Science

Frequently, we use words like rock and stone. Often, these two words are assumed to be synonymous. However, they are *not* synonymous.

Rock

Rocks, as they occur in the earth's crust, are composed of one or more minerals. When rocks are composed of two or more minerals, they are heterogeneous rocks. We can call them rocks only as long as they exist in nature, the earth. Minerals, the formation of rocks, and the three types of rocks were discussed earlier.

Stone

Stone is rock that has been taken from the earth and used to benefit someone. Stone is often used in harmony with nature, as in a stone wall.

Broken Stone

Broken stone is rock that has been broken during the process of winning. It consists of large irregular fragments. Its size may exceed that of boulders. Boulders exceed 256 mm (10 in.) in diameter (Table 13-1). Broken stone is used as it comes from the quarry. Comminution processes are usually not used to produce broken stone.

Crushed Stone

Crushed stone consists of small irregular fragments with sharp edges that result from comminution. The name is derived from the fact that it is stone that has been crushed. Crushed stone sizes range from about 4 mm (0.2 in.) to 200 mm (7.9 in.) in diameter. Pebbles range from 4 mm (0.2 in.) to 64 mm (2.5 in.), while cobbles or cobblestone range from 64 mm (2.5 in.) to 256 mm (10 in.). A further comparison of sizes can be found in Table 13-1.

Gravel

Gravel and crushed stone are not synonymous. However, they are sometimes mixed together when used as coarse aggregate for concrete. This practice contributes to the confusion and misuse of the two terms.

Gravel consists of small irregular fragments of rock with rounded edges. It is a natural form of rock found on and in the earth's crust. The various weathering processes over many years produce gravel. For example, the wave

Table 13-1. Typical Sizes of Crushed and Broken Stone

Clastic Particle	Size Limit*
Boulder	Above 256 mm diameter	} Broken stone
Cobble	64 mm–256 mm
Pebble	4 mm–64 mm	} Crushed stone / Gravel
Granule	2 mm–4 mm
Sand	1/16 mm–2 mm	} Fines
Silt	1/256 mm–1/16 mm
Clay	1/256 mm or less

* Wentworth Scale (based on average particle size).
Note: 1 mm = 0.039 in.

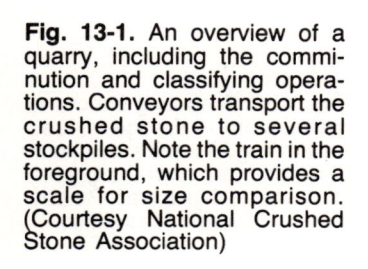

Fig. 13-1. An overview of a quarry, including the comminution and classifying operations. Conveyors transport the crushed stone to several stockpiles. Note the train in the foreground, which provides a scale for size comparison. (Courtesy National Crushed Stone Association)

action along a lake shore produces gravel. Also, the glaciers produced gravel and left it behind in large deposits. These deposits are often referred to as gravel beds. A floating sand and gravel plant are shown in Fig. 13-2. Gravel usually varies in size from about 5 mm (0.2 in.) to 30 mm (1.2 in.). For further comparison of sizes, see Table 13-1.

Raw Materials

The major raw materials used by the crushed and broken stone industry are:

1. Limestone
2. Basalt
3. Granite

The other raw materials, such as marble, quartzite, and sandstone, are grouped together under the heading of miscellaneous rock.

The approximate quantity of raw materials used each year is:

1. Limestone—70%, or more
2. Basalt—10%, or more
3. Granite—6%, or more
4. Miscellaneous rock—10%, or more

Limestone

Limestone is a sedimentary rock. Its chemical composition is calcium carbonate ($CaCO_3$). Actually, the industry uses the term "limestone" to mean *limestone family*. This term includes all the chemical types of limestone. The typical chemical forms of limestone are presented in Table 13-2. The number of chemical forms of limestone helps us understand why this industry uses so much of this material each year.

Table 13-2. Typical Chemical Forms of Limestone

Rock Name	Type of Rock	Chemical Composition
Limestone	Sedimentary	$CaCO_3$
Dolomite	Sedimentary	$CaCO_3 \cdot MgCO_3$
Limestone and chert*	Sedimentary	$CaCO_3$; SiO_2
Dolomite and chert*	Sedimentary	$CaCO_3 \cdot MgCO_3$; SiO_2

* Chert is formed from silica that has been removed from silicate minerals by chemical weathering. This silica is carried by water and eventually settles out. Thus, it may appear as banding or intrusion in limestone deposits.

Basalt

Basalt is a dark-colored, dense, and fine-grained igneous rock. It may even appear black. This rock is primarily composed of two minerals—pyroxene and plagioclase feldspar. The trade name of basalt is "traprock."

Granite

Granite is an igneous rock. Its color varies extensively, due to the different minerals present. The color may vary from black and white (a salt and pepper effect) to various shades of red and brown.

Miscellaneous Rock

A variety of rock is included in miscellaneous rock. The major rocks used include:

1. Sandstone
2. Marble
3. Quartzite
4. Slate

Fig. 13-2. A floating sand and gravel plant. A dredge is used for winning. Barges are used to store and transport the gravel. (Courtesy National Ready Mixed Concrete Association)

Crushed and Broken Stone Industry / 49

Sandstone is a sedimentary rock. Marble, quartzite, and slate are all classified as metamorphic rocks. Actually, *marble* is metamorphized limestone or dolomite. *Quartzite* was originally sandstone. *Slate* was shale before metamorphism occurred. Refer to Unit 5 for more information about basalt, granite, and other rocks.

Procurement

Procuring raw materials for the Crushed and Broken Stone Industry is diagrammed in Fig. 13-3. Prospecting, developing, and winning were introduced and discussed earlier. Prospecting and developing processes discussed earlier are used by this industry. Winning is usually limited to:

1. Superficial type—either open-pit or quarry (see Fig. 13-1)
2. Removal and use of overburden if applicable
3. Drilling and blasting to loosen rock from deposit (Fig. 13-4, 13-5)

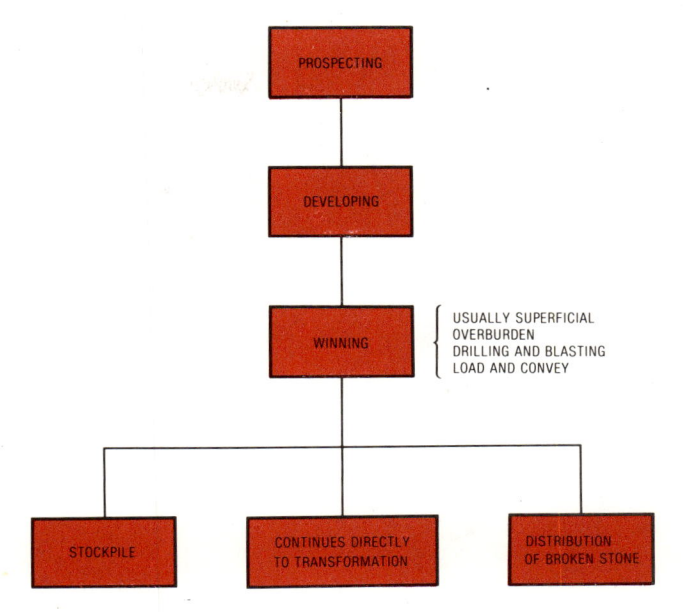

Fig. 13-3. A typical vertical flow sheet for procurement of raw materials used by the crushed and broken stone industry.

Fig. 13-4. Pneumatic drills being used to drill holes for explosives or "shots." (Courtesy National Crushed Stone Association)

Fig. 13-5. A "shot" of explosives blasting rock loose from the face of a quarry. Note the variety of sizes that occur with blasting. (Courtesy National Limestone Institute, Inc.)

50 / *Modern Industrial Ceramics*

Fig. 13-6. Loading broken stone from a recent "shot" with a power shovel. The stone will be taken to the primary crusher. There it will be transformed into crushed stone. (Courtesy National Crushed Stone Association)

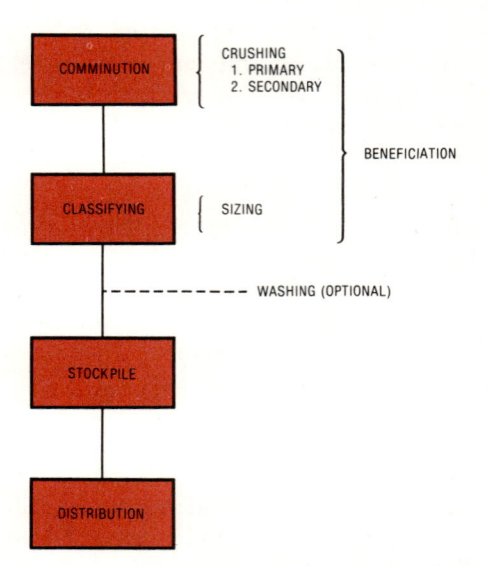

Fig. 13-7. A typical vertical flow sheet for the transformaion of procured rock by the crushed and broken stone industry. Crushed stone is the end product that is distributed to the consumer.

4. Loading and conveying of blasted rock (Fig. 13-6) to:
   a. Stockpile
   b. Transformation processes
   c. Distribution of broken stone to consumer

Transformation

Transforming procured rock is diagrammed in Fig. 13-7. Comminution is usually performed in two steps. Primary crushing (Fig. 13-8) may be done by a jaw crusher or a gyratory crusher. Secondary crushing may be done by a jaw crusher or a hammermill.

Classifying

Classifying the crushed stone into various particle sizes is usually done after the two crushing steps. Heavy steel rods, welded into a grid, and perforated plate steel are used for this operation. Several screens are stacked together with the coarse ones at the top. Belt conveyors are used to move crushed stone during the transformation processes.

Washing

Washing is a separation process. It is done to remove dust particles. Dust particles or *fines* result from the

Fig. 13-8. Feeding broken stone into the primary crusher. This plant uses a jaw crusher for primary crushing. Secondary crushing and classifying operations usually follow primary crushing. (Courtesy National Ready Mixed Concrete Association)

transformation processes. The *fines* range from very small particles to dust-size particles. Particles 4 mm (0.2 in.) or smaller are usually considered fines.

Washing may or may not be done. It is frequently done when the stone is to be used as coarse aggregate for concrete. *Fines* are a by-product of this industry.

Stockpiling

Classified and washed crushed stone is stored in outdoor stockpiles (Fig. 13-9). Stockpiling permits a plant to continue its sales during inclement weather or when breakdowns occur. Distribution is done primarily with various sizes of dump trucks.

Utilization

Products of this industry are used in both manufacturing and construction.

Manufacturing

The major use of broken stone in manufacturing is to produce crushed stone. Crushed stone made from limestone or dolomite may be used in manufacturing for:

1. Flux—iron and steel industry
2. Portland cement—a source of calcium oxide (CaO)
3. Lime—a source of calcium oxide (CaO)
4. Glass—a source of calcium oxide (CaO)

Precast concrete products are made from Portland cement, fine and coarse aggregates, and water. Crushed stone of various sizes is used for the coarse aggregate. Precast concrete products include brick, blocks, pipe, beams, and steps.

Construction

The chief use of broken stone (Fig. 13-10) in construction, or on site, is for *riprap*. Riprap is usually a

Fig. 13-9. Conveyors feeding classified crushed stone onto stockpiles. Note that each stockpile contains a different size of crushed stone. (Courtesy National Crushed Stone Association)

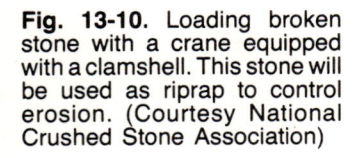

Fig. 13-10. Loading broken stone with a crane equipped with a clamshell. This stone will be used as riprap to control erosion. (Courtesy National Crushed Stone Association)

mixture of particle sizes. It is used for docks, seawalls, and jetties. River and canal banks are lined with riprap to reduce and/or prevent erosion. It is also used under bridges to prevent the growth of vegetation.

Coarse aggregate for concrete is a major use of crushed stone in construction. Think of all the concrete highways, streets, sidewalks, dams, and buildings that are built of concrete. Crushed stone is used for the bed or foundation of highways, streets, and railroads. The trade name for crushed stone used in this manner is road stone or road metal (Fig. 13-11).

Other Uses

Fines are used for agricultural lime if they are from the limestone family. Also, they are used for chicken grit and to sand icy highways.

These are only some of the many ways in which we utilize crushed and broken stone. Each of these uses benefits all of us.

Disposition

This industry does not have disposition problems. In fact, it does not dispose of anything! All of the products produced are sold, even the dust particles (fines). Thus, reclaiming and discarding are not applicable. If any of the products from this industry had to be disposed of, they could simply be returned to their place of origin—the earth.

Fig. 13-11. A stockpile of a larger size of crushed stone. This stone will be used for a roadbed. (Courtesy National Crushed Stone Association)

Unit 14: Dimension Stone Industry

The products of this industry are quite different from those produced by the crushed and broken stone industry. These products are also used differently. The procurement and transformation processes also differ extensively. Thus, these industries are viewed differently.

Material Science

Dimension stone includes the natural blocks and slabs of quarried stone. The blocks and slabs are cut to definite shapes and sizes.

Raw Material

The major raw materials used by this industry are:
1. Granite
2. Marble
3. Limestone
4. Sandstone
5. Slate

The type of rock, composition, color, and typical source of these raw materials can be found in Table 14-1.

Procurement

Procuring raw materials for the dimension stone industry is diagrammed in Fig. 14-1. The prospecting and developing processes used by this industry are the same as those discussed in Unit 8—*Procurement*.

Winning is called *quarrying* for this industry. Quarrying may be either superficial or subterranean (Fig. 14-2).

Cutting the rock into large blocks is the first step in winning. Then the blocks are removed from the quarry and transported to the mill, stockpile, or to a consumer. Most of the blocks enter a stockpile. They are then sold or transformed at the mill.

Cutting processes used in the quarry include:

1. Wire saw
2. Channeling machine
3. Drilling and light blasting (Fig. 14-3)
4. Drilling and broaching (Fig. 14-3)
5. Jet piercing (Fig. 14-4)

These processes are listed in Table 14-2. The blocks vary in size, but they may be 3 ft × 3 ft × 6 ft (0.9 × 0.9 × 1.8 meters).

The huge blocks are removed from the superficial quarry by a crane or derrick (Fig. 14-5). A crane is a movable machine. They are usually either self-propelled or mounted on a heavy-duty truck. The derricks are stationary machines. They usually rotate, and the boom can be raised and lowered.

Transporting the blocks to the stockpile, mill, and/or consumer is done by trucks and the railroads. Some firms have their own private railroads to convey the blocks from the quarry to their mill.

Transformation

Transformation of the blocks into products is diagrammed in Fig. 14-6. Cutting is the major transformation process used to make dimension stone products.

Fig. 14-2. An overall view of a granite quarry. This is superficial winning. Note the five derricks used to hoist the large blocks of granite out of the quarry. (Courtesy The Barre Granite Association)

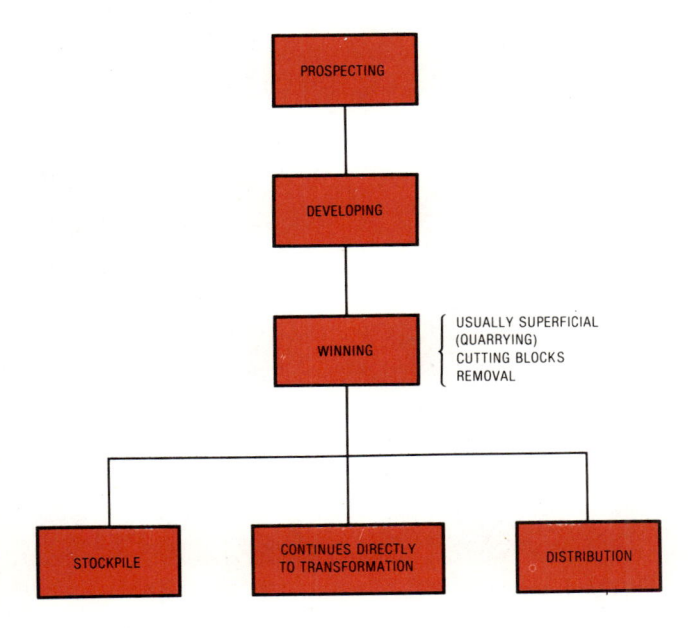

Fig. 14-1. A typical vertical flow sheet for procurement of raw materials used by the dimension stone industry.

54 / Modern Industrial Ceramics

Table 14-1. Raw Materials Used by the Dimension Stone Industry

Name	Type of Rock	Composition: Mineral or Chemical	Color(s)	Typical Source(s)
Granite	Igneous	Hornblende, Feldspar(s), Quartz, Mica(s)	Black and white, and red to brown	Vermont
Marble	Metamorphic	Calcite ($CaCO_3$) or dolomite ($CaCO_3 \cdot MgCO_3$) (Metamorphized limestone)	White, red to brown, black, green, and mixed colors	Universal
Limestone (or dolomite)	Sedimentary	Calcite ($CaCO_3$) Dolomite ($CaCO_3 \cdot MgCO_3$)	Gray	Universal
Sandstone	Sedimentary	Sand size particles usually high in silica/quartz (SiO_2)	Red, blue, gray, white, and shades of these	Northeast U.S.
Slate	Metamorphic	Hydrous alumina silicate ($Al_2O_3 \cdot 2SiO_2 \cdot 2H_2O$) (metamorphized shale)	Red, green, black, and shades of these	Vermont

Fig. 14-3. One method of quarrying the granite blocks involves drilling a series of holes into the stone. Then the web is removed by light blasting or broaching. The worker in the foreground is holding a drill tip. (Courtesy The Barre Granite Association)

Fig. 14-4. In jet piercing, a flame is used to cut the blocks from the quarry. This involves a mixture of oxygen and fuel oil. The tremendous heat generated causes the granite to flake away, resulting in a smoother, but wider cut. (Courtesy The Barre Granite Association)

Dimension Stone Industry / 55

Table 14-2. Cutting Processes Used in Winning Dimension Stone

Cutting Process	Description	Top View
Wire saw	1. Continuous twisted wires carry abrasive grain and water. 2. Abrasive grain does the cutting. 3. Resembles a band saw except usually more than two wheels are used.	Narrow slot with little waste
Channeling machine	1. One or more chisels cut with a chopping action. 2. Chisel(s) resemble a cold chisel.	Wider slot with more waste
Drilling and light blasting	Seldom used due to lack of control and limited, if any, knowledge of hidden rock faults.	1. Drill series of holes. 2. Blast web out between drilled holes.
Drilling and broaching	Broach resembles a flat or blunt chisel	1. Drill series of holes. 2. Broach out web between drilled holes.
Jet piercing	1. Used only on granite. 2. Heat, 4000°F (2182°C), causes a flaking action to occur. 3. Heat produced by burning fuel oil and oxygen.	Fairly smooth cut

Fig. 14-5. The giant blocks weighing 20 to 30 tons (18.1 to 27.2 metric tons) are hoisted from the quarry, using large wood or steel derricks. (Courtesy The Barre Granite Association)

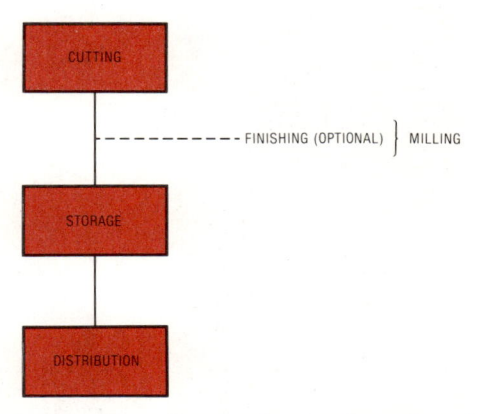

Fig. 14-6. A typical vertical flow sheet for the transformation of rough blocks of stone into dimension stone products. Some products, such as stone sills and curbstones, require only cutting. Other products require extensive finishing. These include memorial stones and statues.

56 / *Modern Industrial Ceramics*

Cutting the straight or flat surfaces may be done with a:

1. Diamond saw (Fig. 14-7)
2. Gang saw
3. Wire saw (Fig. 14-8)

Further details concerning each of these processes is provided in Table 14-3. These processes make a narrow cut with little waste.

Curves are cut by grinding away the stone with a silicon carbide wheel. Arcs, scrolls, and similar curves are cut in this manner.

The finishing of dimension stone is technically very fine cutting. Abrasive grain and water are used with a rotating head on a flat-surface polishing machine to produce a very smooth or glassy surface (Fig. 14-9). The abrasive grain is reduced in size as the finishing process progresses, until the degree of gloss desired is obtained. Silicon carbide grinding wheels are also used for this process.

Lettering, flowers, and similar figures are carved into the stone. This is usually done after the finish process is completed. Sandblasting, pneumatic chisels, and hand tools are used if carving is required. These processes are shown in Figs. 14-10, 14-11, and 14-12. Typical products that are carved include memorial stones, cornerstones, and statues.

The completed products may be stored or immediately distributed. Distribution is done by trucks and railroads.

Utilization

Products of this industry are used in both manufacturing and construction. See Unit 10—*Utilization* for definition of these terms.

Fig. 14-7. A diamond saw is used for the more precise cuts. The spray surrounding the blade is the water used to cool both the saw and the stone. (Courtesy The Barre Granite Association)

Manufacturing

The dimension stone products are manufactured products. Many are primary products used to make secondary products. Some dimension stone products are sold without complete finishing (Fig. 14-13). For example, these blocks are completed by a local firm as memorial stones. Sandblasting (Figs. 14-10 and 14-11) is usually used to complete the finishing.

Table 14-3. Cutting Processes Used to Transform Dimension Stone

Cutting Process	Description	Top View
Diamond saw	1. Resembles a circular saw without teeth. 2. Steel disc with a diamond matrix bonded to perimeter. 3. Types: a. Solid rim b. Slotted rim 4. Diameters: 4 in. to 48 in. (10.2 to 121.9 cm).	Narrow cut with little waste.
Gang saw	1. Resembles a series of parallel hacksaws pushed by pitman arm. 2. Several blades cut concurrently through a block. 3. Blade is a rectangular piece of steel. 4. Blade types: a. Plain b. Diamond matrix bonded to cutting edge. 5. Abrasive grit with water does cutting when a plain blade is used.	Blades may be adjusted to vary width of cuts.
Wire saw	1. Continuous twisted wires carry abrasive grain and water. 2. Abrasive grain does the cutting. 3. Resembles a band saw except usually more than two wheels are used.	Narrow slot with little waste.

Dimension stone as a primary product is used in the manufacture of other products such as:

1. Pool tables
2. Billiard tables
3. Blackboards
4. Tabletops
5. Lamp bases
6. Bookends

Dimension stone as a primary product is also used in construction. The construction uses of dimension stone include:

1. Walls
    a. Solid—garden and retaining wall
    b. Veneer—building interior and exteriors
2. Floors
3. Steps
4. Curbing
5. Fireplaces—mantles, facing, and hearths
6. Patios

Disposition

When disposition of dimension stone is necessary, it is frequently reclaimed. If a stone building is demolished, the stone is often put to new uses. This is a form of

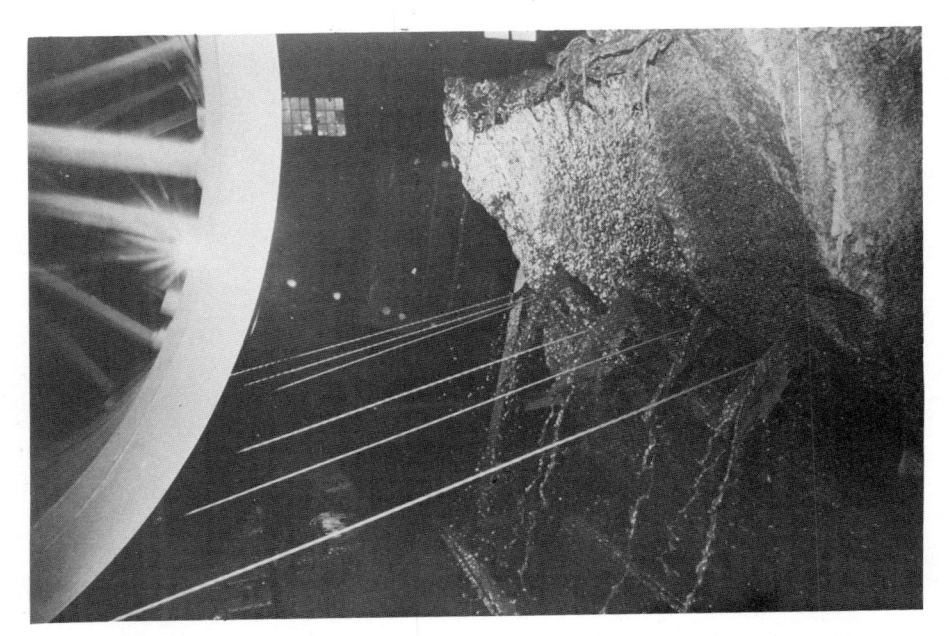

Fig. 14-8. Quarry blocks are taken to a saw plant where they are sawed into slabs of various thicknesses. Six wire saws are being used here to make parallel cuts. (Courtesy The Barre Granite Association)

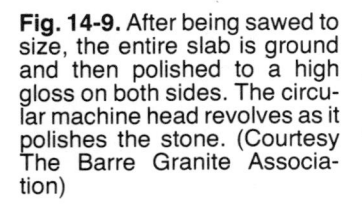

Fig. 14-9. After being sawed to size, the entire slab is ground and then polished to a high gloss on both sides. The circular machine head revolves as it polishes the stone. (Courtesy The Barre Granite Association)

58 / *Modern Industrial Ceramics*

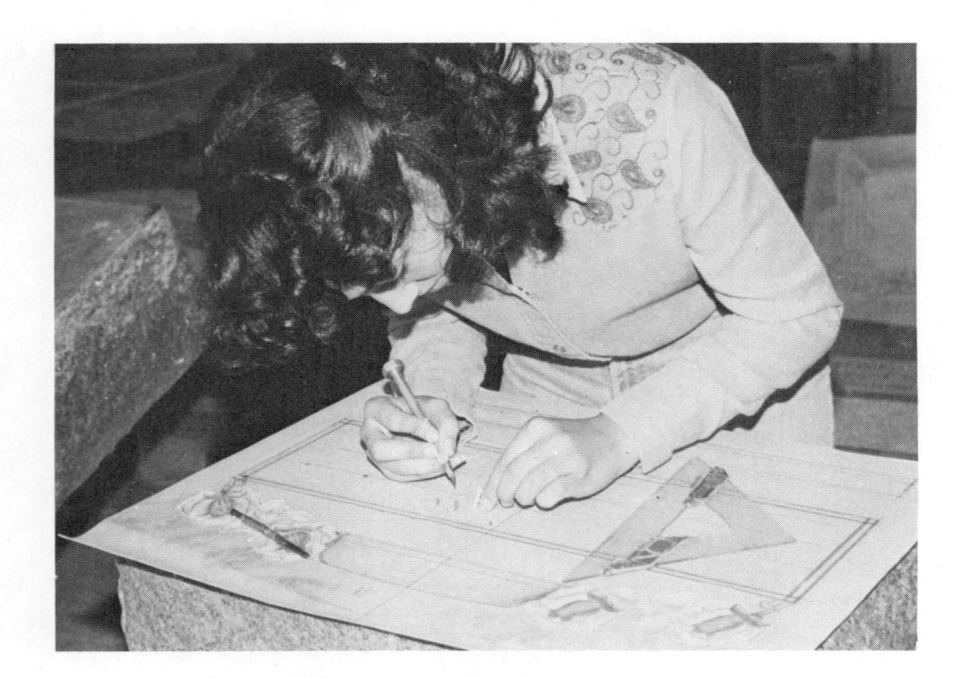

Fig. 14-10. A rubber stencil is applied to the face of a monument to prepare it for sandblasting. Then the design is cut by hand through the stencil.

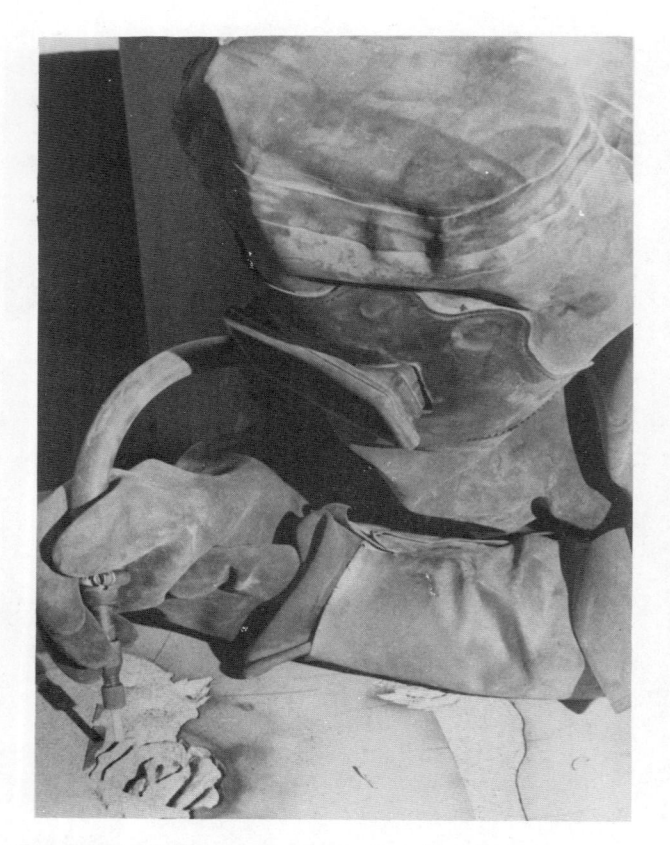

Fig. 14-11. Years of experience guide the hand of the sandblast carver. His skill and judgment determine the depth and shape of the flowers and other symbols on the monument. Note the protective clothing. (Courtesy The Barre Granite Association)

Fig. 14-12. Full and bas-relief are sculptured by hand, using a pneumatic chisel. Hand tools are also used.

Fig. 14-13. Inspecting a stone prior to shipping from the mill. A local firm will do the sandblasting needed to complete this monument. (Courtesy The Barre Granite Association)

reclaiming or reusing. For example, the stone veneer walls may become tabletops or fireplace facing. Dimension stone also may be reclaimed by crushing. Then, it could be used as crushed stone. These uses were discussed in Unit 13.—*Crushed and Broken Stone Industry.* Blocks of used dimension stone are also used like broken stone—for riprap.

Unfortunately, some used dimension stone is discarded. However, this may be beneficial as fill.

Summary

Dimension stone has a fine reputation for quality. Therefore, it is frequently used in construction of public buildings. However, manufactured materials constantly compete with dimension stone. The ceramic products also compete with dimension stone. Concrete, brick, and glass are the ceramic products which provide extensive competition.

Unit 15: Natural Gems

There are two groups of gems. Gems are either natural or synthetic. Natural gems are usually earthy, inorganic, nonmetallic materials. They may have been subjected to high temperatures during their formation.

This unit is limited to the natural gems. They are another form of stone, in a broad sense, called *gemstones*. Actually, all gems are minerals (except organic gems).

Material Science

A general definition for gems is something that is prized for great beauty and/or perfection. Gemstones must have three qualities:

1. Beauty or splendor
2. Durability or hardness
3. Be rare or not abundant

There are three types of gemstones:

1. Precious gems
2. Semiprecious gems
3. Organic gems

Precious gems must have the three qualities listed previously. There are only four precious gems. They are diamonds, rubies, sapphires, and emeralds. These gems are listed with their color and chemical composition in Table 15-1.

Table 15-1. Precious Gems

Name	Color	Chemical Composition
Diamond	Clear (usually colorless)	Carbon (C)
Ruby	Red	Corundum (Al_2O_3)
Sapphire	Blue	Corundum (Al_2O_3)
Emerald	Green	Beryllium aluminum silicate ($Be_3Al_2(SiO_3)_6$)

Semiprecious gems must have the same three qualities as precious gems. However, these gemstones are neither as hard nor as rare as the precious gems. Only about 100 of 1,500 different minerals have all the qualities required in gems. Garnet, opal, quartz, topaz, and turquoise are examples of semiprecious gems. Further information about these gems is provided in Table 15-2.

Organic gems are not ceramic. However, the general definition for gems does not exclude the organic gems nor even the synthetic gems. Organic gems have two possible origins—animal and vegetable. Examples of each type of origin are:

Table 15-2. Semiprecious Gems

Mineral	Gemstone Name	Color
Garnet	Almondine	Violet red
	Pyrope	Wine red
	Rhodolite	Red-purple
Opal	Opal	Colorless, gray or black with varicolored fire
Quartz	Amethyst	Purple
	Agate	Varicolored bands
	Jasper (Bloodstone)	Red, dark green with red
	Rock crystal	Colorless
	Rose	Pink
	Tiger eye	Bands of gold and brown
Topaz	Topaz	Colorless, sherry, gold, blue, red
Turquoise	Turquoise	Sky blue to green

1. Animal origin
    a. Pearl
    b. Coral
2. Vegetable origin
    a. Amber
    b. Jet

Procurement

Procurement of the natural gems is diagrammed in Fig. 15-1. Prospecting, especially for semiprecious gems, is

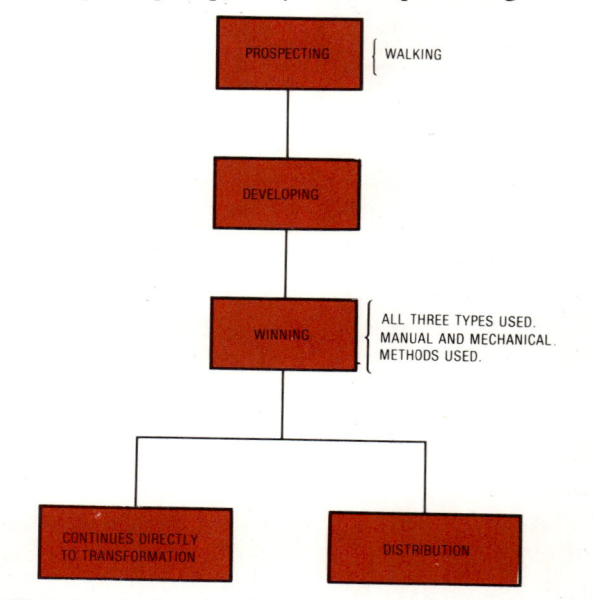

[Flow chart: PROSPECTING (WALKING) → DEVELOPING → WINNING (ALL THREE TYPES USED. MANUAL AND MECHANICAL METHODS USED.) → CONTINUES DIRECTLY TO TRANSFORMATION / DISTRIBUTION]

Fig. 15-1. A typical vertical flow sheet for the procurement of natural gems. The majority of natural gems are procured by one person or company and sold to another who does the transformation.

frequently done by walking. The type and method of winning varies extensively. All three types of winning are used by the industry.

The type of gem dictates the type of winning. For example, pearls are won by subaqueous winning, while diamonds are usually won by superficial winning. The extent of the deposit and the value of the gems may dictate the method of winning. For example, quartz is usually dug by hand with a pick and shovel. In contrast, diamonds are usually dug on a large scale with heavy industrial machines.

Transformation

Transformation of the natural gems is diagrammed in Fig. 15-2. The amount of transformation varies extensively, depending on the type of gem. Some gems, such as pearls, may only require classifying. Beneficiation processes, comminution, and separation are used occasionally. Most of the gems are subjected to cutting processes. These processes include:

1. Faceting
2. Grinding and polishing
3. Tumbling

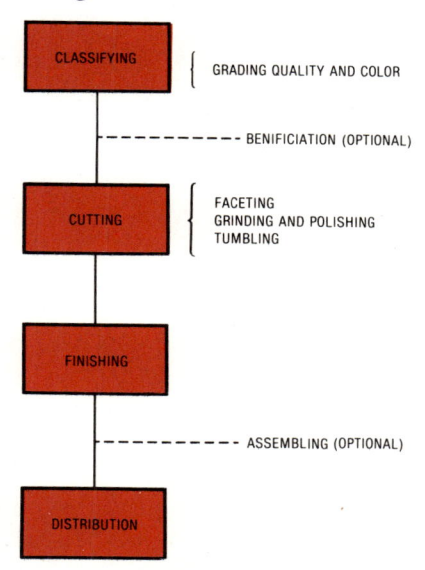

Fig. 15-2. A typical vertical flow sheet for the transformation of natural gems.

All the precious gems and some semiprecious gems are *faceted*. Faceting involves grinding and polishing a flat surface. An exterior surface of a faceted gem is composed of a series of these flat surfaces. The surfaces are ground at different angles. Each surface reflects light, which gives the faceted gem its "sparkle." The process is done with a *facetor*. This is a device which holds the gem at different angles for the grinding and polishing operations. Typical faceted gem cuts are shown in Fig. 15-3. Engagement rings usually have a brilliant cut diamond in their settings. The brilliant cut has 58 facets.

The semiprecious gems are often finished by cutting, grinding, and polishing into *cabochons*. Cabochons are gemstones that are cut convex and highly polished. The typical procedure for making a cabochon is:

1. Cutting slab
2. Lay out shape with template
3. Cut perimeter with trim saw
4. Attach to dopping stick
5. Grind perimeter and convex
6. Polish
7. Remove from dopping stick
8. Attach finding or setting

A variety of cabochon cuts are shown in Fig. 15-4.

Tumbling the semiprecious gems produces *baroques*. Tumblers vary in capacity from 3 pounds (1.4 kilograms) to 40 or more pounds (18.1 kilograms). They resemble ball mills (see Fig. 9-4). Baroques have smooth irregular surfaces produced by tumbling in abrasive grain and water.

Utilization

The natural gem products are manufactured products. Some products (for example, diamonds as loose abrasives) are used as they are manufactured. However, most of them are primary products used in the manufacture of other products. Some examples are:

1. Diamonds
    a. Saws
    b. Drills
    c. Grinding wheels
    d. Coated abrasives
    e. Phonograph needles
    f. Jewelry
2. Ruby and sapphire
    a. Bearings for watches and clocks
    b. Jewelry
3. Emerald
    a. Jewelry
4. Semiprecious and organic gems
    a. Jewelry

Disposition

When disposition of natural gems becomes necessary, they are reclaimed if possible. For example, the gems are removed from jewelry and placed in new settings.

Discarding is necessary when diamond saws, drills, etc. wear out. However, many of these products can be reclaimed for the metal they contain. Thus, the natural gems do not create any real disposition or ecological problems.

62 / *Modern Industrial Ceramics*

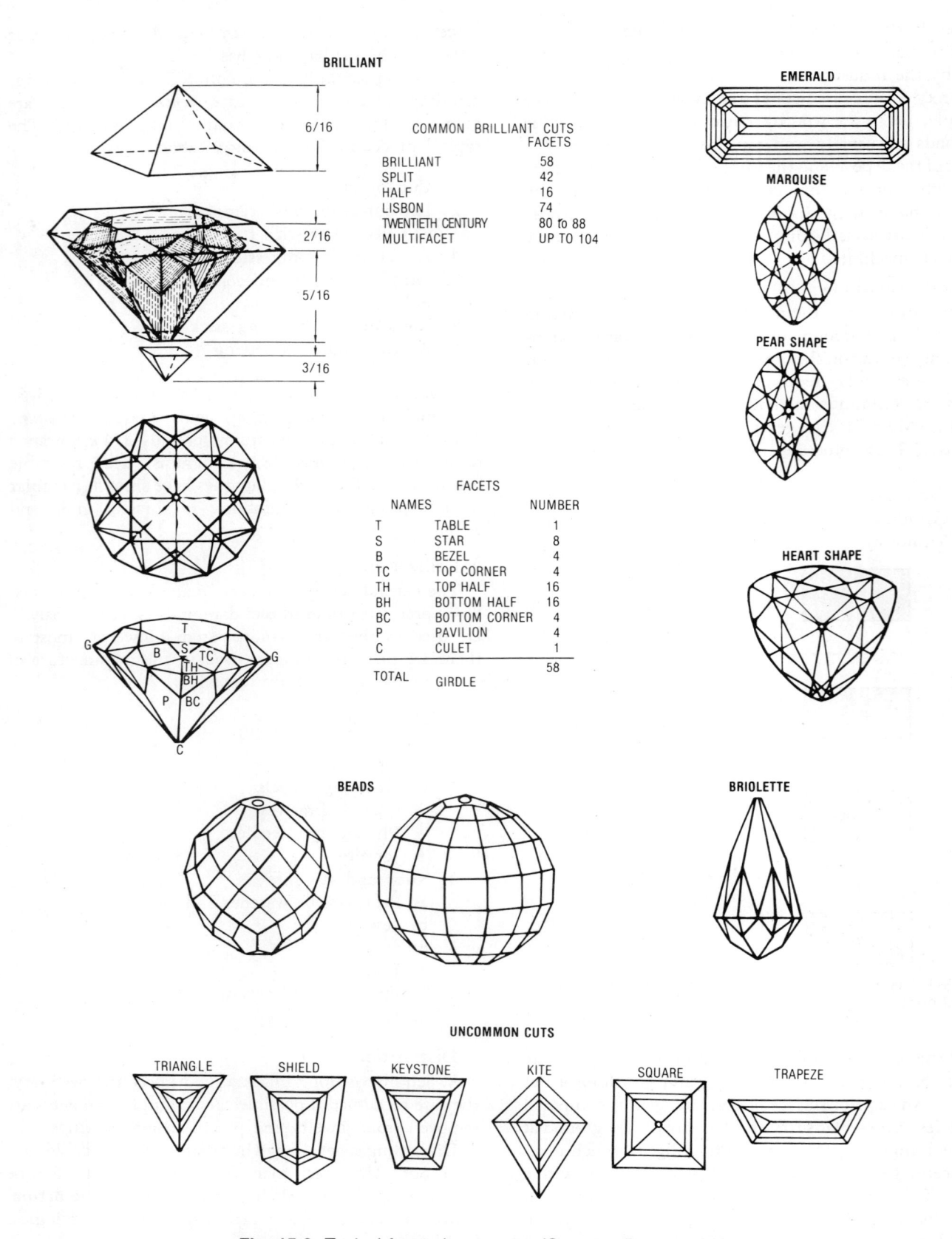

BRILLIANT

COMMON BRILLIANT CUTS	
	FACETS
BRILLIANT	58
SPLIT	42
HALF	16
LISBON	74
TWENTIETH CENTURY	80 to 88
MULTIFACET	UP TO 104

FACETS		
	NAMES	NUMBER
T	TABLE	1
S	STAR	8
B	BEZEL	4
TC	TOP CORNER	4
TH	TOP HALF	16
BH	BOTTOM HALF	16
BC	BOTTOM CORNER	4
P	PAVILION	4
C	CULET	1
TOTAL	GIRDLE	58

EMERALD

MARQUISE

PEAR SHAPE

HEART SHAPE

BEADS

BRIOLETTE

UNCOMMON CUTS

TRIANGLE   SHIELD   KEYSTONE   KITE   SQUARE   TRAPEZE

Fig. 15-3. Typical faceted gem cuts. (Courtesy Bureau of Mines)

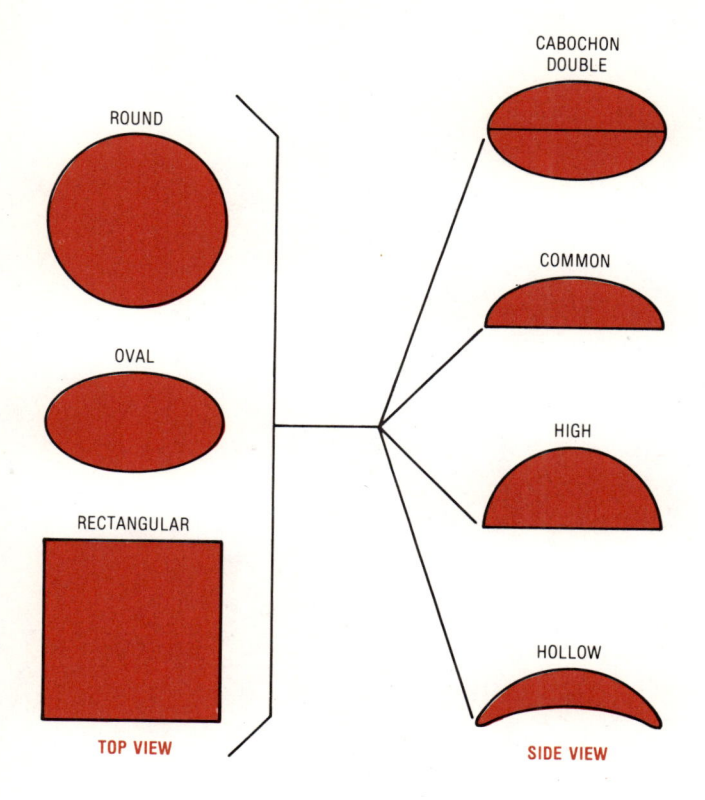

Fig. 15-4. Typical cabochon cuts. (Courtesy Bureau of Mines)

SECTION V

Refractory Industry

Unit 16: Material Science

Refractories provide a means for containing heat. This allows man to attain high temperatures for heat treating in the ceramic, metal, chemical, petroleum, and utilities industries. High temperatures, as previously defined, begin at 540°–590°C (1000°–1100°F). Frequently, refractories are used to retain heat which exceeds 1649°C (3000°F).

Refractories are important to several industries. Many of the ceramic industries require high temperatures. The metal, chemical, petroleum, and utilities industries also require high temperatures.

Definition

The technical definition of *refractoriness* is *the ability of a material to withstand high temperatures*. Technically, a refractory material will withstand 1482°C (2700°F) or higher. Ceramists describe this temperature as *Cone 19* (C/19). *Cone 19* refers to a pyrometric cone. These are discussed in the Section VI—*Kiln Industry*. We are interested in the refractoriness of only inorganic nonmetallic materials.

Materials

The major raw or natural materials used to manufacture refractory products are:

1. Diatomite
2. Chrome ore
3. Fireclay
4. Kaolin
5. Silica
6. Zircon

Other natural materials, rocks and minerals, are used to manufacture synthetic refractory materials. Some of these are:

1. Bauxite
2. Dolomite
3. Limestone
4. Mica
5. Sand
6. Magnesia (from sea water) and magnesite
7. Zircon

The major synthetic materials used to manufacture refractory products include:

1. Alumina
2. Calcium oxide
3. Ceramic fibers
4. Dolomite (dead burnt)
5. Graphite
6. Magnesia
7. Magnesite (dead burnt)
8. Mullite
9. Silicon carbide
10. Tricalcium aluminate
11. Vermiculite
12. Zirconia

Types of Refractories

All refractories, in a broad sense, may be broken down into types by their chemical nature. *Chemical nature* refers to the chemical reaction of the refractory at high temperatures.

This is important when a material being heat-treated touches the refractory. This occurs in rotary kilns, glass tanks, blast furnaces, and similar furnaces. Many materials become chemically active at high temperatures. Then, they will be eaten away if an acid material is in contact with a basic material. To prevent this, an acid should be used with an acid or a basic brick with a basic slag. See Table 16-1 for examples.

Table 16-1. Chemical Nature of Refractories*

Refractory Group	Example of Refractory Material
Acid	Silica
Basic	Calcium oxide
Neutral	Mullite or graphite
Super	Silicon carbide

* Chemical reaction at high temperatures

The *American Society for Testing Materials (ASTM)* has adopted classifications for the commonly used heavy refractories (Table 16-2). The heavy refractory brick or traditional refractories included in Table 16-2 are classified according to material. "Heavy," as used here, means a hard, dense product.

The term *fireclay brick* is included in Table 16-2. Fireclay brick and firebrick are identical. The word *firebrick* is frequently misused. Often, it is used as a common name for all refractories. This usage is incorrect. Firebricks are a refractory product made from fireclay. Fireclay is a natural clay that is resistant to high temperatures.

Firebrick may be either dense or porous. Firebricks that have a porous texture are called insulating firebrick (IFB). Insulating firebrick are made of fireclay mixed with a combustible material, such as sawdust. This mixture is formed, dried, and fired. The sawdust will burn out during firing, leaving small voids or pores. The porous structure increases the insulating efficiency of the firebrick. Thus, they are usually used to insulate kilns.

Products

The variety of refractory products produced can be confusing. However, they may be divided into three groups:

1. Heavy refractories
2. Insulating refractory materials
3. Castable refractories

The heavy refractories include standard (Fig. 16-1) and special shapes. The standard shapes include bricks or straights.

Insulating refractory materials are the insulating firebrick (IFB) and refractory ceramic fibers (Table 16-3). Insulating firebrick are used for both backup insulation and hot-face insulation.

The castable refractories are the mortars, plastics, concretes, and coatings. These are made in both powdered and wet or premixed forms.

Table 16-2. Classification of Heavy Refractory Brick (Heavy or Traditional Refractories)

Fireclay brick:
   Pouring pit (PCE* below 15)
   Low-duty (PCE 15)
   Medium-duty (PCE 29)
   Semisilica (min SiO_2, 72%)
   High-duty (PCE 31½):
     Regular
     Spall-resistant
     Slag-resistant
     High-fired
   Superduty:
     Regular
     High-fired
Kaolin (high-grog, high-fired)
High alumina:
   50% Al_2O_3 (PCE 34)
   60% Al_2O_3 (PCE 35)
   70% Al_2O_3 (PCE 36)
   80% Al_2O_3 (PCE 37)
   85% Al_2O_3
   90% Al_2O_3
   99% Al_2O_3 (Al_2O_3, 97% min)
     1. Sintered grain plus bond
     2. Fused grain plus bond
     3. Fusion cast
Silica:
   Conventional (0.5 to 1.0% Al_2O_3, TiO_2, and alkalies)
     Regular
     Hot-patch (more spall-resistant)
   Superduty (0.2 to 0.5% Al_2O_3, TiO_2, and alkalies)
     Regular
     Hot patch (more spall-resistant)
   Lightweight (lower thermal conductivity)
Basic:
   Magnesia
   Magnesia-chromite
   Chromite-magnesia
   Chromite
   Forsterite (2 $MgO \cdot SiO_2$)
   Dolomite (CaO, MgO)
     Fired, silicate-bonded
     Fired, direct-bonded
     Tar-bonded
     Fired, tar-impregnated
     Fusion cast
     Steel cased
Carbon:
   Carbon
   Graphite
Special:
   Zirconia
   Zircon
   ZrO_2-SiO_2-Al_2O_3 (fused glass-tank blocks)
   Silicon carbide
     Clay-bonded
     Frit-bonded
     Nitride-bonded
     Oxynitride-bonded
     Recrystallized
   Acidproof brick (dense, resistant to acids)

* Pyrometric cone equivalent

Material Science / 67

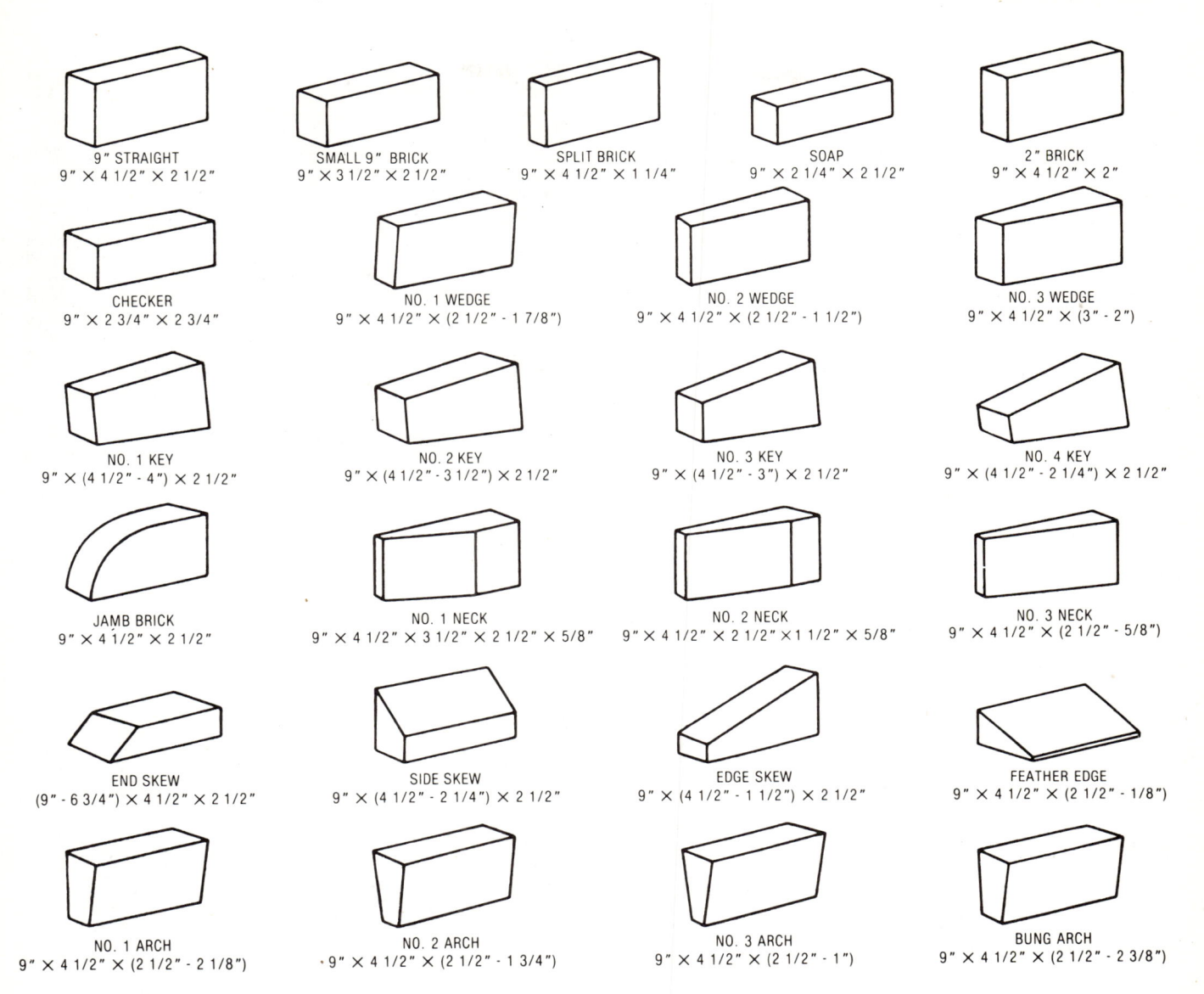

Fig. 16-1. Standard 9-in. refractory shapes manufactured in the United States. A refractory brick or straight is 2½ × 4 × 9 in. (6.4 × 10.2 × 22.9 cm).

Shapes shown (with dimensions):
- 9" STRAIGHT — 9" × 4 1/2" × 2 1/2"
- SMALL 9" BRICK — 9" × 3 1/2" × 2 1/2"
- SPLIT BRICK — 9" × 4 1/2" × 1 1/4"
- SOAP — 9" × 2 1/4" × 2 1/2"
- 2" BRICK — 9" × 4 1/2" × 2"
- CHECKER — 9" × 2 3/4" × 2 3/4"
- NO. 1 WEDGE — 9" × 4 1/2" × (2 1/2" - 1 7/8")
- NO. 2 WEDGE — 9" × 4 1/2" × (2 1/2" - 1 1/2")
- NO. 3 WEDGE — 9" × 4 1/2" × (3" - 2")
- NO. 1 KEY — 9" × (4 1/2" - 4") × 2 1/2"
- NO. 2 KEY — 9" × (4 1/2" - 3 1/2") × 2 1/2"
- NO. 3 KEY — 9" × (4 1/2" - 3") × 2 1/2"
- NO. 4 KEY — 9" × (4 1/2" - 2 1/4") × 2 1/2"
- JAMB BRICK — 9" × 4 1/2" × 2 1/2"
- NO. 1 NECK — 9" × 4 1/2" × 3 1/2" × 2 1/2" × 5/8"
- NO. 2 NECK — 9" × 4 1/2" × 2 1/2" × 1 1/2" × 5/8"
- NO. 3 NECK — 9" × 4 1/2" × (2 1/2" - 5/8")
- END SKEW — (9" - 6 3/4") × 4 1/2" × 2 1/2"
- SIDE SKEW — 9" × (4 1/2" - 2 1/4") × 2 1/2"
- EDGE SKEW — 9" × (4 1/2" - 1 1/2") × 2 1/2"
- FEATHER EDGE — 9" × 4 1/2" × (2 1/2" - 1/8")
- NO. 1 ARCH — 9" × 4 1/2" × (2 1/2" - 2 1/8")
- NO. 2 ARCH — 9" × 4 1/2" × (2 1/2" - 1 3/4")
- NO. 3 ARCH — 9" × 4 1/2" × (2 1/2" - 1")
- BUNG ARCH — 9" × 4 1/2" × (2 1/2" - 2 3/8")

Table 16-3. General Types of Insulating Refractory Materials

Type	Weight, psf	Use Limit, °F
Insulating firebrick (IFB)	varies	3300 +
Diatomaceous, block with asbestos and lime	23	1800-1900
Fibers:		
Slag wool block	15-20	1500-1700
Slag wool blanket	10	800-1000
Glass wool blanket	3	800-1000
Silica-alumina wool block	12-20	2000-2300
Silica-alumina wool blanket	3-8	2000-2300
Silica-alumina wool paper	8-10	2000-2300
Silica-alumina wool loose	3-10	2000-2300
Vermiculite:		
Block	19	1500-1600
Loose	10	1500-1600
Kaolin-gypsum	30	1600
Foamed glass	10	1000

Unit 17: Procurement

Procurement of raw materials for the refractory industry is diagrammed in Fig. 17-1. The prospecting and developing processes previously discussed are used by this industry. Many of these companies have a continuous prospecting program. Their objective is to find new sources of raw materials to replace deposits as they are depleted.

All three types of winning or mining may be used by this industry. However, superficial winning is the most commonly used method (Fig. 17-2). This type of winning is the least expensive of the three types. Some subterranean or underground mines are used. The method used least often is subaqueous winning. There are a few raw materials found underneath the surface of the water used by the refractory industry.

A different approach to subaqueous winning is used by the refractories industries. This method involves pumping sea water to a plant. Sea water contains magnesia, which is magnesium oxide (MgO). Magnesia is removed from the water. Then it is made into refractories.

Raw materials are stockpiled after the winning process (Fig. 17-3). This insures a continuous plant operation.

Fig. 17-2. Superficial winning of bauxite in Alabama. A crawler-type tractor with a front-mounted scoop is used for digging and loading. Trucks transport the bauxite to the plant. (Courtesy Harbison-Walker Refractories, Dresser Industries, Inc.)

Fig. 17-3. Stockpiling raw materials insures continuous plant operation. Here a clamshell-equipped crane unloads fireclay. (Courtesy A. P. Green Refractories Co.)

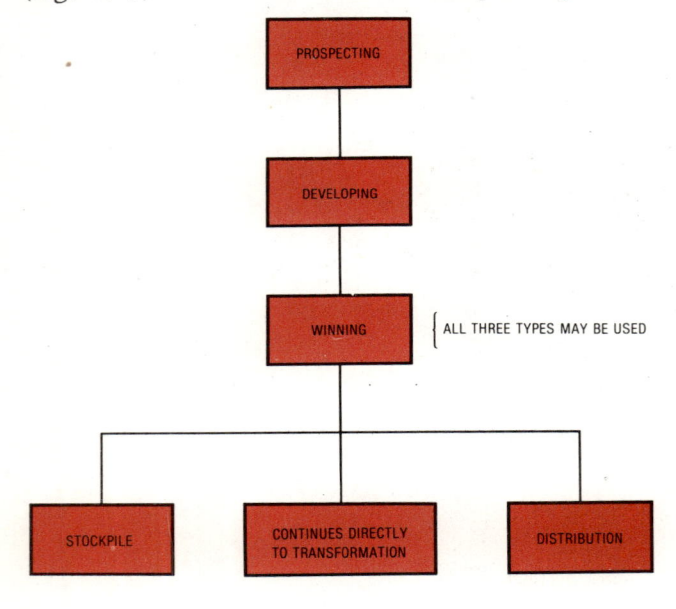

Fig. 17-1. A typical vertical flow sheet for procurement of raw materials used by the refractory industry.

Unit 18: Transformation

Typical transformation of materials into refractory products is diagrammed in Figs. 18-1 and 18-2. Heavy refractories, such as the bricks and shapes in Fig. 16-1, are the traditional products of this industry. We will use these as examples to simplify our discussion of transformation.

Beneficiation

Beneficiation of raw materials into usable form is the first transformation process. Raw materials such as fireclay and kaolin usually require comminution. This includes crushing and grinding into a powder. Then the powder is classified into particle sizes by sieves.

Some raw materials such as bauxite and dolomite require special processing before or during beneficiation. This usually involves a heat-treating process. For example, the heat drives off chemically combined water from bauxite. The heat removes carbon dioxide from the dolomite.

Batching

Batching is the ceramic term for mixing. The two types of batching are dry and wet. Dry batching is necessary if two or more powdered materials are to be combined. Then wet-mixing is done by adding water. Usually, water is added to assist the forming process.

Machines are used for batching. These include mixers, wet pan, pug mill (Fig. 18-3), and blunger. The amount of water, when added, determines the machine used.

Forming

Forming is used to shape materials without the removal or addition of material. The major forming methods used by the refractories industry are *pressing* and *extruding*. Slip casting, hand molding, and some other forming methods are occasionally used.

Pressing

Pressing (Fig. 18-4) is frequently used, since many refractory materials have little or no plasticity. The pressing methods include:

1. Dry pressing
2. Impact and vibrating pressing

Dry pressing usually involves a mix with 10% water or less. The water is a temporary binder and makes the mix granular. The mix is fed into a recessed steel die. Then it is compressed into the shape of the die. The pressed shape is ejected from the die and the cycle repeated. The compression is usually achieved by hydraulic pressure.

Impact and vibrating pressing causes the particles in the mix to pack into a dense uniform mass. This shaping is also done in a steel die.

The preceding presses may produce one or more shapes each cycle. The number produced in each cycle depends on the size of the press and size of the product being formed.

Extrusion

Extrusion forming is usually limited to clay in the plastic state. Powdered clay becomes plastic when less than 50% water, by weight, is added. Plasticity allows the clay to be formed and retain its shape when the forming force is removed.

The clay is batched by pugging and de-airing. Then the extrusion machine forces the plastic clay through a die. It emerges as a horizontal column. Then the column is cut to length. A wire cutter that works similar to an egg cutter is used.

Other Forming Methods

Some special refractory shapes are made by slip casting and hand molding. Powder clay, other refractory material(s), water, and an electrolyte are mixed together into a liquid. This mixture or *slip* is poured into a porous plaster mold. The mold absorbs the water and shapes the

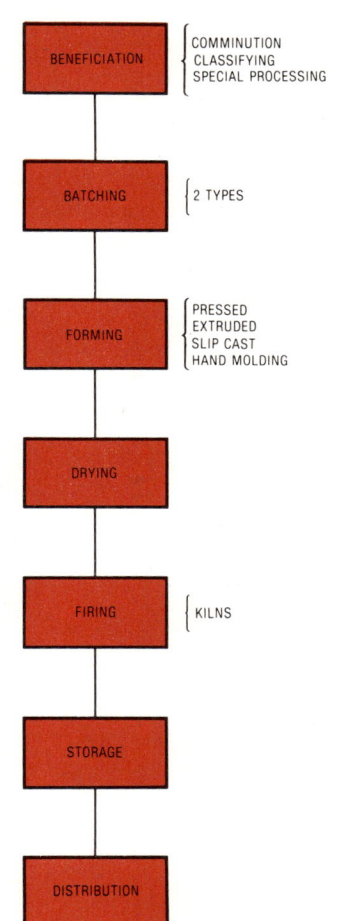

Fig. 18-1. A typical vertical flow sheet for the transformation of materials into refractory products, such as heavy refractories and insulating firebrick. Powdered castable refractory products would require only batching, bagging, and storage.

70 / *Modern Industrial Ceramics*

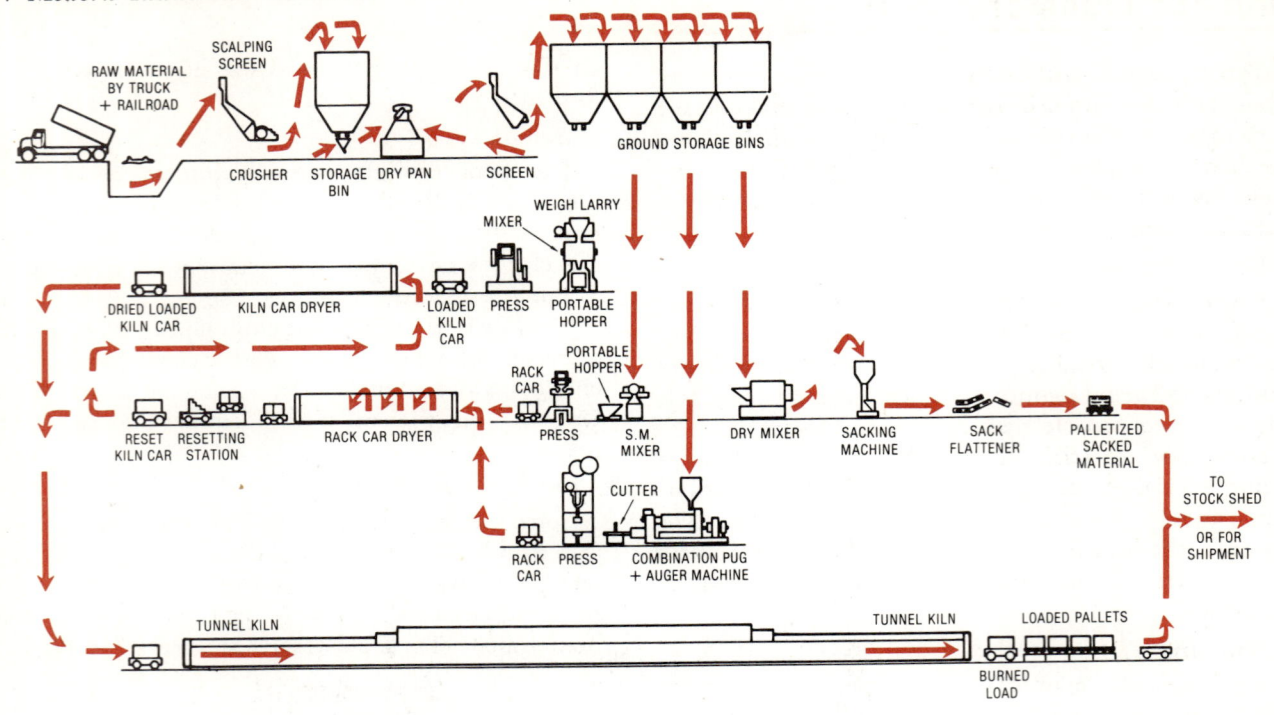

Fig. 18-2. Pictorial manufacturing flow sheet for heavy refractories, insulating firebrick (IFB), and powdered castable refractories. (Courtesy The Refractories Institute)

product. Then the shape is removed from the mold.

Hand molding involves pressing the plastic clay into a mold. The mold may be made of porous plaster to absorb the water. Then the shape is removed from the mold.

Drying

Many refractory products require drying after forming. The purpose of *drying* is to remove most of the water added to assist the forming process. Air or *force drying* is used. Force drying is faster. This is usually done in a heated chamber. Temperature and humidity in the chamber are controlled. A product does not have permanent strength after drying. This is achieved by firing (Fig. 18-5).

Firing

Firing is done in kilns at high temperatures to give the product permanent strength. Firing is a controlled heat-treating process. This process may require 24 or

Fig. 18-3. Fireclay and water are wet-mixed into a plastic state by a pug mill. This process precedes extrusion. (Courtesy A. P. Green Refractories Co.)

Fig. 18-4. Forming refractory bricks or straights by pressing. This machine produces two bricks in each pressing cycle. (Courtesy A. P. Green Refractories Co.)

Fig. 18-5. Air drying refractory shapes on a plant floor before firing. (Courtesy Harbison-Walker Refractories, Dresser Industries, Inc.)

more hours. The firing temperature ranges from 2000° to 3200°F (1093° to 1760°C). Periodic and continuous kilns are used for firing. Both are discussed in the following Section VI—*Kiln Industry*. *Periodic kilns* are often used for small quantities or special refractories.

Continuous kilns are used for large volume or mass production. Specifically, the tunnel kiln with cars is used (Fig. 18-6). For example, refractory bricks are stacked on cars. Then a continuous series of cars is moved through the kiln. While in the kiln, they pass through preheating, firing, and cooling zones. After firing, the refractory bricks are transferred to pallets.

The refractory products may be stored in a warehouse or immediately distributed. Distribution is done by trucks and railroads.

Refractory Products

The three refractory products groups previously discussed are:

1. Heavy refractories
2. Insulating refractory materials
3. Castable refractories

The preceding transformation processes are used to manufacture *heavy refractories.* Insulating firebricks (IFB), an insulating refractory material, are also manufactured using the preceding processes.

Refractory ceramic fibers, another *insulating refractory* material, require different transformation processes (Fig. 18-7). Basically, these fibers are manufactured by dry batching raw materials, such as alumina, silica, and borax glass. These materials are melted at a temperature of 3600°F (1982°C) in an electric furnace. The molten material is poured into a jet of steam. This produces a fluffy bulk fiber. The fiber ranges up to 10 microns or 10,000 millimicrons in diameter and 1½ inches (3.8 cm) in length. Fibers may be used in bulk or consolidated into wool, blanket, or similar products. All these products are lightweight and usually flexible.

Castable refractories also require different processing. Powdered castable refractories are dry batched, bagged (Fig. 18-8), and stored for distribution (see Fig. 18-2). Wet, or *premixed,* castable refractories are dry and wet batched, packed in containers, and stored for distribution.

Fig. 18-6. A continuous tunnel kiln with a car exiting. These are basic refractory brick. (Courtesy General Refractories Company)

72 / *Modern Industrial Ceramics*

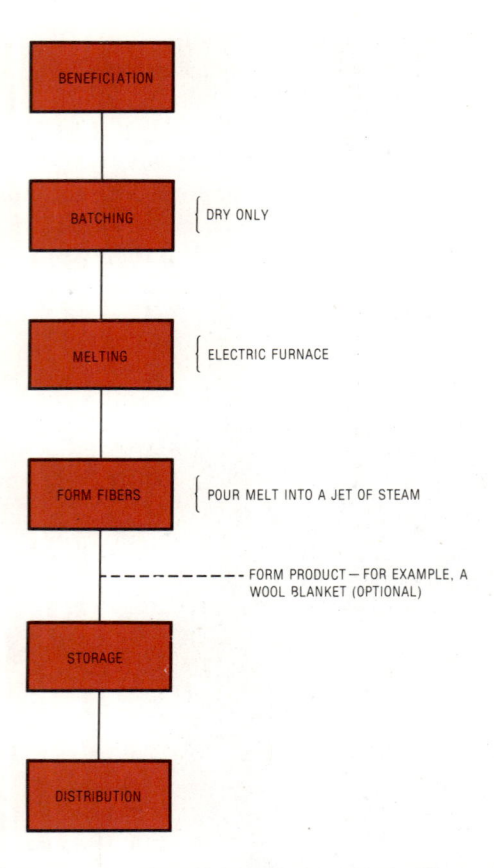

Fig. 18-7. Typical vertical flow sheet for the manufacture of refractory ceramic fibers. The finished product is lightweight and usually flexible.

Fig. 18-8. Packaging a dry castable refractory in 100-lb paper bags. The bags are stacked on pallets for storage and shipping. (Courtesy General Refractories Company)

Unit 19: Utilization

Refractory products are used in both manufacturing and construction. There are many applications for these products. Some of them are used in our homes and others are used in industry. The major consumers of refractory products are the ceramics, chemicals, metals, petroleum (Fig. 19-1), and utilities industries.

Fig. 19-1. A 5-in. thickness castable refractory lining being applied to a catalytic cracking unit at a petroleum refinery plant. The castable refractory is being applied by a pneumatic gun. (Courtesy General Refractories Company)

Fig. 19-2. A front-loading electric periodic kiln. Insulating firebricks (IFB) are used for the interior of the kiln. A special refractory shape holds the kiln elements. (Courtesy American Art Clay Co., Inc.)

Manufacturing

Refractory products are manufactured products. Many are primary products used to make secondary products. For example, firebricks are used to insulate the smaller movable periodic kilns (Fig. 19-2). Your school may have one or more of these kilns. They are used to fire porcelain enamels and pottery. Furnaces and some hot-water tanks for the home have refractory fireboxes.

Kiln furniture (Fig. 19-3) is a manufactured refractory product used inside kilns, including shelves, posts, and stilts. These products are used by hobbyists, schools, and industries.

Construction

Insulating firebrick, refractory brick, and other shapes are used in industry to build the large industrial kilns and furnaces. These are usually built on site, due to their large size.

The continuous-tunnel kiln introduced in Unit 18 is an example. Some other types of kilns built on site for use by the ceramic industries include:

1. Rotary kiln (Fig. 19-4)
2. Vertical kiln
3. Glass tank

Fig. 19-3. A shuttle-type periodic kiln with transfer car used to fire kiln furniture. The kiln furniture on the kiln car includes saggars, setters, and shelves. (Courtesy Bickley Furnaces, Inc.)

74 / *Modern Industrial Ceramics*

Fig. 19-4. The interior of a rotary kiln is lined with refractory brick. Note its size. This kiln slowly rotates when operating. The Portland cement, gypsum, and lime industries use this type of continuous kiln. (Courtesy Harbison-Walker Refractories, Dresser Industries, Inc.)

Fig. 19-5. A metal refractory anchor is attached to a stainless steel J-bolt that is bolted to the metal shell of a soaking pit at a steel mill. Workers ram the plastic refractory around the anchors to hold them firmly in place. (Courtesy General Refractories Company)

The large furnaces used by the metal industries are built on site (Fig. 19-5). Blast furnaces (Fig. 19-6), open-hearth, and the basic oxygen furnace are typical of these furnaces. The large chimneys for industrial plants are also built of refractory brick and shapes.

In the home, refractory bricks are used to line the interior of a fireplace. A hidden refractory product is used in the chimney. This is the flue liner.

Fig. 19-6. Blast-furnace troughs are lined with graphitic plastics and ramming mixes to provide a longer service life. They are quickly resurfaced back to original lines with the same material, as shown. (Courtesy General Refractories Company)

Unit 20: Disposition

Disposition of refractory products occurs after manufacturing and when they are broken or worn out. All the refractory products are solids (Fig. 20-1). When disposed of, they are considered solid waste. Disposition of these products is accomplished either by reclaiming or by discarding.

Reclaiming

Products are reclaimed either by reusing or by recycling. Both types of reclaiming are recommended.

Reuse

Some of the refractory products are designed to be reused. This is true for element holders and kiln furniture, such as stilts, posts, and shelves. These products are not disposed of until they are either broken or contaminated. Then they can only be recycled, if possible, or discarded.

Most of the refractories removed from a kiln that is to be rebuilt are worn out. These have to be discarded. Occasionally, some of them can be reused in the rebuilt kiln.

Recycle

Plate pins are used in *saggars* to support glazed plates when fired. A small amount of glaze flows onto the plate pins during firing. After firing, the pins are removed and inspected. The pins are recycled by cutting off the glazed end. When the pins become too short they can be substituted for stilts.

Kiln building and repairing produces odd pieces of unused firebrick (Fig. 20-2). These can be crushed, sized, and used as grog.

Pieces of insulating firebrick can also be recycled as trivets. These are used to hold pieces of metal during the firing of porcelain enamels.

Broken kiln posts and shelves can be recycled (Fig 20-3). These are simply cut down in size. A diamond saw and coolant are used to cut these hard, dense products.

Unusable broken kiln furniture, unused pieces, contaminated, and worn out refractories can be recycled. This can be used as riprap and hard fill.

Discarding

High temperature and abrasion during use cause many refractory products to eventually break down. This is easily observed with insulating firebrick, since they are softer than refractory brick.

Discarding is the easy way to dispose of these refractory products. This has been the traditional practice. However, all of these products are solids. Many are hard, dense solids. Thus, they can be put to beneficial use—riprap and hard fill, instead of being discarded.

Fig. 20-2. These odd pieces of unused insulating firebrick can be recycled. They can be crushed, sized, and used as grog.

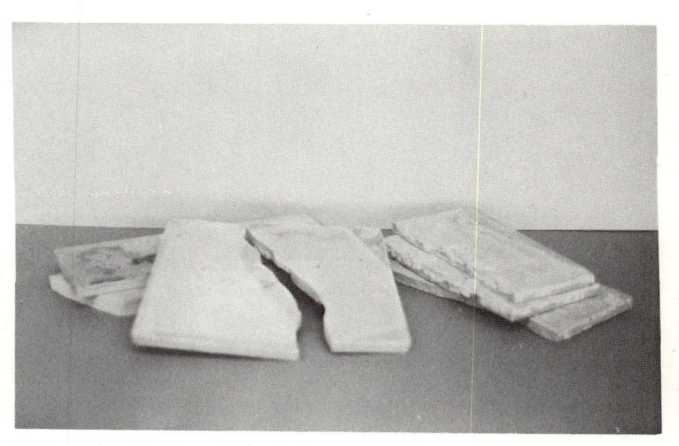

Fig. 20-3. Broken kiln shelves can be recycled. They are simply cut down in size.

Fig. 20-1. All the refractory products are solids. These pieces of insulating firebrick, used firebrick, and refractory brick, and element holder can be reclaimed by recycling.

SECTION VI

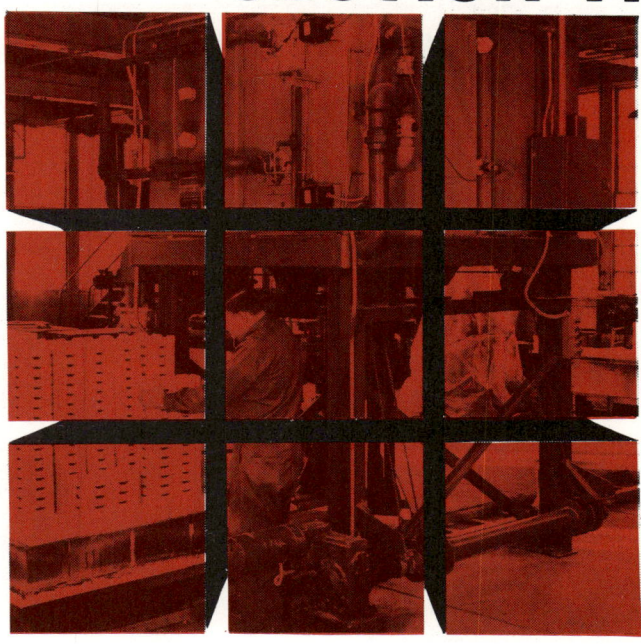

Kiln Industry

Unit 21: Introduction

Kilns are very important to the ceramic industries. They provide the means for high-temperature heat treatment of ceramic products.

Definition of Kiln

A *kiln* is a refractory-lined chamber. This chamber is heated to fire or heat-treat ceramic products.

All the kilns have the following basic components:

1. Refractory chamber
2. Heat source and controls
3. Provision for loading and unloading
4. Temperature measuring device (optional)

Use of Kilns

The ceramic industries use kilns for several different heat-treating processes. The major heat-treating processes are listed in Table 21-1. A product example for each process is also included in the table. Each heat-treating process will be defined in the following sections.

Types of Kilns

There are two major types if kilns—*periodic* kilns and *continuous* kilns. The periodic kilns operate on a cycle. The continuous kilns operate constantly. Each will be defined further and illustrated in Units 22 and 23.

Table 21-1. Major Heat-Treating Processes Used By The Ceramic Industry

Heat-Treating Process*	Product Example
1. Calcining	Plaster
2. Burning	Portland cement
3. Firing or form glass bond	Clay bricks
4. Melting	Glass
5. Anneal	Glass
6. Form crystal structure	Silicon carbide

*Processes are listed in the sequence they will be discussed in the following sections.

Unit 22: Periodic Kilns

This unit is an introduction to the discussion of the various types of periodic kilns used by the ceramic industries. The discussion is limited to the major periodic kilns now in use.

Definition

Periodic kilns are designed to operate in cycles. A typical operating cycle includes:

1. Load or stack
2. Fire
3. Cool
4. Unload or unstack

This operating cycle is typical for many of the refractory, clay, glaze, and abrasive products.

The size of periodic kilns varies extensively. They range in size from the size of a toaster to as large as a small building.

Major Types of Kilns

The major periodic kilns in use today are:

1. Electric
2. Updraft
3. Downdraft
4. Muffle
5. Shuttle
6. Glass pot
7. Day tank
8. Kettle
9. Electric resistance furnace
10. Electric arc furnace

A description of each type of kiln and its use follows.

Electric Kiln

The smaller periodic electric kilns are commonly used by schools and hobby ceramics. They are used to fire clay and glaze products. Firing of the porcelain enamels, such as copper enameling, is another use. The larger electric kilns are used by many ceramic industries.

Electric kilns are operated on either 110 volts or 220 volts. Two types of electrical resistance elements are used to heat these kilns. Wire-wound and silicon carbide resistance rods are used. The wire-wound elements are made of either nichrome or kanthal. The kanthal elements are the most common wire-wound elements in use today. Wire-wound elements are similar in appearance to a long stretched-out spring.

A door for loading the electric kilns is found at either the front (Fig. 22-1) or the top (Fig. 22-2).

During firing, the atmosphere inside an electrical kiln is oxidizing. The term *oxidizing* means there is sufficient oxygen present for firing. Lack of oxygen causes some glaze colors to change in color.

Fig. 22-1. A front-loading electric periodic kiln. The wirewound heating elements are inside the refractory holders located on the kiln sides and floor. The temperature inside the kiln is indicated on the pyrometer (lower left front). The pyrometer is connected to a thermocouple that protrudes inside the kiln. (Courtesy American Art Clay Co., Inc.)

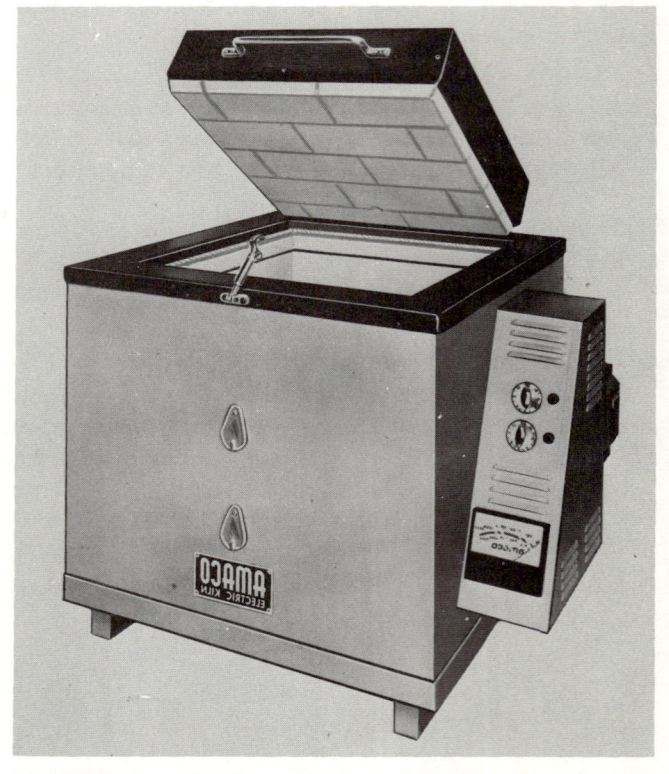

Fig. 22-2. A top-loading electric periodic kiln. Two covered peepholes are located on the front of the kiln. They are used to visually check the kiln interior when firing. (Courtesy American Art Clay Co., Inc.)

Updraft Kiln

The term *updraft* indicates the location of the flue in a combustible-fuel kiln. The flue is located in the top of the kiln. Thus, the draft or flow of heat is vertical, as indicated in Fig. 22-3. The fuel used to heat the kiln may be natural gas, propane, oil, coal, or wood. Gas is the most common fuel used today. The updraft kiln may be either front- or top-loading.

Downdraft Kiln

The *downdraft* kiln is the opposite of an updraft kiln. The flue is located at or in the kiln floor. Thus, the heat rises to the kiln ceiling and then flows downward to the flue, as indicated in Fig. 22-4. The same combustible fuels are used as with the updraft kiln. Usually, the downdraft kiln is a front-loading kiln.

Updraft and downdraft kilns are used when a reduction firing is needed. During firing, the normal kiln atmosphere is oxidizing. The flow of air into the kiln can be reduced. This causes the kiln atmosphere to become short of oxgyen, which is reduction. A reduction atmosphere causes unique color changes to occur with some clay and glazes.

Muffle Kiln

This kiln has a muffle or extra refractory wall inside its chamber. A space between the refractories and muffle permits the heat and fumes from the combustible fuel to flow upward, without contacting the ware being fired. The muffle protects the ware from any contamination (Fig. 22-5).

Saggars provide another method of protecting ware from contamination. They are used in combustible-fuel kilns without muffles. A saggar is a box, usually made of fireclay. The ware is placed inside the saggar, and several saggars may be stacked in the kiln (Fig. 22-6).

Shuttle Kiln

A part of the shuttle kiln is shuttled or moved to allow loading and unloading. The movable part of the kiln varies. For example: the entire kiln, except the bottom, moves vertically in an elevator-type kiln; or the bottom moves vertically (see Fig. 22-6); or the kiln bottom and front are a car on rails that is pulled outward. Another variation allows the kiln car bottom to be moved through when doors are opened at each end (Fig. 22-7). This type of kiln has at least two kiln cars that allow one to be unloaded and restacked while the other loaded car is inside the kiln for firing. Shuttle kilns are heated with either electricity or a combustible fuel.

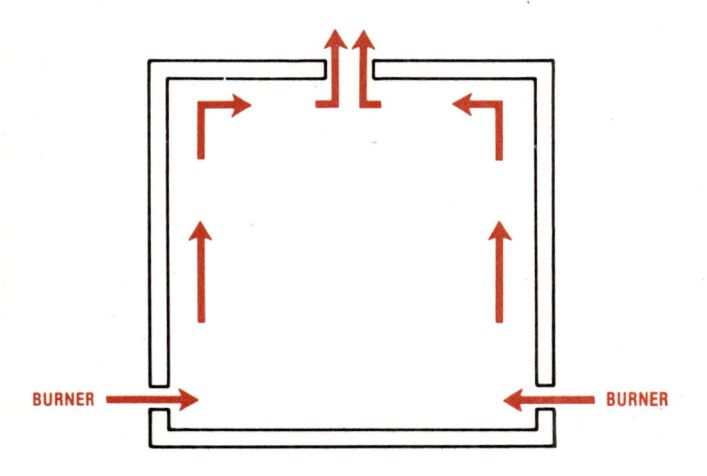

Fig. 22-3. Full-section elevation drawing of a typical updraft kiln. This kiln is heated by combustible fuels. The arrows indicate the flow of heat and fumes inside the kiln.

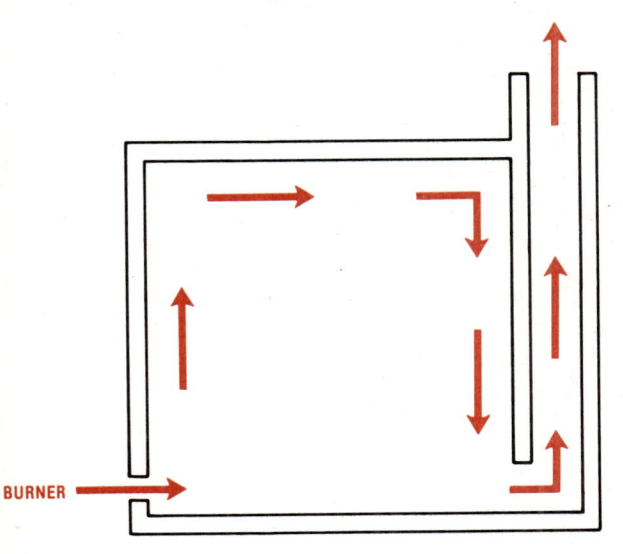

Fig. 22-4. Full-section side elevation drawing of a typical downdraft kiln. This kiln is heated by combustible fuels. The arrows indicate the flow of heat and fumes inside the kiln.

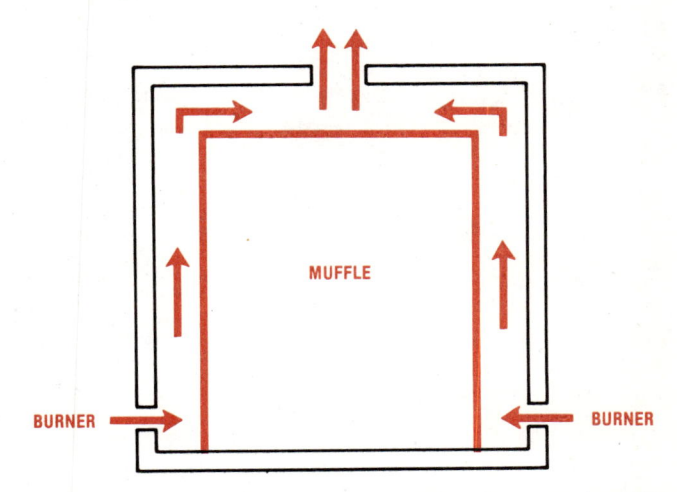

Fig. 22-5. Full-section elevation drawing of a typical muffle kiln. Ware is stacked inside the muffle before firing. The arrows indicate the flow of heat and fumes. Note how the muffle protects the ware being fired.

80 / *Modern Industrial Ceramics*

Fig. 22-6. A bottom-loading elevator-type shuttle kiln. A winch (right foreground) is used to raise the kiln bottom. Ware to be fired is inside the saggars being stacked on the kiln bottom. (Courtesy Bickley Furnaces, Inc.)

Fig. 22-7. A group of four shuttle kilns with a transfer car system. These kilns are 18 ft. (5.5 meters) in width and 50 ft. (15.3 meters) in length. Each shuttle kiln holds three cars across the width and two cars along the length for a total capacity of six cars. A typical firing cycle takes 36 hours cold-to-cold. (Courtesy Bickley Furnaces, Inc.)

Glass Pot

Glass pots are containers used to melt small or special batches of glass. They are either open or closed containers. The open type of glass pot resembles a crucible. The closed type of glass pot has a roof with an opening at the top of a side wall for charging and removing the glass. Both types of glass pots are placed inside a furnace that provides the heat.

Day Tank

Day tanks are also used to melt batches of glass. Usually, there are gas burners that heat the top surface of the glass.

A day tank holds enough glass for one or more persons to work in a normal day. Another batch of materials is added at the end of the day. These materials are melted overnight. A new batch of glass is ready for the next working day. Thus, the term *day tank* has been applied.

Kettle

A kettle is used to make plaster from gypsum. Its shape resembles a large kettle with a cover. The kettles can hold 10 to 20 tons (9.1 to 18.2 metric tons) of gypsum. Some kettles may be 20 ft (6 meters) or more in height. The heat treating or calcining of gypsum is a low-temperature (340°F or 171.1°C) process.

Electrical Resistance Furnace

The abrasives industry uses this type of furnace to make silicon carbide. The furnace has a long low rectangular or trough-like shape. A large graphite electrode is found in each end for heating by electrical resistance. Refractory brick are used for the bottom, sides, and ends. The top is open to allow release of fumes. A charge is placed inside the furnace, with coke through the center to connect the electrodes. Silicon carbide crystals are formed through the center during the one or more days of heating. Then, the furnace is cooled and emptied before repeating the cycle (Fig. 22-8).

Electric Arc Furnace

This furnace is used to make aluminum oxide. It is shaped like a large crucible. The crucible wall is water-cooled. Graphite electrodes enter the furnace through the open top. An electric arc provides heat to melt the charge. The electrodes are slowly raised until the melt fills the furnace. Then, the charge is cooled to form crystals, dumped, and the process repeated (Fig. 22-9).

Periodic Kilns / 81

Fig. 22-8. Full-section elevation drawing of a typical electrical-resistance furnace. This furnace is used to make silicon carbide. Typical dimensions: height, 8 ft (2.4 meters); width, 8 ft (2.4 meters) and length, 20 ft (6.1 meters).

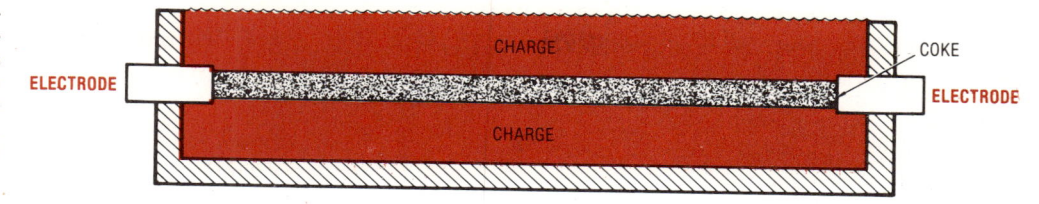

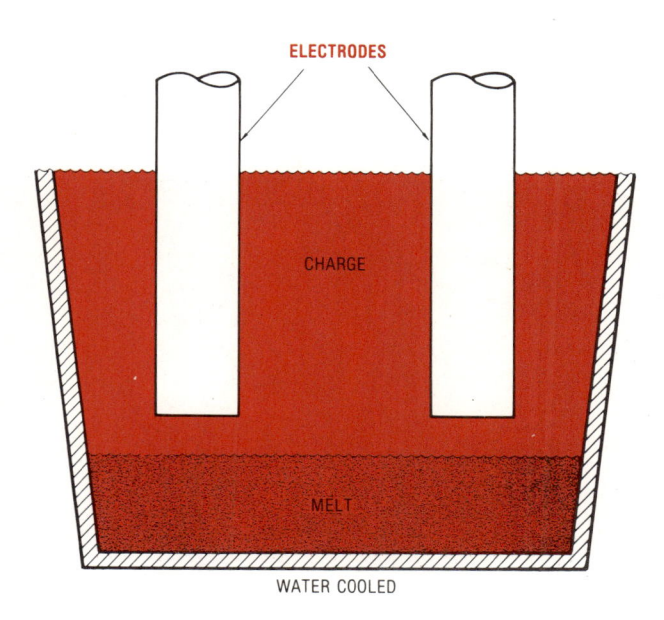

Fig. 22-9. Full-section elevation drawing of a typical electric-arc furnace. This furnace is used to make aluminum oxide. Typical dimensions: height, 6 ft (1.8 meters) and diameter, 10 ft (3.1 meters).

Unit 23: Continuous Kilns

This unit is an introduction to the *continuous* kilns used by the ceramic industries. It is limited to the major types of continuous kilns in use today.

Definition

Continuous kilns operate constantly—24 hours per day, 7 days per week, throughout the entire year. Usually, three zones can be identified through the length of the kiln. They are: (1) preheat; (2) firing; and (3) cooling.

Major Types

The major types of continuous kilns in use today are:

1. Tunnel
2. Vertical
3. Rotary
4. Glass tank
5. Lehr

Tunnel Kiln

Tunnel kilns are used to fire refractory, clay, glaze, porcelain enamel, and abrasive products. Thus, they are used by several different ceramic industries.

Tunnel kilns are built in two shapes—straight-line and circular. The straight-line is the more common type. The length of the straight-line kiln varies from 6 to 200 ft (1.8 to 61 meters) or more. They are usually heated by either electricity or natural gas.

The ware being fired passes through three zones: (1) preheat; (2) firing; and (3) cooling (Fig. 23-1).

Ware is moved through a tunnel kiln by one of three

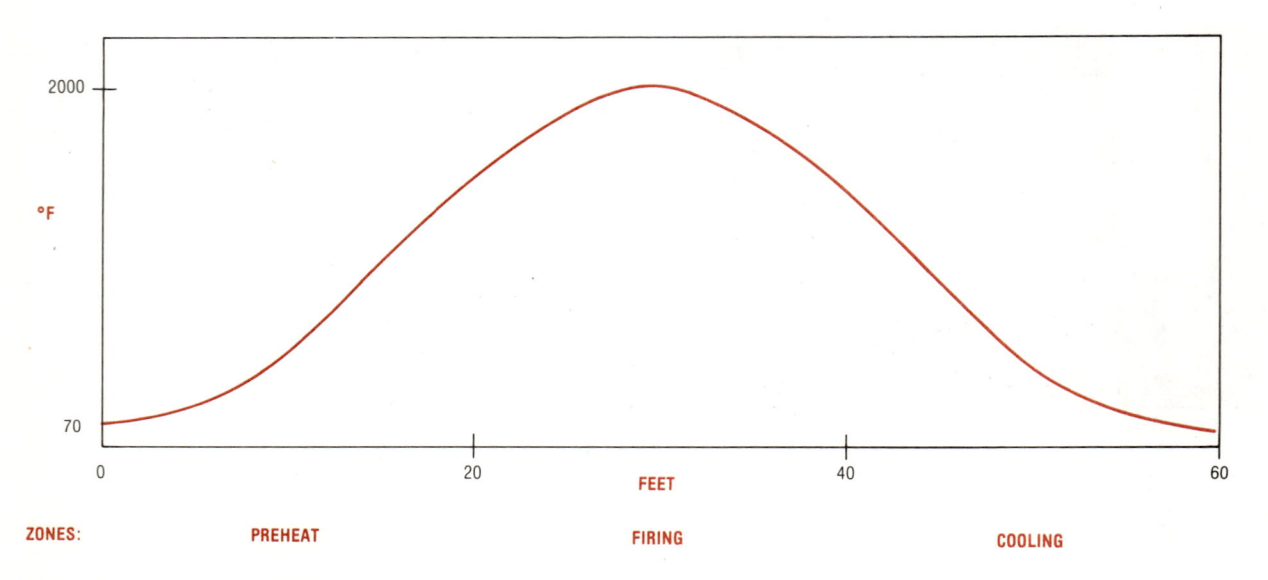

Fig. 23-1. A typical firing curve for tunnel kilns. Note the three zones.

Fig. 23-2. An electrically heated laboratory-size roller-hearth tunnel kiln. The rollers seen outside the kiln continue through its length. Ceramic rollers convey the setters. Ware is placed on each setter, as shown as the kiln entrance.

methods—roller hearth, pusher slab, and kiln cars. The smaller tunnel kilns use the roller-hearth and pusher-slab methods.

Roller hearth—Ware is placed on kiln shelves or setters. A continuous line of setters is carried through the kiln by ceramic rollers (Fig. 23-2).

Pusher slab—Ware is placed on kiln shelves or slabs that rest on two rails. A continuous line of slabs is pushed through the kiln (Fig. 23-3).

Kiln car—Ware is stacked on a refractory car top. Saggars or kiln shelves and posts may be used to increase the height of the load. A continuous line of cars is slowly rolled through the kiln. The cars are usually propelled by a hydraulic pusher. Kiln cars move on wheels and rails similar to those used for railroad cars (Fig. 23-4).

Vertical Kiln

The vertical or shaft kiln is used to make lime. Usually, these kilns are built with a steel exterior, refractory lining, and cylindrical shape. Square and rectangular cross sections have also been used. The dimensions range from 10 to 24 ft (3 to 7.3 meters) in diameter and 35 to 75 ft (10.7 to 22.9 meters) in height. Natural gas and fuel oil are usually used for heat.

Crushed raw material, such as limestone, is fed into the top of the kiln. The material is first preheated in the top portion. Then, as the material moves downward inside the kiln, it is calcined. Cooling takes place at the bottom of the kiln. Then, the lime is discharged (Fig. 23-5).

Fig. 23-3. A laboratory-size, pusher-slab tunnel kiln. Each slab carries one tile. Note the pusher mechanism outside the kiln entrance. (Courtesy R. T. Vanderbilt Company, Inc.)

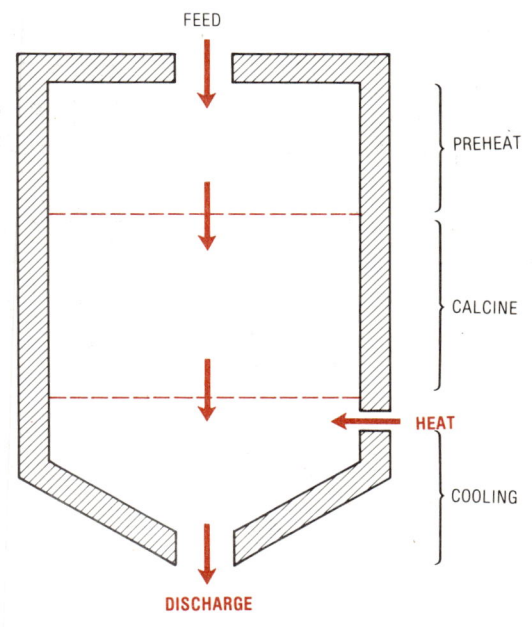

Fig. 23-5. Simplified full-section drawing of a typical vertical or shaft kiln. This kiln usually has a steel exterior and refractory lining. Each of the three zones is full of material that slowly moves downward.

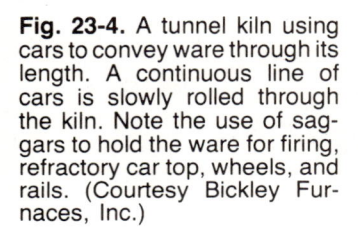

Fig. 23-4. A tunnel kiln using cars to convey ware through its length. A continuous line of cars is slowly rolled through the kiln. Note the use of saggars to hold the ware for firing, refractory car top, wheels, and rails. (Courtesy Bickley Furnaces, Inc.)

Rotary Kiln

The major use of the rotary kiln is to burn several materials used to make Portland cement. It is also used for calcining when making plaster and lime.

This kiln is a steel cylinder with refractory lining. Dimensions range up to 10 ft (3 meters) or more in diameter and 300 ft (91.4 meters) or more in length. It is inclined to the horizontal, usually ½ in. (1.3 cm) per ft (30.5 cm). The entire kiln slowly rotates (usually 1 rpm) when operating. It turns on large gears (Fig. 23-6).

The raw materials are fed into the high end of the cylinder. The inclination and rotation cause the feed to slowly move downward through the kiln. The burner is at the lower end of the kiln. Thus, the heat rises and travels upward through the length of the kiln. Combustible fuels are used to provide the heat.

Preheating occurs in the higher portion of the kiln. Burning occurs toward the lower end where the temperature is highest. Cooling takes place outside the kiln after discharge.

Glass Tank

A glass tank is used to melt raw materials into glass. The tank resembles a covered Olympic-size swimming pool (Fig. 23-7). Glass tanks are built of refractory bricks that have a chemical nature similar to glass. This is necessary, since the glass is in contact with many of the refractories.

Heat is supplied by side burners with an exhaust on the opposite side. The flames flow across the top surface of the glass. The direction of the heat flow is reversed about every 30 minutes. Natural gas or oil is the fuel.

Batched raw materials are fed into one end called a *doghouse*. The batch becomes molten in the melting section of the tank, which is the largest part of the tank. A bridge and throat prevent floating material from passing into the fining area. The glass is then fed through the forehearth to the forming machines (see Fig. 23-7).

Lehr

A lehr is used to anneal glass. Nearly all glass needs to be *annealed* after forming. Annealing is a controlled cooling process.

The lehr resembles a straight-line tunnel kiln. However, there are three ways in which lehrs differ from the tunnel kilns. One of these differences is the conveyance. Usually, continuous stainless steel mesh belts carry the glassware through the lehr. Another difference is that a lower temperature is used. The usual maximum temperature is about 1000° F (538°C). The third difference is that glassware enters the lehr while the glassware is very hot. Thus, a lehr is very hot at its entrance. It gradually cools throughout its length (Fig. 23-8).

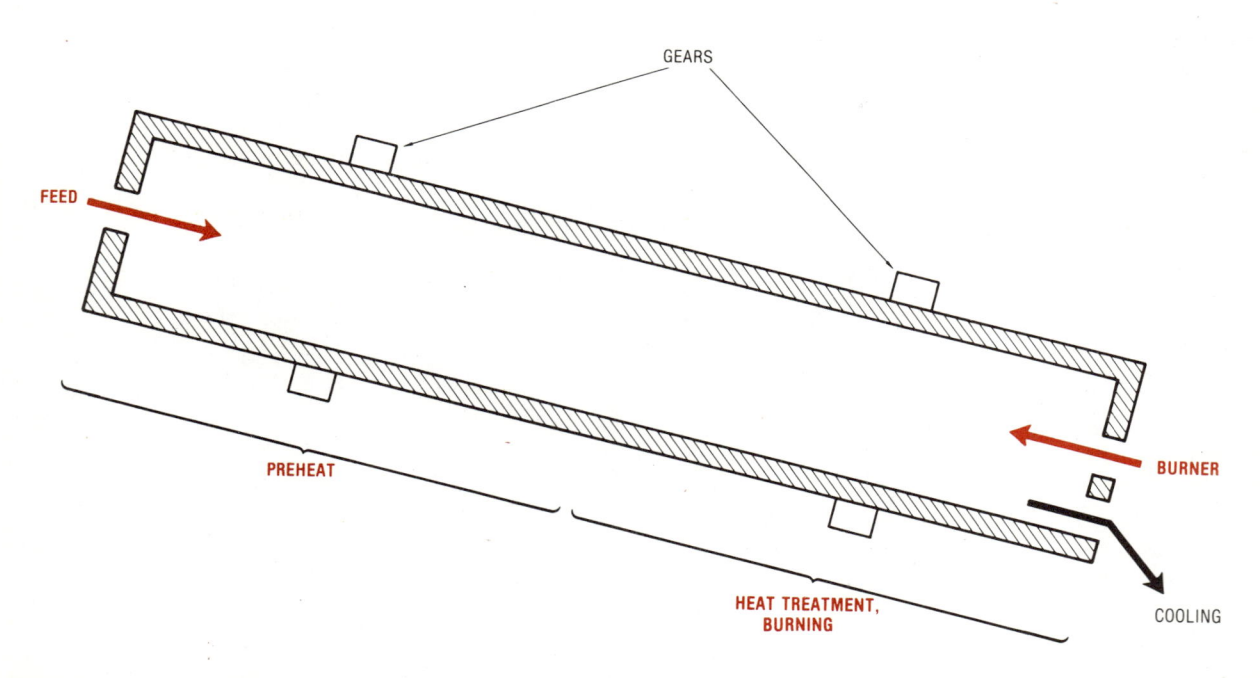

Fig. 23-6. Simplified full-section drawing of a rotary kiln. This kiln is a huge steel cylinder with a refractory lining. It is inclined toward the horizontal and slowly rotates on large gears. Typical specifications: diameter, 10 ft (3.1 meters); length, 300 ft (91.5 meters); inclined ½ in. (12.7 mm); and rotation, 1 rpm.

Continuous Kilns / 85

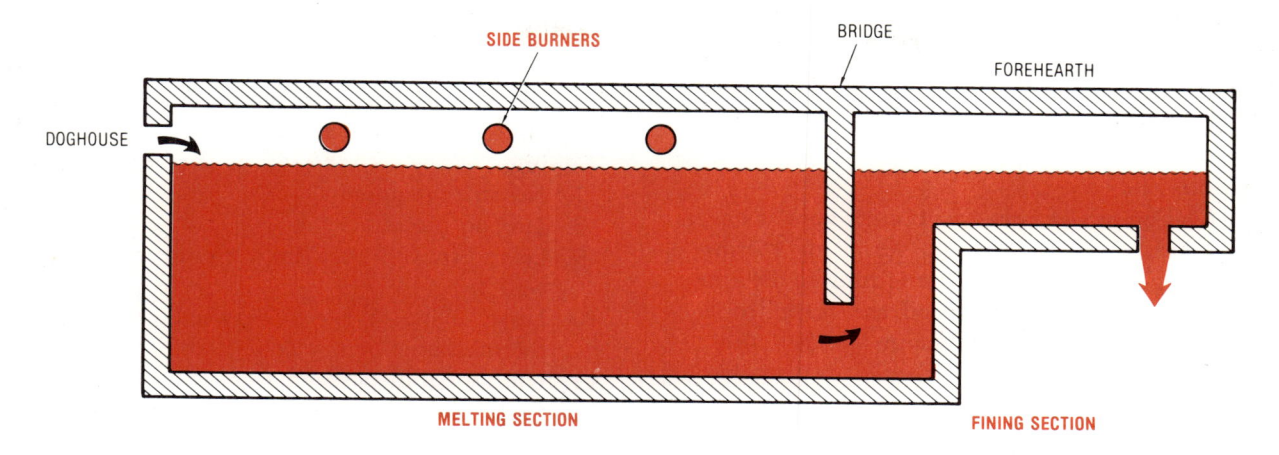

Fig. 23-7. Simplified elevation drawing of a glass tank. It resembles a covered Olympic-size swimming pool.

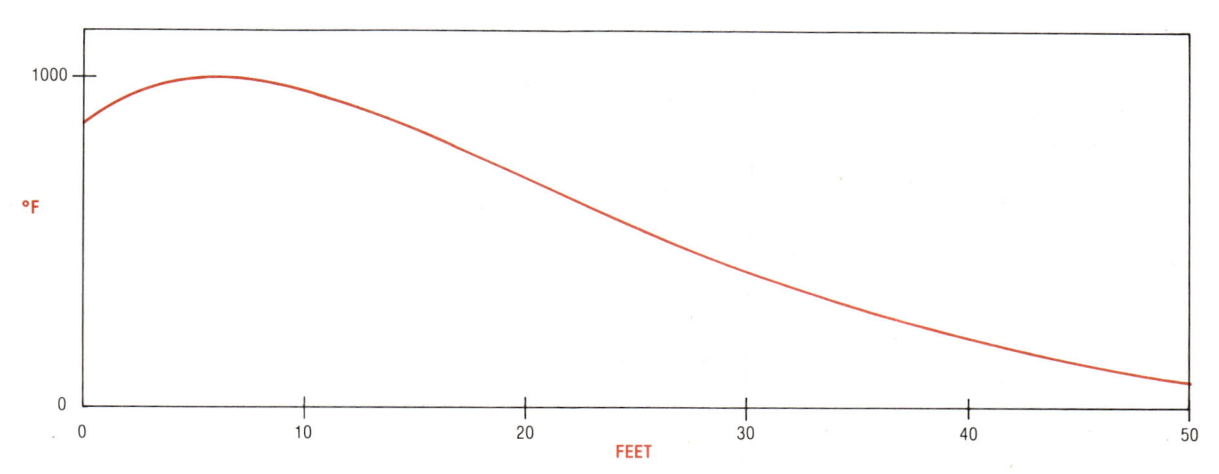

Fig. 23-8. A typical cooling curve for a lehr. The horizontal axis in feet represents the length of the lehr.

Unit 24: Kiln Operation

This unit is concerned with the general procedures for periodic kilns and with temperature measurement. Refer to the kiln's service manual for the specific procedures concerning element replacement, maintenance, and repair work.

Definitions

Kiln wash is a protective refractory coating applied only to the top surface of the kiln shelves. This coating prevents excess glaze or drippings from fusing to the kiln shelf. A kiln wash can be scraped off and reapplied as needed. It is usually made of 50% kaolin and 50% flint, with enough water added to achieve a paint-like consistency, and it is applied with a brush.

Technically, *greenware* is any unfired clayware. However, the term is usually used to describe the dry unfired clay that is ready for firing.

Bisque firing is the first firing of dry clay ware. The usual maximum temperature is *Cone 07*. This temperature leaves the fired clay porous and easier to glaze.

Glaze firing is usually the second firing of the clay ware. It is often done at a higher temperature, such as *Cone 04*, than the bisque fire. The purpose of the higher temperature is to cause the glaze to melt, forming a glassy coating. Also, it causes the clay to become vitreous or glassy.

Preparing for Stacking

The general procedure for preparing a periodic kiln prior to stacking should include the following:

1. Inspect the kiln interior for dust and loose material.
2. Vacuum if necessary.
3. Inspect the kiln shelves for loose kiln wash and glaze.
4. Clean and recoat, if necessary.
5. Use a kiln shelf on four ½-in. (12.7 mm) posts to protect bottom of kiln.
6. Kiln shelf size should be 1 in. (25.4 mm) less than the kiln's interior width and length, to allow the heat to circulate.
7. Test the heat source: for example, open the kiln door; turn on each electric kiln element separately; listen for a hum; and cautiously feel the warmth. *Do not touch the hot elements!*

Stacking for Bisque Firing

The general procedure for stacking a periodic kiln for bisque firing includes:

1. Wash your hands.
2. Handle greenware carefully. Do not pick it up by handles. Support it from underneath.
3. Select only *dry* greenware. If in doubt, place the ware against your cheek. If it feels cool, the ware is damp and not ready for firing.
4. Separate greenware into groups with similar height.
5. Shorter pieces should be placed on the lower shelves with the taller pieces on the top shelves.
6. Greenware may touch and be stacked two pieces high, if the second piece is smaller.
7. No stilts required.
8. Lids on pieces.
9. Leave 1 in. (2.54 cm) between greenware and kiln wall.
10. Leave 2 in. (5.1 cm) between greenware and thermocouple or any control device.
11. Use four kiln posts to support each shelf.
12. Assemble a three cone plaque.
13. Place the cone plaque so it can be seen through the peephole.

Stacking for Glaze Firing

The general procedure for stacking a periodic kiln for glaze firing includes:

1. Wash hands and repeat as needed to prevent transfer of glaze to another piece.
2. Handle glazed pieces carefully; do not pick up by handles, and support from underneath.
3. Inspect and blow off dust, if necessary.
4. Separate glazed pieces into groups with similar height.
5. Shorter pieces should go on the lower shelves with taller pieces on top shelves.
6. Use correct size of stilt to provide stable support. Stilts raise a glazed piece above the kiln shelf. This prevents fusing of the piece to the shelf.
7. Pieces do not touch. Allow ½ in. between pieces.
8. Leave 1 in. (2.54 cm) between pieces and kiln wall.
9. Leave 2 in. (5.1 cm) between pieces and thermocouple or any control device.
10. Use four kiln posts to support each shelf.
11. Assemble a three-cone plaque.
12. Place cone plaque so it can be seen through peephole.

The results of the preceding steps can be seen in Fig. 24-1.

Firing Procedure

The general firing procedure for an electric periodic kiln is:

1. If applicable, set kiln shutoff or other control device.
2. Set up a kiln log for time, temperature, and other details.
3. Lock the kiln door.

4. Open the peephole. This allows moisture to escape from the kiln. Close it when the kiln reaches 1000°F (538°C).
5. Turn on the kiln.
    a. If the kiln has only two toggle switches, turn on the bottom element; or.
    b. If the kiln has selector switches—low, medium, and high, turn on "low" for first hour, "medium" for the second hour, and "high" until the kiln reaches temperature.
6. Take periodic temperature readings if equipped with a pyrometer, such as every half hour, and record on log.
7. Watch the cone plaque as the kiln temperature approaches shutoff.
8. Allow kiln to cool at its own rate, usually overnight. Cooling time can be shortened by opening the peephole and kiln door. Do this with caution.
9. Unstack the kiln when the ware can be handled with bare hands.

Temperature Measurement

Temperature is the degree of heat. Special instruments are necessary to measure the high temperatures inside kilns. These are either direct- or indirect-reading instruments.

Indicating Pyrometer

The indicating pyrometer or thermocouple pyrometer is a direct-reading instrument. It is composed of two parts—a *thermocouple* and a *pyrometer*.

A thermocouple is made of two wires of different metals. The usual combinations are either *alumel* and *chromel* or *platinum and 10% rhodium platinum*. A thermocouple has hot and cold junctions. The two wires are twisted and welded together at the hot junction. The hot junction projects 1 to 2 in. (2.54 to 5.1 cm) inside the kiln. Each wire at this end of the thermocouple is insulated inside porcelain tubing. The cold junction is where the two wires attach to the back of the pyrometer (Fig. 24-2).

A pyrometer is a millivoltmeter. The scale has been calibrated to read in degrees Fahrenheit and/or Celsius.

The thermocouple and pyrometer operate on the principle of dissimilar metals. When the hot junction is heated, a slight electric current is generated. The current is measured on the pyrometer.

Recording Pyrometer

A recording pyrometer is similar to the indicating pyrometer. It has a thermocouple and a pyrometer with a pen. The pen records a line corresponding to the temperature on special graph paper. Thus, the term *recording* pyrometer. Two types of paper are used. When a roll of paper is used, it is called a *strip chart*. If a round disc is used, it is a *circular chart*.

Optical Pyrometer

The optical pyrometer is held by a person who aims it through an open peephole. This can be done from several feet away from the kiln. There is no thermocouple.

Optical pyrometers are constructed with three basic parts—telescope or optical system, lamp bulb, and

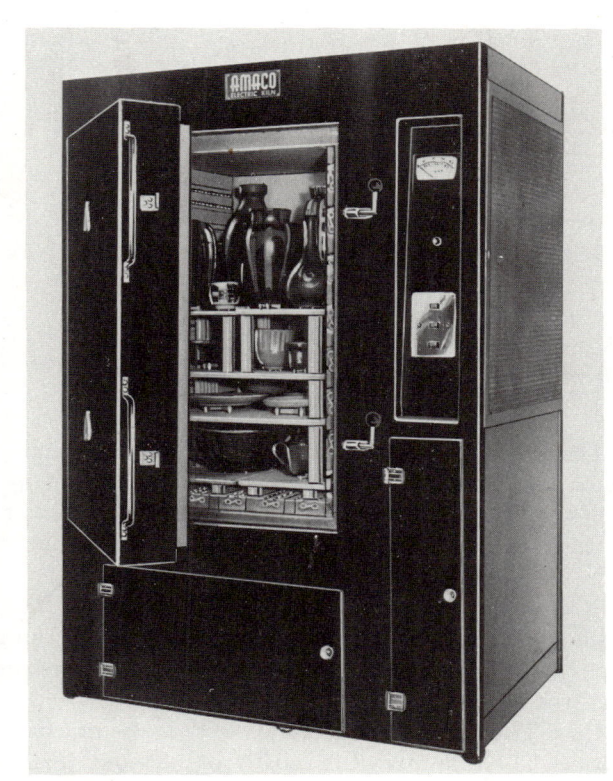

Fig. 24-1. A front-loading electric periodic kiln stacked for glaze firing. The shorter pieces are placed on the lower shelves and the taller pieces are placed on the top shelf. Note the use of stilts and kiln posts. (Courtesy American Art Clay Co., Inc.)

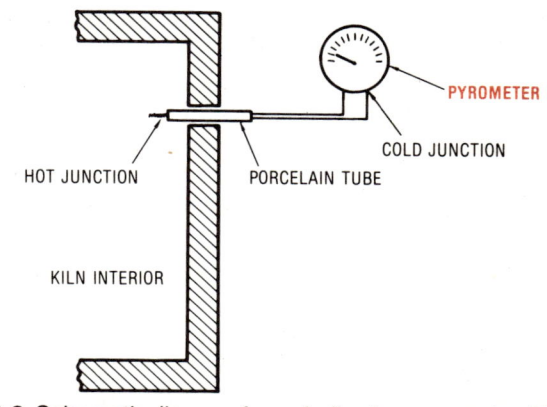

Fig. 24-2. Schematic diagram for an indicating pyrometer. The two wires are twisted and welded together only at the hot junction. This end projects into the kiln 1 to 2 in. (25.4 to 50.8 mm).

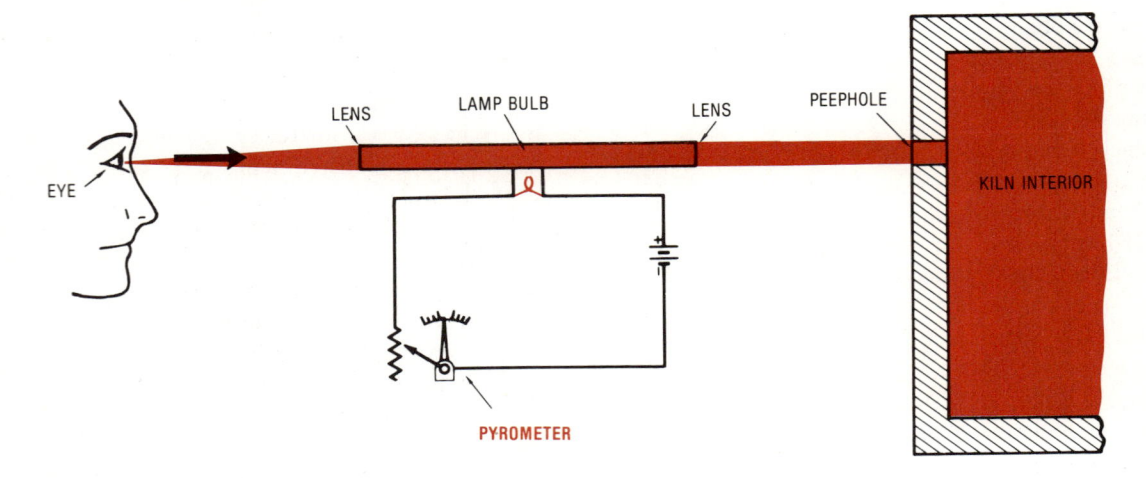

Fig. 24-3. Schematic diagram for an optical pyrometer. It is aimed through the peepholes at the kiln interior. This may be done from a reasonable distance.

pyrometer that measures the current to the lamp. While aiming the optical system, the current to the filament in the lamp bulb is adjusted until its intensity is equal to that of the kiln interior. Then the pyrometer is read (Fig. 24-3).

Pyrometric Cones

Pyrometric cones are indirect-reading instruments. They are made of clay, flint, feldspars, and similar materials. Their composition is similar to clay and glaze ware.

The basic shape of the pyrometric cones is a three-sided pyramid. The cones incline 8° from the vertical (Fig. 24-4). There are two sizes of cones. The large cones are 2½ in. (6.35 cm) in height. The small cones are 1⅛ in. (2.86 cm) in height.

The *cone number* is on the bending face of the large cones. On the small cones, the number is on a side next to the bending face. The large cone numbers range from *Cone 022* (1112°F, or 600°C) to *Cone 01* and from *Cone 1* to *Cone 42* (theoretically 3660°F, or 2016°C), as shown in Table 24-1. Small cones are available in *Cone 022 to 01* and from *Cone 1 to 20*. Table 24-1 includes cones *C/022* to *C/12* (cone number is abbreviated *C/*).

How They Work—The pyrometric cones deform progressively as the heat inside a kiln increases. They react to heat and to time. The large cones are usually heated at the rate of 270°F (132°C) per hour. The small cones are usually heated at the rate of 540°F (282.2°C) per hour.

Cones are selected in relation to the maturing temperature of the clay or glaze to be fired. Usually, three pyrometric cones are used for each firing. One cone is heated to deform at the maturing temperature of the clay or glaze, a second cone is heated to deform at the next higher temperature, and a third cone is heated to deform at the next lower temperature.

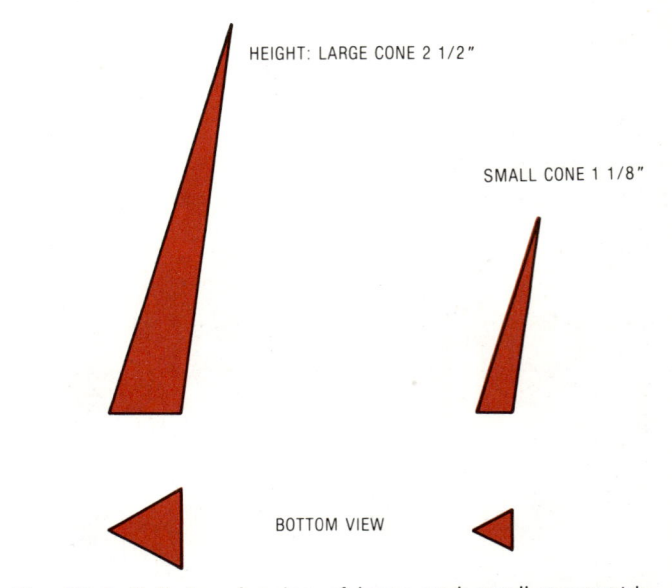

Fig. 24-4. Full-size drawing of large and small pyrometric cones. Note their shapes in the two views.

The lower-temperature cone bends first. Its bending warns you that the maturing temperature is approaching. The kiln is turned off when the maturing temperature cone bends halfway downward. The third cone is a guard cone that bends only if the kiln is overfired.

When more than three cones are used, the extra cone(s) act as additional warning cones. Each additional warning cone would be one cone number lower in temperature than the preceding cone.

How They are Assembled—Cones are assembled in a plaque. The term *plaque* is usually pronounced "pack." The plaque is placed in the kiln so it can be seen through the peephole (Fig. 24-5).

One or more additional cone plaques may be used in other parts of the kiln to check the distribution of heat.

Kiln Operation / 89

This is useful when firing large kilns, such as kilns over 1 cu ft (28.3 cu decimeters) in capacity. At least one cone plaque should be used for each firing. When only one plaque is used, it should be placed so it can be seen through the peephole.

A cone plaque may be assembled so that all three cones bend right or left (Fig. 24-6). Plaques can be made from plastic clay, about ¾ in. × 1 in. × 4 in. (1.9 cm × 2.54 cm × 10.2 cm) and dried before use. Commercially made plaques for either large or small cones can be purchased (Fig. 24-7). The commercial plaques are made so the cones bend only to the left.

Always check your work after assembling a plaque. All the cones should be inclined in the same direction. The cone numbers are in line or on identical surfaces. All the cone numbers are in sequence.

The sequence of the cone numbers (abbreviated C/) for a C/07 bisque firing is C/08, C/07, C/06. These cones are bending left. The sequence is C/06, C/07, C/08, if bending right. A C/04 glaze firing would be C/05, C/04, C/03 to bend left. The sequence would be C/03, C/04, C/05 to bend right.

Cones and plaques can be fired only once. They provide a permanent record of a firing. Thus, a cone plaque should be used for every firing. This is true, even when the kiln is equipped with an indicating pyrometer.

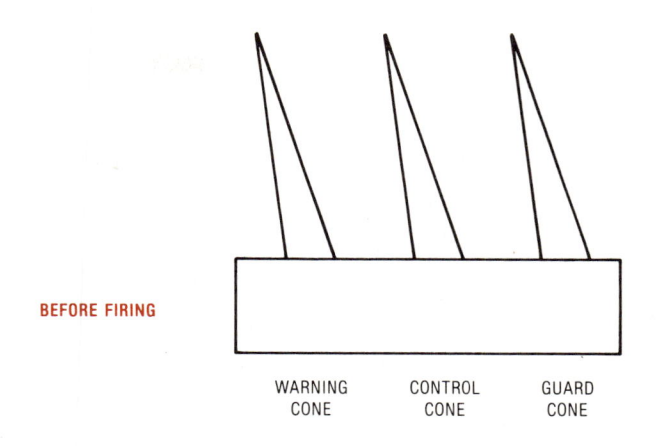

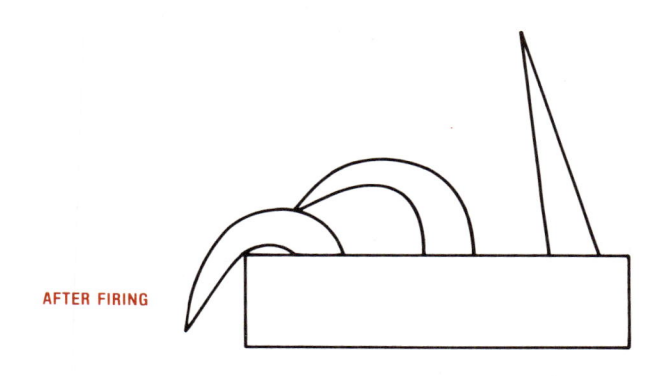

Fig. 24-6. Front view of a three-cone plaque bending toward the left.

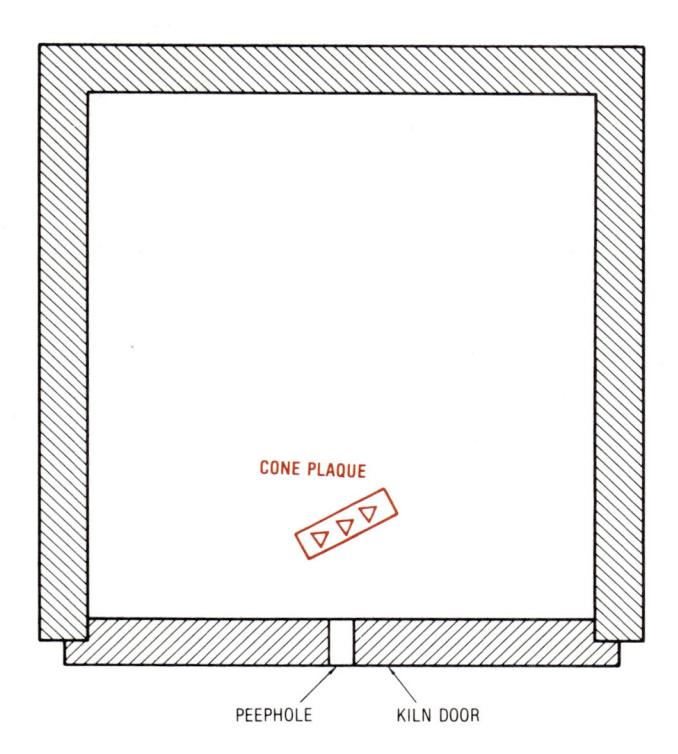

Fig. 24-5. Full-section top view drawing of a periodic kiln with a cone plaque placed inside the peephole. The plaque will bend to the left. It is set at a slight angle (about 30°) to the peephole. This allows the entire plaque to be viewed.

Fig. 24-7. Commercial plaques for large cones (top) and small cones (bottom). The cones bend only toward the left in these plaques. The large cones are inserted through the bottom of the plaque with the cone number parallel to the one smooth interior surface (left-hand side of triangular opening above). The small cones are inserted from the side of the plaque with the cone number visible. Large cone plaques are available in 2- and 4-hole modes, as shown. Small cone plaques are available in 3- and 4-hole (shown) modes. (Courtesy The Edward Orton Jr. Ceramic Foundation)

90 / *Modern Industrial Ceramics*

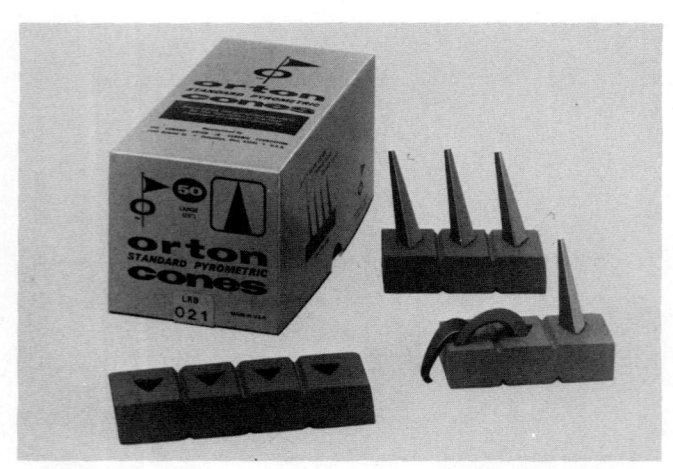

Fig. 24-8. Large cones (2½ in., or 6.35 cm) cover the entire series from *Cone Nos. 022 to 01* and *Nos. 1 to 42*. *Cone Nos. 022 to 016* contain lead compounds. *Cone Nos. 010 to 01* and *Nos. 1 to 3* contain iron oxide. These cones contain organic binders. They are packed in boxes of 50 cones. The cones require mounting in plaques, as shown, or in clay pats. Note the cone plaques for before and after firing. (Courtesy The Edward Orton Jr. Ceramic Foundation)

Fig. 24-9. Small cones (1⅛ in., or 2.86 cm) are made in *Cone Nos. 022 to 01* and *Nos. 1 to 20* from the same composition of materials as the large cones. However, they are about half the size of the large cones. They are packaged in blister packs of 10 cones and in boxes of 50 cones. These cones contain organic binders. They require mounting in plaques, as shown, or in clay pats. Note the cone plaques for before and after firing. (Courtesy The Edward Orton Jr. Ceramic Foundation)

The large cone plaques (Fig. 24-8) and small cone plaques (Fig. 24-9) are shown before and after firing. Also, additional information concerning each size cone is shown.

Self-supporting cones (Fig. 24-10) are now available. These cones are 2½ in. (6.35 cm) in height. They eliminate the need for plaques, which reduces cost and time.

Fig. 24-10. Self-supporting cones (2½ in., or 6.35 cm) are new and do not require a holder. They are available in *Cone Nos. 022 to 01* and *Nos. 1 to 23*. These cones contain organic binders. They are packed in blister packs of 5 cones and in boxes of 25 cones. Note the cones for before and after firing. (Courtesy The Edward Orton Jr. Ceramic Foundation)

Table 24-1. Temperature Equivalents for *Orton* Standard Pyrometric Cones (As Determined at the National Bureau of Standards)

CONE NUMBER	LARGE CONES 60°C / 108°F		LARGE CONES 150°C / 270°F		CONE NUMBER	SMALL CONES 300°C / 540°F	
022	585°C.	1085°F.	600°C.	1112°F.	022	630°C.*	1165°F.*
021	602	1116	614	1137	021	643	1189
020	625	1157	635	1175	020	666	1231
019	668	1234	683	1261	019	723	1333
018	696	1285	717	1323	018	752	1386
017	727	1341	747	1377	017	784	1443
016	764	1407	792	1458	016	825	1517
015	790	1454	804	1479	015	843	1549
014	834	1533	838	1540	014	870*	1596
013	869	1596	852	1566	013	880*	1615
012	866	1591	884	1623	012	900*	1650
011	886	1627	894	1641	011	915*	1680
†010	887	1629	894	1641	†010	919	1686
09	915	1679	923	1693	09	955	1751
08	945	1733	955	1751	08	983	1801
07	973	1783	984	1803	07	1008	1846
06	991	1816	999	1830	06	1023	1873
05	1031	1888	1046	1915	05	1062	1944
04	1050	1922	1060	1940	04	1098	2008
03	1086	1987	1101	2014	03	1131	2068
02	1101	2014	1120	2048	02	1148	2098
01	1117	2043	1137	2079	01	1178	2152
1	1136	2077	1154	2109	1	1179	2154
2	1142	2088	1162	2124	2	1179	2154
3	1152	2106	1168	2134	3	1196	2185
4	1168	2134	1186	2167	4	1209	2208
5	1177	2151	1196	2185	5	1221	2230
6	1201	2194	1222	2232	6	1255	2291
7	1215	2219	1240	2264	7	1264	2307
8	1236	2257	1263	2305	8	1300	2372
9	1260	2300	1280	2336	9	1317	2403
10	1285	2345	1305	2381	10	1330	2426
11	1294	2361	1315	2399	11	1336	2437
12	1306	2383	1326	2419	12	1355	2471

CONE NUMBER	LARGE CONES 150°C / 270°F				CONE NUMBER	P.C.E. CONES 150°C / 270°F	
12	1306°C.	2383°F.	1326°C.	2419°F.	12	1337°C.	2439°F.
13	1321	2410	1346	2455	13	1349	2460
14	1388	2530	1366	2491	14	1398	2548
15	1424	2595	1431	2608	15	1430	2606
16	1455	2651	1473	2683	16	1491	2716
17	1477	2691	1485	2705	17	1512	2754
18	1500	2732	1506	2743	18	1522	2772
19	1520	2768	1528	2782	19	1541	2806
20	1542	2808	1549	2820	20	1564	2847
23	1586	2887	1590	2894	23	1605	2921
26	1589	2892	1605	2921	26	1621	2950
27	1614	2937	1627	2961	27	1640	2984
28	1614	2937	1633	2971	28	1646	2995
29	1624	2955	1645	2993	29	1659	3018
30	1636	2977	1654	3009	30	1665	3029
31	1661	3022	1679	3054	31	1683	3061
31½					31½	1699	3090
32	1706	3103	1717	3123	32	1717	3123
32½	1718	3124	1730	3146	32½	1724	3135
33	1732	3150	1741	3166	33	1743	3169
34	1757	3195	1759	3198	34	1763	3205
35	1784	3243	1784	3243	35	1785	3245
36	1798	3268	1796	3265	36	1804	3279
37	ND	ND	ND	ND	37	1820	3308
38	ND	ND	ND	ND	38	1850*	3362
39	ND	ND	ND	ND	39	1865*	3389
40	ND	ND	ND	ND	40	1885*	3425
41	ND	ND	ND	ND	41	1970*	3578
42	ND	ND	ND	ND	42	2015*	3659

* Temperatures approximate. See Note 3. N.D.—not determined. P.C.E.—Pyrometric Cone Equivalent (small cone only)
† Iron-free (white) are made in numbers 010 to 3. The iron-free cones have the same deformation temperatures as the red equivalents when fired at a rate of 60 Centigrade degrees per hour in air.

Notes:
1. The temperature equivalents in this table apply only to Orton Standard Pyrometric Cones, *when heated at the rates indicated, in an air atmosphere.*
2. Temperature Equivalents are given in degrees Centigrade (°C.) and the corresponding degrees Fahrenheit (°F.). The rates of heating shown at the head of each column of temperature equivalents were maintained during the last several hundred degrees of temperature rise.
3. The temperature equivalents were determined at the National Bureau of Standards by H. P. Beerman (See Journal of the American Ceramic Society, Vol. 39, 1956), with the exception of those marked (*).
4. The temperature equivalents are not necessarily those at which cones will deform under firing conditions different from those under which the calibrating determinations were made. For more detailed technical data, please write the Orton Foundation.
5. For reproducible results, care should be taken to insure that the cones are set in a plaque with the bending face at the correct angle of 8° from the vertical, with the cone tips at the correct height above the top of the plaque. (Large Cone 2"; small and P.C.E. cones 15/16")

September 1964

(Courtesy The Edward Orton Jr. Ceramic Foundation)

SECTION VII

Gypsum Industry

Unit 25: Material Science

This section introduces the first of three major *hydrosetting* materials. *Hydrosetting* means that a powdered material will chemically *set* (change into a homogeneous solid) when water is added.

Hydrosetting Materials

The three hydrosetting materials are plaster, lime, and Portland cement. Manufacturing hydrosetting materials can be expressed by a simple word equation:

$$\text{raw material(s)} \xrightarrow{\Delta} \text{primary product}$$

Adding water to the preceding word equation results in the following:

$$\text{raw material(s)} \xrightarrow{\Delta} \text{primary product} + \text{water} \longrightarrow \text{homogeneous solid}$$

The equation provides an overview of this section. Since the equation begins with raw materials, this is the first topic discussed.

Raw Materials

The major raw material used by this industry is *rock gypsum*. It is a sedimentary rock. Chemically, it is hydrous calcium sulfate ($CaSO_4 \cdot 2H_2O$). The water is chemically combined with the calcium sulfate.

Physically, gypsum is fairly soft. It has a hardness of 2 on Moh's scale of mineral hardness (Table 25-1). You can scratch it with your fingernail. Usually, it is white in color.

Minor raw materials used by the gypsum industry include:

1. Selenite
2. Satin spar
3. Alabaster
4. Gypsite
5. Anhydrite

Selenite is the crystal form of gypsum. It is a clear mineral. *Satin spar* is a white fibrous variety of the mineral. *Alabaster* is a massive form of gypsum. *Gypsite* is a mixture of gypsum and white sand. *Anhydrite* is calcium sulfate ($CaSO_4$).

Table 25-1. Moh's Scale of Mineral Hardness*

No.	Mineral	Common Objects**
1	Talc	
2	Gypsum	
		2.5 fingernail**
3	Calcite	
		3-3.5 Copper coin(penny)**
4	Fluorite	
5	Apatite	
		5.5 common glass ** or knife blade
6	Feldspar (orthoclase)	
		6.5 Steel file
7	Quartz**	
8	Topaz	
9	Corundum	
10	Diamond	

* A scratch test or a measure of the mineral's resistance to a mechanical force.
** Basic items frequently used for hardness testing.

Unit 26: Procurement

Procurement of raw materials for the gypsum industry is diagrammed in Fig. 26-1. The prospecting and developing processes previously discussed are used by this industry.

Prospecting in the United States has revealed two major deposits. The first deposit starts in southwestern Texas. It extends northward through central Kansas, Iowa, and into southern Michigan. From Michigan, the deposit extends southeast through northern Ohio into western New York. The second deposit starts in California's Imperial Valley. It extends northward through Nevada and Utah, with a small portion extending into Montana.

Superficial (Fig. 26-2) and subterranean winning are both used. The depth of the deposit determines the type of winning used. For example, in western New York the deposit is about 88 feet (26.8 meters) below the surface. Thus, subterranean or underground winning is used. Some subterranean winning processes can be seen in Figs. 26-3, 26-4, and 26-5.

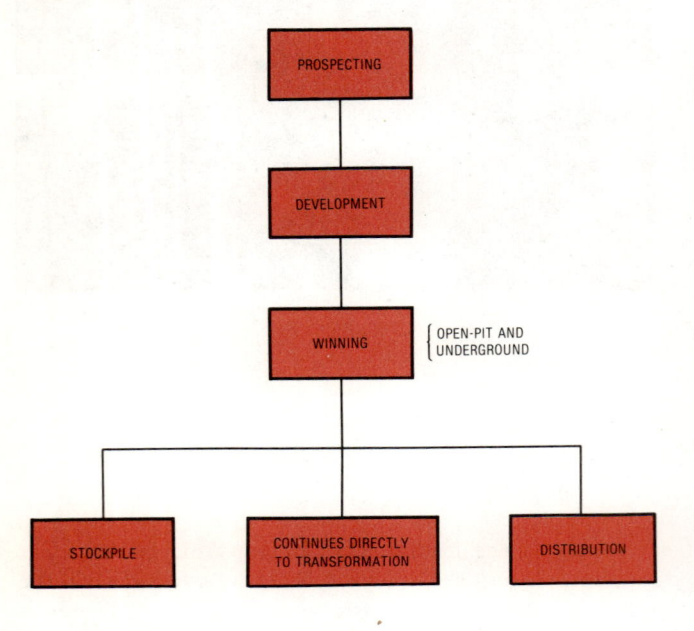

Fig. 26-1. A typical vertical flow sheet for procurement of raw materials used by the gypsum industry.

Fig. 26-2. Blasting in an open-pit mine produces these large irregular rock gypsum fragments. They are easily broken with a pneumatic chisel. (Courtesy United States Gypsum Company)

Fig. 26-3. Drilling holes for explosives in an underground mine. Note the hard hat and battery light safety equipment. (Courtesy Gold Bond Building Products, Division of National Gypsum Company)

Procurement / 95

Fig. 26-4. Blasting often produces large irregular rock fragments as in this underground mine. Their size is being reduced so machines can load the rock. (Courtesy Gold Bond Building Products, Division of National Gypsum Company)

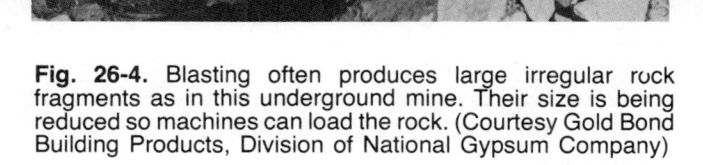

Fig. 26-5. A self-propelled machine loading loose rock in an underground mine. The small railroad cars will transport the rock to the mine shaft. There it will be conveyed to the surface. (Courtesy Gold Bond Building Products, Division of National Gypsum Company)

Unit 27: Transformation

Transformation of gypsum into plaster is diagrammed in Fig. 27-1. Plaster is manufactured in the form of a powder.

Beneficiation

Beneficiation prepares the raw material for the heat-treating process. Both comminution and classifying processes are used. Comminution involves primary and secondary crushing. Primary crushing reduces the raw material to 3 in. (7.6 cm) or less. Secondary crushing reduces it to ½ in. (1.3 cm) or less. Fine grinding further reduces the raw material. Jaw crushers and hammermills are frequently used for crushing. Dry pans and rod mills are often used for grinding.

Calcining

Calcining involves the use of heat in a kiln, but not to the point of fusion, to cause a chemical and/or physical change. This heat-treating process is done in one of three units:

1. Kettle
2. Rotary kiln
3. Autoclave

The kettle and the rotary kiln are the most frequently used processes.

Kettle

The kettle is a periodic kiln. It requires 1 to 4 hours at 340°F (171°C) for calcining, depending on its size. The kettle is a covered vertical cylinder 10 to 12 feet (3 to 3.7 meters) in diameter. The kettles may be 20 or more feet (6 meters) in height, and they can hold 10 to 20 tons (9.1 to 18.1 metric tons) of gypsum.

The kettle is fed through a door at the top. The gypsum is stirred by agitators while being heated from the bottom. After calcining, the contents are dumped into cooling bins.

Rotary Kiln

The rotary kiln operates continuously. It requires 45 minutes at 340°F (171°C) for calcining. The typical rotary kiln used by this industry is 12 to 15 feet (3.7 to 4.6 meters) in diameter and 150 feet (45.7 meters) in length. It is a rotating cylinder that is slightly inclined to the horizontal (Fig. 27-2). This kiln was discussed previously in Section VI.

Autoclave

The autoclave is a periodic process. It is a closed cylinder similar to a pressure cooker that uses steam at

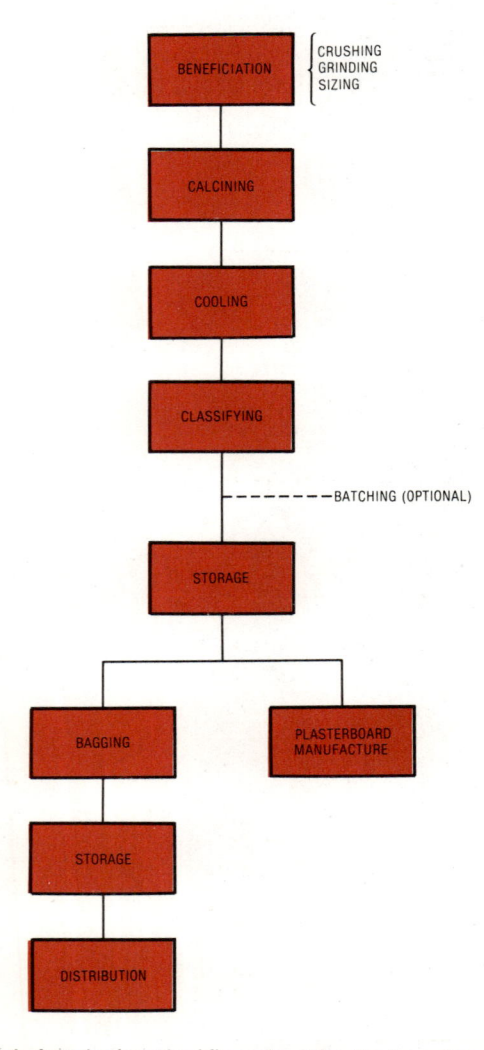

Fig. 27-1. A typical vertical flow sheet for the transformation of gypsum into plaster. The final product, plaster, is in the form of a powder.

Fig. 27-2. Exterior view of a rotary kiln, looking toward the lower end. The burner that fires the kiln is inside the building at the lower end. This continuous kiln is a very large rotating cylinder that is slightly inclined to the horizontal. To appreciate its size, compare the kiln with a railroad car. (Courtesy United States Gypsum Company)

250°F (121°C) to produce special plasters. The time required for calcining varies, depending on the plaster to be produced.

Purpose of Calcining

The purpose of calcining is to drive off some or all of the chemically combined water. This water is called the *water of crystallization*. When three-fourths (¾) of the water of crystallization is driven off, *plaster of Paris* is produced. When *all* of the water is driven off, dead, burnt plaster called *Keene cement* is produced.

The calcining process is expressed by this word equation:

$$\text{gypsum} \xrightarrow{\Delta} \text{plaster} + \text{water} \uparrow$$

Usually, only three-fourths (¾) of the chemically combined water is driven off. This is clearly shown by the chemical equation:

$$CaSO_4 \cdot 2H_2O \xrightarrow{\Delta} CaSO_4 \cdot \tfrac{1}{2}H_2O + 1\tfrac{1}{2}H_2O \uparrow$$

This equation is incorrect, since we cannot have less than one molecule. Thus, the balanced chemical equation is:

$$2(CaSO_4 \cdot 2H_2O) \underset{\longleftarrow}{\xrightarrow{\Delta}} 2CaSO_4 \cdot H_2O + 3H_2O \uparrow$$

The lower arrow means that this equation is reversible without the use of heat. Or plaster plus water yields gypsum.

Physically, plaster is changed from a powder back to gypsum, a homogeneous solid, when water is added. Plaster is the only known material that sets by the addition of only water. It reverts to the original natural material, gypsum. This was shown in the preceding chemical equation.

This is the reason plaster is called a hydrosetting material. *Hydrosetting* means that a powdered material will chemically set or change into a homogeneous solid when water is added. Hydrosetting materials are usually in powder form. A powder is easier to mix with water.

Plasterboard

Most of the plaster manfactured each year in the United States is used to make plasterboard. There are three types of plasterboard:

1. Wallboard (interior)
2. Lath (interior)
3. Sheathing (exterior)

Plasterboard is basically a type of sandwich. A gypsum core is sandwiched between two layers of heavy paper. The core is composed of plaster, water, additives, adhesive, foaming agent, and accelerator. Each layer of paper is composed of a face, filler, and bond liner.

Plasterboard manufacturing is diagrammed in Fig. 27-3. The width of plaster wallboard is 4 ft (1.2 meters).

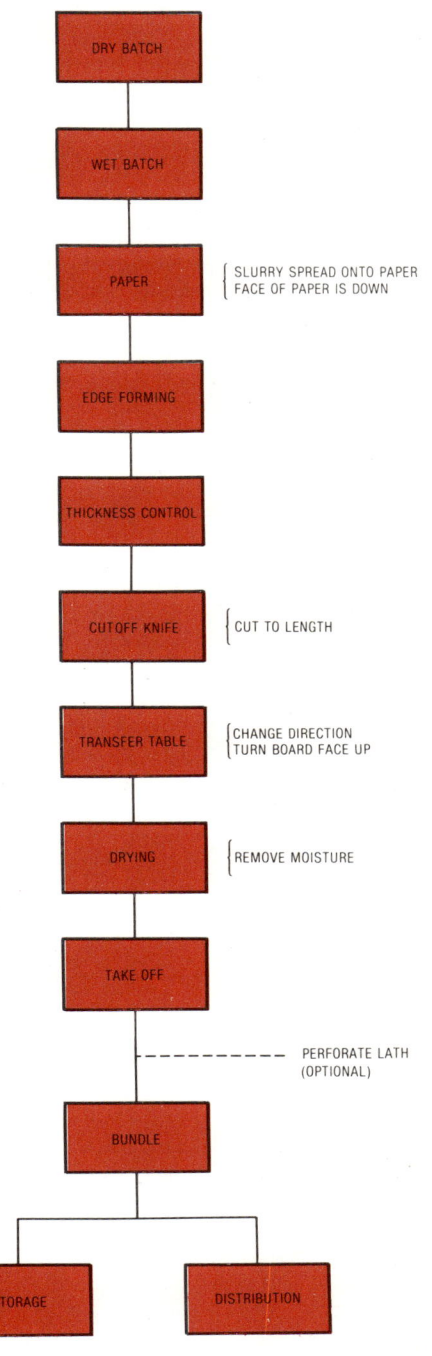

DRY BATCH
↓
WET BATCH
↓
PAPER — SLURRY SPREAD ONTO PAPER / FACE OF PAPER IS DOWN
↓
EDGE FORMING
↓
THICKNESS CONTROL
↓
CUTOFF KNIFE — CUT TO LENGTH
↓
TRANSFER TABLE — CHANGE DIRECTION / TURN BOARD FACE UP
↓
DRYING — REMOVE MOISTURE
↓
TAKE OFF
┆
┆ PERFORATE LATH (OPTIONAL)
↓
BUNDLE
↓
STORAGE   DISTRIBUTION

Fig. 27-3. A typical vertical flow sheet for the manufacture of plasterboard.

98 / *Modern Industrial Ceramics*

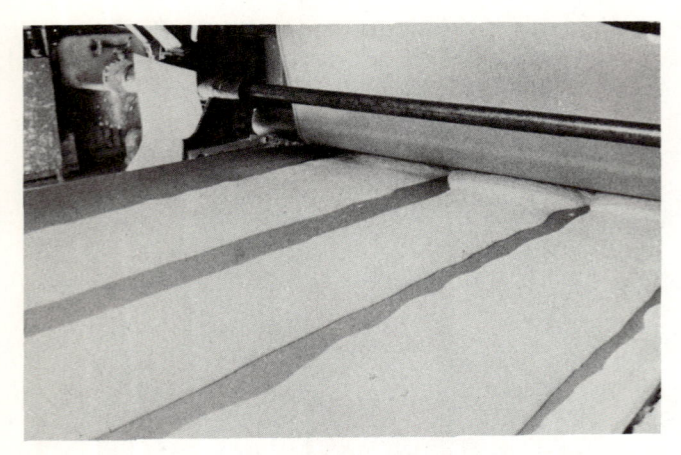

Fig. 27-4. Edge forming and thickness control on the plasterboard machine are shown here. The slurry for the gypsum core has already been spread onto the face liner. The vertical sheet of paper is the backing from an overhead roll. (Courtesy United States Gypsum Company)

Thus, the board-making machine is a long, narrow machine. All the operations through thickness control (Fig. 27-4) occur at one end of the machine. A moving conveyor belt supports the continuous sheet while the gypsum core sets (Fig. 27-5). Several hundred feet of belt movement are required to allow time for setting. An accelerator, raw gypsum, is mixed with the plaster used in the core. This shortens the chemical setting time.

The floor plan of a typical plasterboard manufacturing plant is diagrammed in Fig. 27-6. The floor plan shows the relationship of the board-making machine, transfer table, drier, and bundling. The transfer table turns each piece of plasterboard. Then it feeds them into a drier that removes excess moisture. After drying, the new plasterboard is bundled. The pieces are assembled into bundles of two pieces with the board faces together. Then each bundle is taped at the ends. Storage or immediate distribution completes transformation.

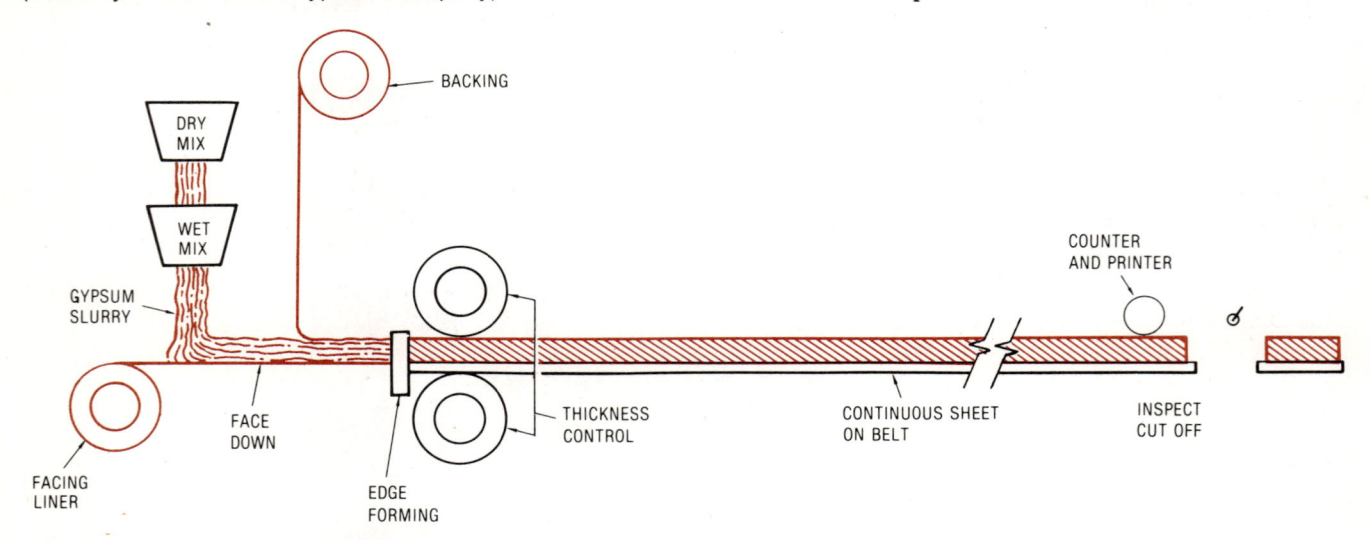

Fig. 27-5. Elevation drawing of a typical plasterboard machine.

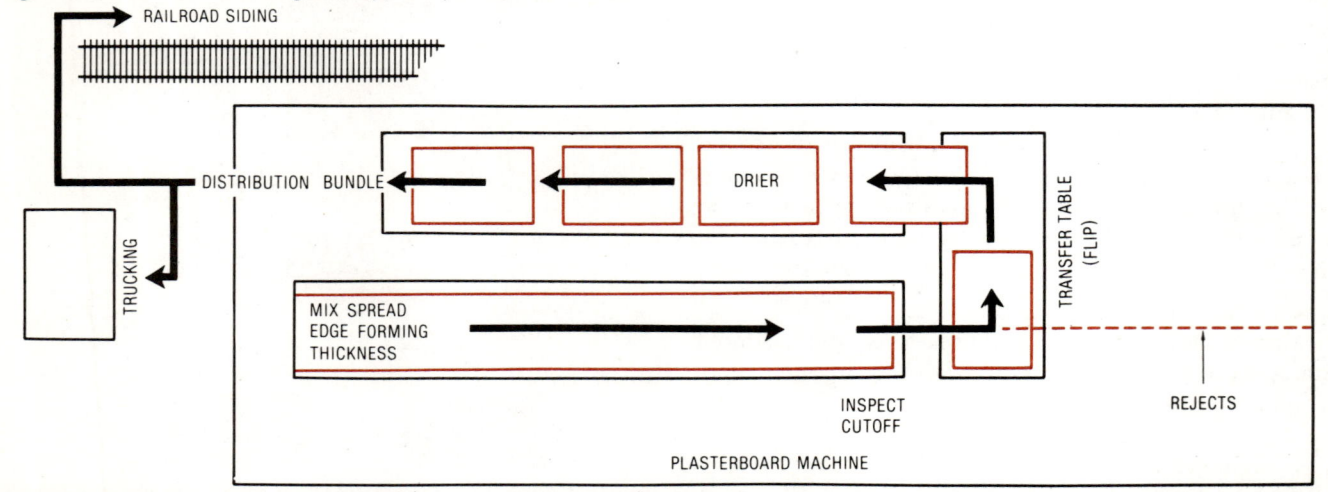

Fig. 27-6. Floor plan of a typical plasterboard manufacturing plant. Note the long narrow plasterboard machine. These machines range from 525 to 880 feet in length.

Unit 28: Utilization

Products of the gypsum industry are used in agriculture, manufacturing, and construction. Manufacturing and construction were previously defined.

Agriculture

Rock gypsum, not calcined, is ground and used for land plaster. Land plaster is sprinkled on peanut plants while they are in bloom. Another agricultural use is in mushroom beds. Rock gypsum is mixed with manure to make mushroom beds.

Manufacturing

There are three major types of manufactured gypsum products—rock gypsum, plaster, and plasterboard.

Rock Gypsum

Rock gypsum, not calcined, is used to make Portland cement. The gypsum acts as a retarder in the Portland cement. This cement is a hydrosetting material that is discussed later.

Raw rock gypsum acts as an accelerator when it is added to plaster and water. It is used in the gypsum core of plasterboard products to shorten its chemical setting time. An accelerated set allows the plasterboard machine, discussed in Unit 27, to operate faster.

Most of the rock gypsum is calcined to make plaster.

Plaster

Plaster is in a powdered form. There are many types of plaster that include:

1. Plaster of Paris
2. Pottery plaster
3. Hard dense plasters
4. Wall plasters (usually contain additives)
5. Keene cement or waterproof plaster

Plaster is used to make models and molds. These are used by several industries including ceramics, polymer, and automobiles. Plaster models and molds used by the ceramic industries are discussed in Section VIII.

Small amounts of plaster are used in a variety of other products. These include aspirin, chalk, matches, and toothpaste.

The major use of plaster is in plasterboard.

Plasterboard

Plaster wallboard is available in several thicknesses—⅜, ½, ⅝ in. (0.95, 1.3, 1.6 cm); one width—4 ft (1.2 meters); several lengths—8, 10, 12, 14, 16 ft (2.4, 3.0, 3.7, 4.3, 4.9 meters); and several face surfaces. The face surface may be a plain paper that can be wallpapered or painted. Prefinished wallboard may have paper, vinyl, or thin wood veneer surfaces.

Plaster lath and sheathing are hidden from view inside a wall, since they are covered with other materials. Plaster lath is the base for plastered interior walls. The usual size of lath is ⅜ × 16 × 48 in. (0.95 × 41 × 122 cm). Plaster sheathing is the base for exterior siding. The usual size is ½ in. × 4 × 8 ft (12.7 mm × 1.2 × 2.4 meters).

Plaster wallboard is used in prefabricated homes, housing modules, and mobile homes. All of these uses are manufactured secondary products.

Construction

Both plaster and plasterboard are used extensively in construction. The trade name for the plaster wallboard used on interior walls is *dry wall*. This wallboard is installed dry, thus the name (Figs. 28-1 and 28-2). Joints of the plain unfinished wallboard are usually *taped*. The wallboard joint compound and tape make a smooth continuous wall surface.

Wet Wall

The trade name for the plaster used for interior walls is *wet wall*. This means that the wall plaster is applied wet. Actually, it is in a heavy viscous state when applied to the wall. Plaster lath is usually used as the base for the wet walls (Fig. 28-3). Wet walls are usually applied in two coats or layers. The brown coat is applied first to the lath

Fig. 28-1. Installing plaster wallboard. Commercial buildings have metal framing, while wood framing is usually used in homes.

Fig. 28-2. Nailing plaster wallboard to wood framing. Metal corner bead will be nailed over this corner. Only the bead will show after joint compound has been applied.

Fig. 28-3. Installing plaster lath over metal framing for a commercial building, such as a school. Lath is the base for a plastered wall. This lath will be covered with at least two coats of wall plaster. (Courtesy United States Gypsum Company)

(Fig. 28-4). Then, a finish coat is applied. This second coat is usually white. Thus, the finish coat is also referred to as the white coat.

Keene Cement

Keene cement is used to make waterproof walls. It is applied by the same procedure as a wet wall.

Also, Keene cement is used to make simulated stone. Colorants may be added to the Keene cement to make the "stone" more realistic.

Fig. 28-4. Applying the first or brown coat for a plastered or "wet" wall over metal lath. The plasterer has a hawk in his left hand to carry plaster. He is using a steel trowel to apply plaster to the lath. (Courtesy United States Gypsum Company)

Unit 29: Disposition

Gypsum products are used in agriculture, manufacturing, and construction. Disposition is unnecessary for agricultural uses of gypsum. Gypsum is a natural material that simply becomes part of the soil.

Gypsum products used in manufacturing can be disposed of by reclaiming or discarding if necessary. Often, there is nothing remaining to dispose of, for example, aspirin or toothpaste.

In construction, the gypsum products become part of the total structure. This includes wall plaster and the plasterboard products. Disposition of buildings is the problem of the construction industry. It is not the responsibility of the gypsum industry.

Reclaiming

Manufactured gypsum products are reclaimed by reusing or recycling. Both of these reclaiming methods are possible and recommended.

Reuse

The whitewares industry uses many plaster molds. Often, they have molds to dispose of that are worn, chipped, or discontinued shapes. These molds can be donated to educational institutions for instructional and student use.

Hobby ceramic shops also use plaster molds. These shops usually sell their molds when there is no longer a demand for the shape produced. They often sell the used molds at a reduced price. This provides them with money to invest in new molds. Thus, the mold reuse cycle continues.

Manufacturing and construction firms often have plasterboard pieces left over when the product or project is completed. These pieces of wallboard, lath, and sheeting can be given to educational institutions. The plasterboard can be used for instructional and student use—for example, testing, models and/or mock-ups, or a small building.

Recycle

Working plaster molds become contaminated with clay and/or deflocculant when extensively used. Eventually the mold becomes unusable. Molds saturated with deflocculant may grow white crystals or "whiskers" on their exterior. These molds can be recycled by gently washing in water. Washing is successful, since the clay and deflocculant are soluble in water.

Broken plaster molds can be recycled by repairing them. The pieces can be reassembled, using a small amount of white glue. The excess glue should be completely removed with a damp cloth.

Technically, worn out or broken molds can be recycled into new plaster. This could be done by washing to remove contaminates, crushing, and calcining.

It is technically possible to recycle all set plaster. That is, to calcine it a second time to make new plaster. Unfortunately, none of the manufacturers appear to be doing this. Set plaster often contains additives, such as the core of the plasterboard, and/or is contaminated. These and economic factors make recycling by recalcining infeasible at this time.

All the plaster molds can be recycled for agricultural uses. Crushing the molds is the only process required. In fact, many other gypsum products could be recycled in this manner.

Discarding

Discarding of many gypsum products continues. The quantity involved and/or geographic location often discourages any reclaiming effort. Although most gypsum products are solids, they cannot be recycled in the same manner as stone or refractories. Gypsum is a soft material (Moh's hardness = 2). Some gypsum products also absorb water. These factors make recycling as riprap infeasible. However, they can be beneficial if used for hard fill.

SECTION VIII

Plaster Mold Industry

Unit 30: Introduction

The plaster mold industry is a manufacturing industry. Technically, they manufacture a secondary product. Their basic material, plaster, is a primary product. This industry is composed of firms that:

1. Manufacture and sell only plaster molds.
2. Manufacture and sell plaster molds along with other products.
3. Manufacture plaster molds for only their own use; for example, manufacturers of fine china dinnerware (Fig. 30-1).

Limitation

This section is limited to the plaster molds used by ceramists. Ceramists that use plaster molds include:

1. Hobby ceramics
2. Ceramic artists or designers
3. Potters
4. Ceramic industries (Fig. 30-2)

Ceramic industries that are major users of plaster molds include the refractory and clay industries. The two major divisions of the clay industries are the structural clay products and the whitewares. Of these two divisions, the whiteware manufacturers are the major users of plaster molds (Fig. 30-3).

Plaster molds are also used by other industries. These industries include the crafts, metals, and polymer industries. The medical field, especially dentistry, also uses plaster molds. Basically, the molds used by the

Fig. 30-1. An industrial casting shop: dumping slip from a double-drain mold (top) and removing the moist clay piece, a cream pitcher, from the opened mold (bottom). These are only two of several steps in the slip-casting process. (Courtesy Syracuse China Corporation)

104 / *Modern Industrial Ceramics*

medical field function the same as those used by ceramists. However, the porosity and density of the mold may be different from those used by ceramists.

Physical Properties

Porosity is the major physical property of the plaster molds used by ceramists. Its ability to absorb water is important for clay forming. Other important physical properties include:

Fig. 30-2. Oval platters cast in solid-type slip-casting molds. These are two-piece solid casting molds with removable spares. A *spare* is an opening in the top of the mold that allows the slip to be poured into the mold; and it is a reservoir for extra slip. The two closed molds show where the removable spares are located. This is an industrial casting shop. (Courtesy Shenango China Products)

Fig. 30-3. Cup handles being removed from a two-piece gang mold in an industrial casting shop. Each casting (solid-type) of this mold produces eight cup handles. Note the number of identical molds and how they are clamped together on a special casting table. (Courtesy Syracuse China Corporation)

1. Plaster does not have a shape of its own when in the liquid state.
2. When in the liquid state, it is easily poured into a form.
3. Plaster faithfully reproduces detail when it chemically sets.
4. Usually, it neither shrinks nor expands when chemically setting.
5. Set plaster is easy to work, due to its softness.

Types of Plaster

Pottery plaster has all of the preceding physical properties. This plaster is preferred for making molds that are used for clay forming (Fig. 30-4). Pottery plaster is sold under several trade names. The major manufacturers include:

1. Georgia Pacific Corp., Gypsum Division
2. National Gypsum Company
3. United States Gypsum Company

Porosity is not required of the models. Thus, the models can be made of any plaster that does not contain additives. Pottery plaster is preferred by many ceramists. Other plasters that can be used for models are:

Fig. 30-4. Cup handles being removed from a complex gang mold in an industrial casting shop. Each casting (solid-type) produces 24 cup handles. Note how the molds are stacked vertically with a removable *spare,* as discussed earlier. (Courtesy Shenango China Products)

1. Plaster of Paris
2. Casting plaster
3. Molding plaster
4. Hard dense plaster
5. Keene cement

The following units discuss processes used by the plaster mold industry. These include batching of plaster, model making (Fig. 30-5), and mold making.

Fig. 30-5. A moldmaker completing a plaster mold for a coffee mug. The model will be used to make a four-piece drain-type slip casting mold. (Courtesy Syracuse China Corporation)

Unit 31: Batching of Plaster

Batching of plaster is clearly expressed by the word equation:

$$\text{plaster} + \text{water} \rightarrow \text{gypsum} + \text{heat} \uparrow$$

The chemical equation for this process is:

$$2CaSO_4 \cdot H_2O + 3H_2O \underset{\Delta}{\rightleftharpoons} 2(CaSO_4 \cdot 2H_2O) + \text{Heat} \uparrow$$

The process is reversible by applying heat. This heat-treating process, calcining that is required to reverse the process, was discussed in Section VII.

General Procedure

The general procedure for batching of plaster includes:

1. Determine the amount of materials (plaster and water)
2. Prepare for batching of plaster
3. Actual batching of plaster process
4. Completion of the batching of plaster process

Only the English units of measure will be used in this procedure. This will help you avoid possible errors that could occur if both English and metric measure were given.

Materials required are:

1. Pottery plaster
2. Form (if expendable or not reusable)
3. Newspaper
4. Paper towels
5. Water

Equipment required is:

1. Safety glasses and lab coat or apron
2. Scales (English pounds and ounces) with scoop and scoop counterweight
3. Fluid measure
4. Form (if reusable type)
5. Mixing bowl (flexible type preferred, such as brass or plastic)
6. Trash can

Amounts of Materials

Determining the amount of materials includes:

1. Determine the volume in cubic inches (in^3) of your form.
   a. Formula
      (1) $V = lwh$ for rectangular solids; or
      (2) $V = \pi r^2 h$, in which $\pi = 3.14$ for cylinders
   b. Note: To simplify your work, treat all shapes as one of the above.
2. Volume divided by 81 (cu in.) equals quarts of water.
   a. Round off to nearest ¼, ½, ¾, or full quart
   b. It is better to have slightly more water than not enough.
   c. There are 81 cubic inches (in^3) in one (1) quart of water
3. For each quart of water use 2 lb 12 oz of pottery plaster.
   a. See Table 31-1 for water–plaster ratios.
   b. *Note:* This will achieve the correct mold porosity.

Table 31-1. Water–Plaster Ratios*

Cubic Inches (in³)	Water	Plaster
10.13	¼ pt = 4 fluid oz	5 oz
20.25	½ pt = 8 fluid oz	11 oz
40.5	1 pt = 16 fluid oz	1 lb 6 oz
60.75	1½ pt = 24 fluid oz	2 lb 1 oz
81	1 qt = 32 fluid oz	2 lb 12 oz
101.25	1¼ qt = 40 fluid oz	3 lb 7 oz
121.5	1½ qt = 48 fluid oz	4 lb 2 oz
141.75	1¾ qt = 56 fluid oz	4 lb 13 oz
162	2 qt = 64 fluid oz	5 lb 8 oz

* Only for pottery plaster when used for molds to achieve correct porosity. Above ratios may also be used for models, although they do not require porosity.

Preparation

Preparation for batching of plaster includes:

1. Safety
   a. Eye safety
   b. Wear lab coat or apron
2. Use clean mixing and measuring equipment
   a. Prevent contamination
   b. Set plaster which is gypsum, acts as an accelerator
   c. Select equipment
      (1) Bowl
      (2) Fluid measure
3. Optional—cover work surface with old newspaper. This makes cleanup easier.
4. Obtain, assemble, and/or prepare form.
5. Weigh plaster in pounds and ounces
6. Water
   a. Measure in fluid ounces—1 qt = 32 oz
   b. Temperature
      (1) Cold—retards set
      (2) Lukewarm—best
      (3) Hot—accelerates set

Batching

The actual batching of plaster process includes:

1. Slowly sift plaster into water (Fig. 31-1)
   a. Do not dump plaster
   b. No lumps
   c. No foreign matter

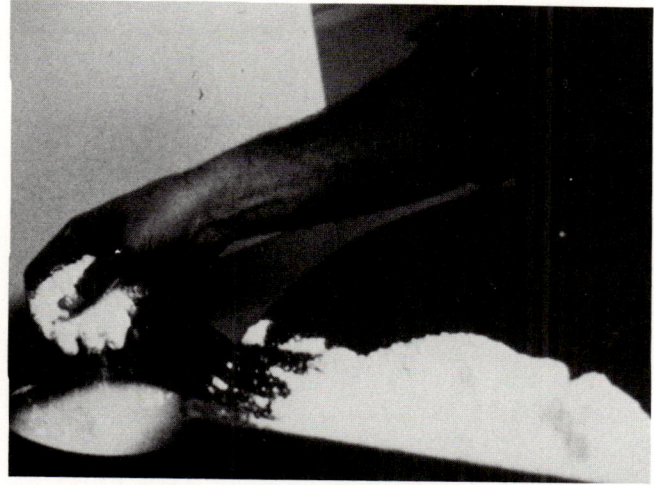

Fig. 31-1. When batching plaster, slowly sift the plaster into the water. Look for and remove any lumps or foreign matter, such as pieces of the paper bag, while sifting the plaster.

2. Slake
   a. Time—2 to 5 minutes or longer
   b. Slake means to allow a material to take on water at its own rate or pace.
3. Mix (Fig. 31-2)
   a. Use one hand—remove any rings
   b. Keep hand submerged to prevent forming air bubbles
   c. Note *feel* of the plaster as it changes condition
   d. Mix until your finger leaves a *slight* trail. The consistency has thickened.

Fig. 31-2. Mix the plaster until your finger leaves a *slight* trail. When the consistency has thickened, it is now ready to pour.

4. Pour
   a. Pour when consistency has thickened and your finger leaves a slight trail
   b. Pour slowly and uniformly—try not to trap air
   c. Vibrate form to release air bubbles if possible
5. Clean up mixing utensil immediately and completely (*Note:* Submerging in water does not stop plaster from setting.)

Completion

Completion of the batching of plaster process includes:

1. Remove form 5 to 10 minutes after chemical set occurs.
2. Remove form carefully.
3. Clean form, if reusable type, and return equipment to storage.
4. Note heat of hydration. This heat results from the chemical reaction.

Unit 32: Model Making

Model making means to manufacture or make a model. A *model* is identical in shape to the final product (Fig. 32-1). It is used for making a mold. The function of a model is to form the interior shape of the mold. This is the part of the mold that actually forms the clay.

The above is true for all types of molds that require a model. Two types of molds do not require a model. They are the *drape mold* and the *jiggering mold*.

Model Materials

Models may be made from several materials. These include:

1. Plaster
2. Clay
3. Synthetic clay
4. Wood
5. Plastics
6. Glass

Pottery plaster is preferred by many ceramists for model making. Actually, any type of plaster, usually without additives, can be used.

Both unfired and fired clay may be used. Unfired clay may be plastic, leatherhard, or greenware. Fired clay may be bisque or glazed. Synthetic clay is manufactured. Basically, it is a mixture of powdered clay and petroleum jelly. It does not air dry. This is an advantage over plastic clay.

Any type of wood may be used for models. The same is true for plastics and glass. Wood and plastics are usually easier to use than glass. *Caution* must be used with glass, since it is breakable and, therefore, potentially dangerous.

Designing Models

A simple model with draft in one direction only is recommended for your first model making experience (see Fig. 32-1). Draft in one direction only means that an angle or taper on the model allows it to be easily withdrawn after pouring plaster over it to make a model. Small models, 1 to 4 in. (2.5 to 10.2 cm) in width and length, are also recommended for the first few experiences.

All clay shrinks when dried and fired. Thus, you may wish to increase the model size to allow for shrinkage. If the percent of shrinkage is known, simply add to the final size desired. Or, use the general rule of allowing 10 percent for shrinkage. This means all dimensions of the final product are increased by 10 percent. For example, a square tile ¼ in. × 4 in. × 4 in. is to be the final product. Calculate additional size to allow for shrinkage as follows:

1. Final thickness

    ¼ in. × 10% = 0.25 × 0.10
    $\qquad\qquad\quad$ = 0.025 in. (for shrinkage)
    0.25 in. + 0.025 in. = 0.275 in.
    $\qquad\qquad\qquad\quad$ = $^9/_{32}$ in. (approx.)

2. Final sides (2)

    4.0 in. × 10% = 4.0 × 0.10
    $\qquad\qquad\quad$ = 10.4 in. (for shrinkage)
    4.0 in. + 0.4 in. = 4.4 in.
    $\qquad\qquad\qquad$ = 4 $^{13}/_{32}$ in. (approx.)

3. Therefore, final tile dimensions to allow for shrinkage are:

    0.275 in. × 4.4 in. × 4.4 in. (for shrinkage)
    $^9/_{32}$ in. × 4$^{13}/_{32}$ in. × 4$^{13}/_{32}$ in. (approx.)

Models for one-piece press molds should be attached to a mount. A *mount* holds the model in position or centered. It also prevents movement when "liquid" plaster is poured over the model. The mount also acts as a handle for

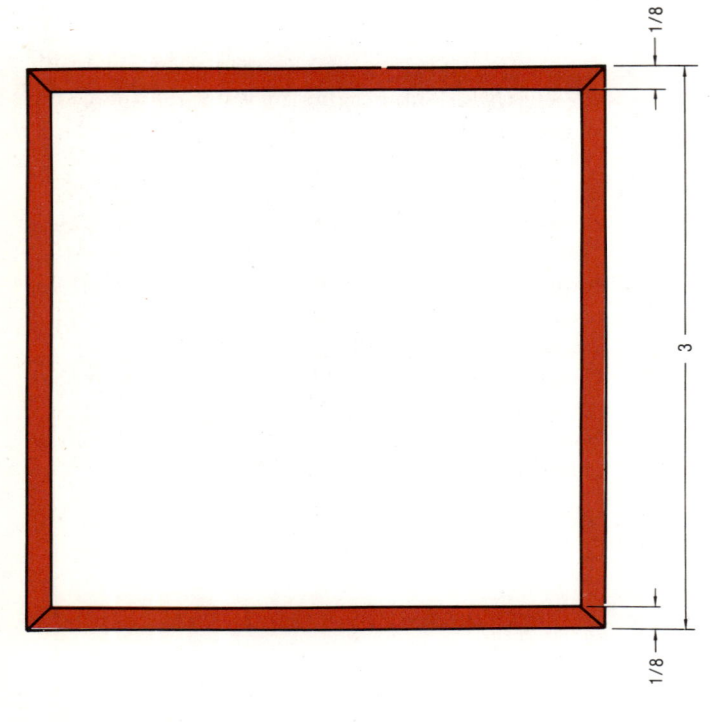

Fig. 32-1. A tile model for a one-piece press mold. Note that this simple model has draft or taper in only one direction. This model was made by direct carving a block or set plaster.

removing the model from the mold after the plaster chemically sets. Dimensions for a mount are thickness not more than 1 in. (2.5 cm); width and length are equal to the dimensions of the mold.

A correctly designed model for a slip-casting mold has a spare and trimshelf. A *spare* is an opening in the top of the mold that allows the *slip* to be poured into the mold and acts as a reservoir for extra slip (Fig. 32-2). The *trimshelf* is part of the spare. It is the flat shelf on the spare (see Fig. 32-2).

The spare and trimshelf can be made of the material used for the model. Often, they can be part of the model and made with it.

Dimensions for a spare are: height, ½ to 1 in. (1.3 to 2.5 cm); and diameter, 1 in. (2.5 cm) larger than the model. A tapered square with trimshelf is recommended (see Fig. 32-2).

The function of a spare and trimshelf in a mold is described in Unit 33—*Mold Making*.

Plaster Model Making Methods

The major methods of plaster model making are:

1. Direct carving (Fig. 32-3)
2. Lathe
3. Turning box (Fig. 32-4)
4. Pin template (Fig. 32-5)
5. Plaster wheel (Fig. 32-6)

Direct Carving

Direct carving means to cut the shape of the model from a piece of set plaster (see Fig. 32-3). This is done with hand tools and/or machines. Woodworking tools and machines can be used to shape, usually by cutting set plaster. A separate set of tools to be used only for plaster work is recommended. Set plaster readily contaminates and dulls the tools. Thus, the standard woodworking machines, such as a circular saw, are *not* often used on set plaster. Another suggestion is to use a hacksaw to cut plaster, since the blades can be replaced easily and inexpensively.

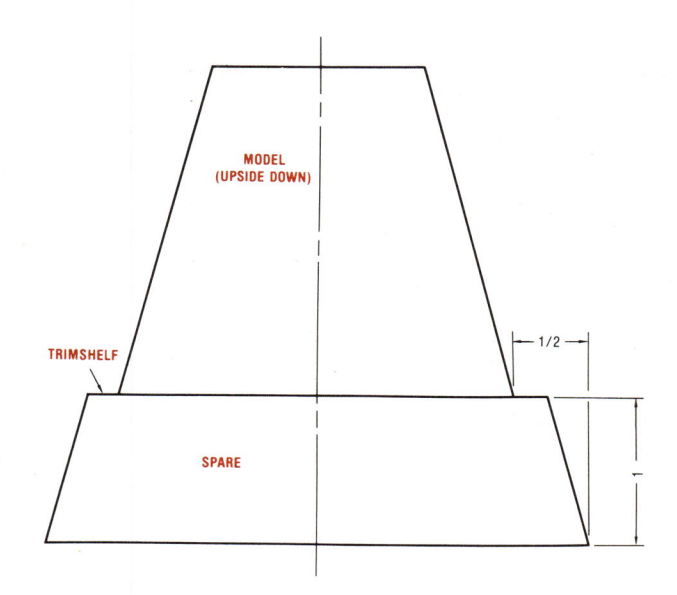

Fig. 32-2. A model for a one-piece drain-type slip casting mold. Note the recommended dimensions for the spare and trimshelf.

Fig. 32-3. Direct-carving a model for a press mold. Note the use of standard tools—*surform* and stencil knife. The remaining tool is a moldmaker's knife.

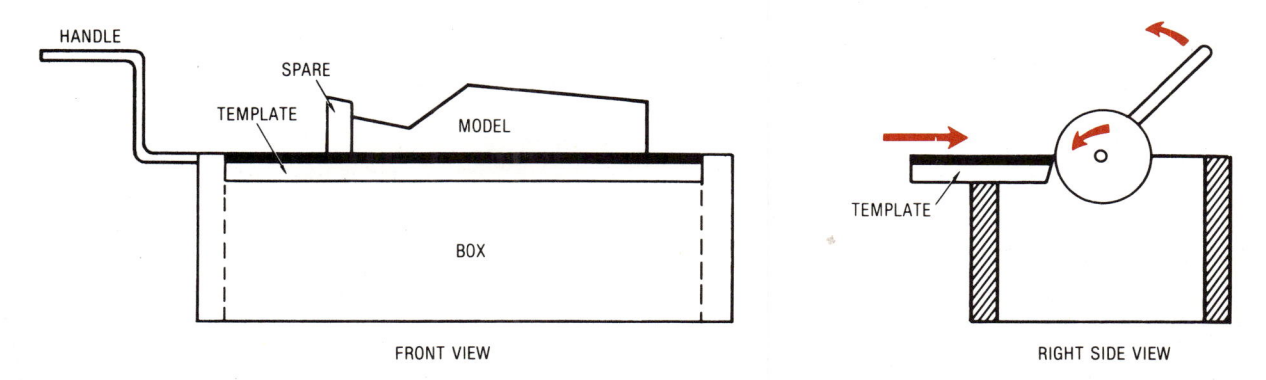

Fig. 32-4. A turning box is a manually powered template lathe. The model revolves around the horizontal axis while the template is stationary.

110 / Modern Industrial Ceramics

All types of shapes are possible when direct carving. The other model making methods are all limited to a cylindrical shape.

Lathe

A lathe is a machine that is used to revolve a cylinder block of set plaster on a horizontal axis. The operator cuts the desired shape with turning tools while the set plaster is revolving. Usually, a woodworking lathe is used. A metalworking or machine lathe could also be used. It is recommended that a cylinder of plaster be cast around a metal rod or tubing. One end of the rod or tubing should project 1 in. (2.5 cm) from the plaster. The opposite end should be flush with the plaster. This arrangement allows the plaster to be safely mounted in the lathe.

A wood dowel, flush with both ends, can also be used. This is mounted between the lathe centers. Usually, models for tall objects, such as vases, are turned on a lathe.

Turning Box

The turning box is a manually powered template lathe, (see Fig. 32-4). The model is built up around a rotating horizontal shaft. Freshly mixed plaster in a viscous state is used to build up the model. A stationary template, usually made of sheet metal, cuts the shape of the revolving model.

The turning box is usually limited to models for tall objects, such as vases. Both the turning box and the pin template are types of template forming.

Pin Template

A pin template is the opposite of a turning box (see Fig. 32-5). The template moves around a vertical axis while the plaster model is stationary.

Built-up soft plaster and a sheet-metal template are also used with this method. Thus, the concept and procedure are basically the same as those for the turning box.

Models for saucers, plates, and bowls can be made by this method. This method is primarily used for thin flat models.

Plaster Wheel

The plaster wheel is a machine that revolves a cylindrical block of plaster in the vertical axis. The operator cuts the desired shape with turning tools while the "soft" plaster is revolving.

Technically, the plaster wheel is a vertical lathe. Actually, it closely resembles a potter's wheel. In fact, a potter's wheel can be used.

A good plaster wheel has a three-sided enclosure that serves as a splash and backboard. It also has a variable-speed electric motor.

Usually, a plaster bat is attached to the plaster-wheel head. This bat will have an undercut or *chum* to hold the model in place (see Fig. 32-6). A model can be removed and replaced on the chum (Fig. 32-7). A chum is a simple chuck that revolves the model with the plaster wheel. It allows the model to be removed by lifting vertically. Technically, a chum is a frustrum of a cone with a key cut into the top surface. The key locks the model to the chum.

Procedures

Molds are discussed in Unit 33—*Mold Making*. General procedures for several different types of molds are provided. These procedures include model making, if applicable.

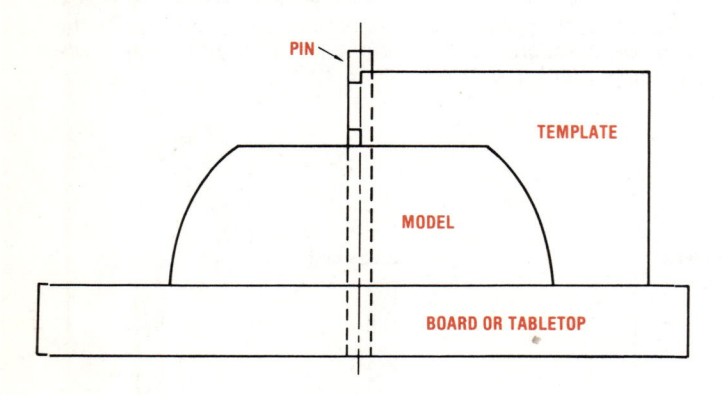

Fig. 32-5. A pin template for a soup bowl. The template revolves around the vertical axis while the model is stationary. Note that the model is shaped while in the upside-down position.

Model Making / 111

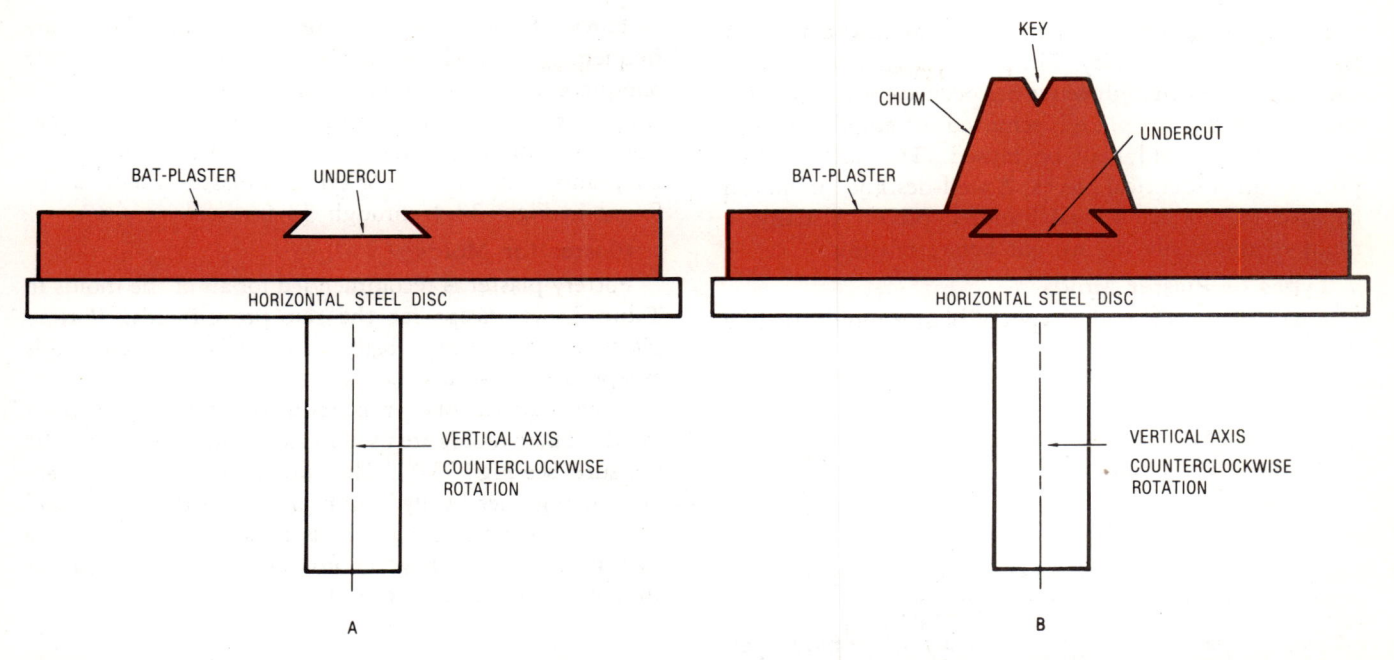

Fig. 32-6. Two plaster wheel bats, including undercut bat for very small diameter, tall, or very large cylindrical models (A); and bat with chum for medium-size cylindrical models (B). The undercut bat locks a model in place until cut off. A chum allows the model to be removed, vertically, and replaced on the plaster wheel when desired. The chum is a simple chuck that makes the model revolve with the plaster wheel.

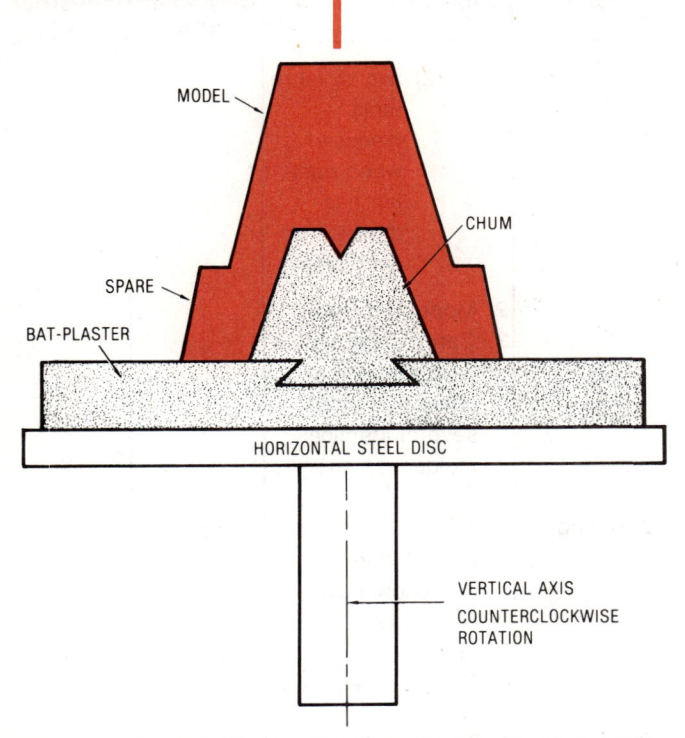

Fig. 32-7. Full-section drawing of a model with spare on a plaster wheel. The model is for a one-piece drain-type slip-casting mold. Note how the chum allows the model to be removed and replaced.

Unit 33: Mold Making

Mold making means to manufacture or make a mold. A *plaster mold* forms the clay. The plaster absorbs water from the clay. When enough water has been absorbed or when the clay is firm, the piece is removed, dried, cleaned up, and fired. A mold may be reused. This allows mass production of identical pieces. A well-designed mold can be used to produce several hundred pieces. Usually, a mold requires only drying out before reusing.

Types of Plaster Molds

There are five (5) major types of plaster molds used by ceramists. They are:

1. Drape molds
2. Press molds
3. Slip casting molds
4. Jiggering molds
5. Ram-press molds

Three of these (slip casting, drape, and press) are frequently referred to as the "simple molds." Only one-piece molds are "simple molds." Molds can be either simple or complex, depending on their design. Information concerning each mold is provided in Table 33-1. Diagrams of the five major types of plaster molds can be found in Figs. 33-1 through 33-5.

Plaster for Molds

Pottery plaster is recommended for all of the molds in Table 33-1, except for the ram-press molds. Pottery plaster is used, since porosity is required of all these molds except the ram-press molds.

A hard dense plaster is required for the ram-press molds. These molds are used in a machine with hydraulic pressure (see Fig. 33-5). The two halves of the mold are pressed together by hydraulic pressure—thus, the need for a strong plaster. Ram-press molds simply force the clay into shape. They do not absorb any significant amount of water from the clay.

Table 33-1. Types of Plaster Molds (used by ceramists)

Type	Details	State of Clay When Formed	Industrial Usage
Drape	1. Type: only one 2. No model 3. Manual process 4. Limited to draft in only one direction 5. Used by potters	Plastic	No
Press	1. Types: one- and two-piece 2. Model required 3. Manually press plastic clay into mold	Plastic	Yes (limited usage)

Slip casting	1. Types
	  a. Drain cast
	    (1) one-piece mold-draft in one direction
	    (2) two or more mold pieces—when casting has draft in two directions
	  b. Solid cast
	    (1) one-piece mold-draft in one direction
	    (2) two or more mold pieces—when casting has draft in two directions
	2. Model required
	3. Can manually pour or use in a machine that holds several identical molds
		Liquid (specifically casting mold)	Yes

Jiggering	1. Types
	  a. Flat ware
	  b. Hollow ware
	2. No model
	3. Machine process
	4. Template required
		Plastic	Yes

Ram press	1. Types: only one 2. Model required 3. Machine process 4. Two-piece mold only required	Plastic	Yes

Mold Making / 113

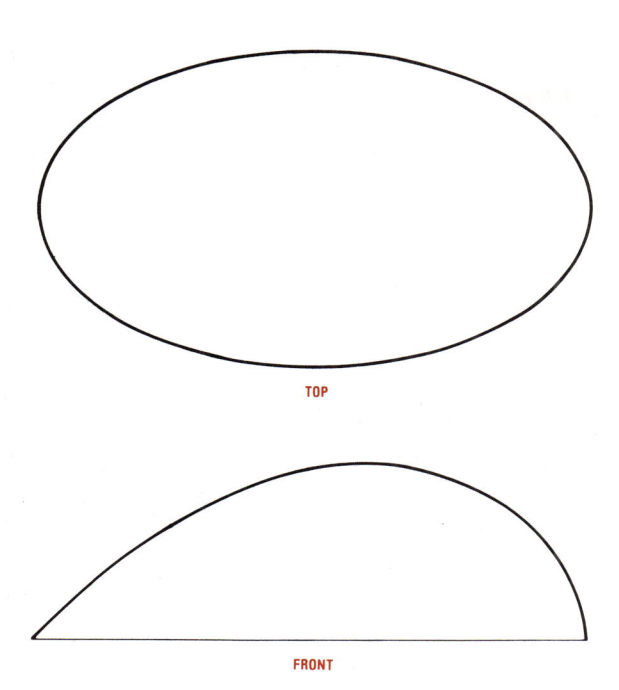

Fig. 33-1. Front and top views of a typical drape mold. Identical pieces are difficult to produce on this type of mold.

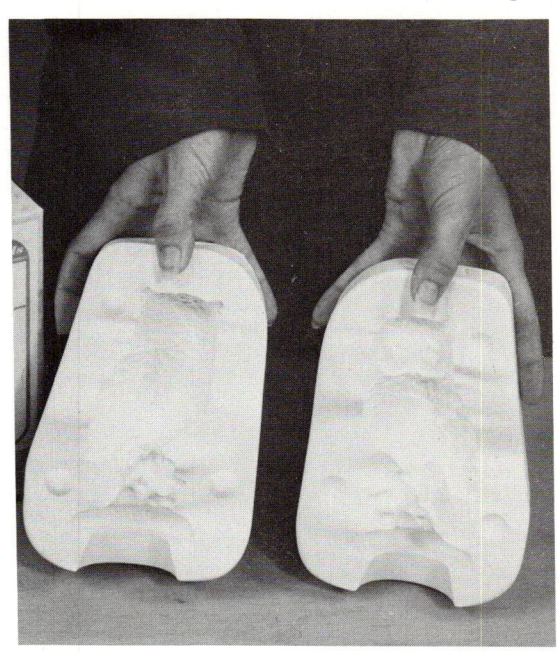

Fig. 33-3. A two-piece drain-type slip casting mold. The three knobs are joggles that align the mold when it is assembled. A *fettle* line will occur on the owl where the two pieces of the mold join. (A *fettle* is a parting line or seam.) The opening at the bottom is the spare. The owl will be upside down when the slip is poured into the assembled mold via the spare. (Copyright © Duncan Ceramic Products, A Division of Duncan Enterprises)

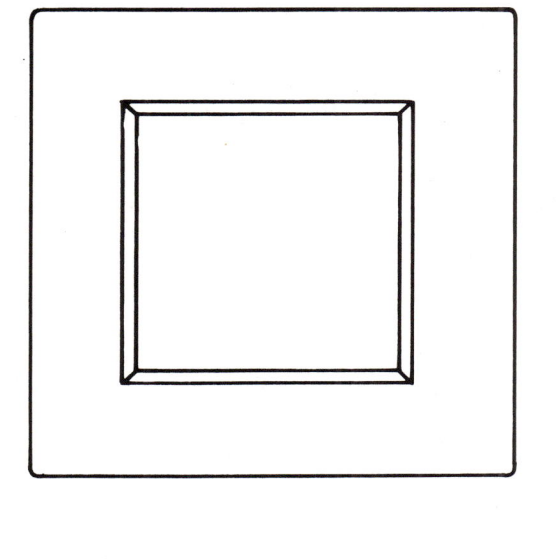

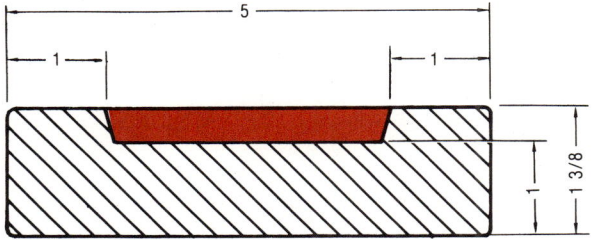

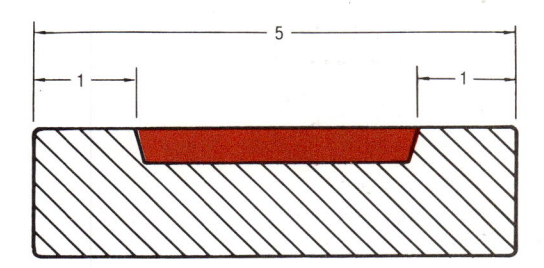

Fig. 33-2. A one-piece press mold for a square tile. Note that the mold is 1 inch larger than the tile to provide adequate absorption and strength.

114 / *Modern Industrial Ceramics*

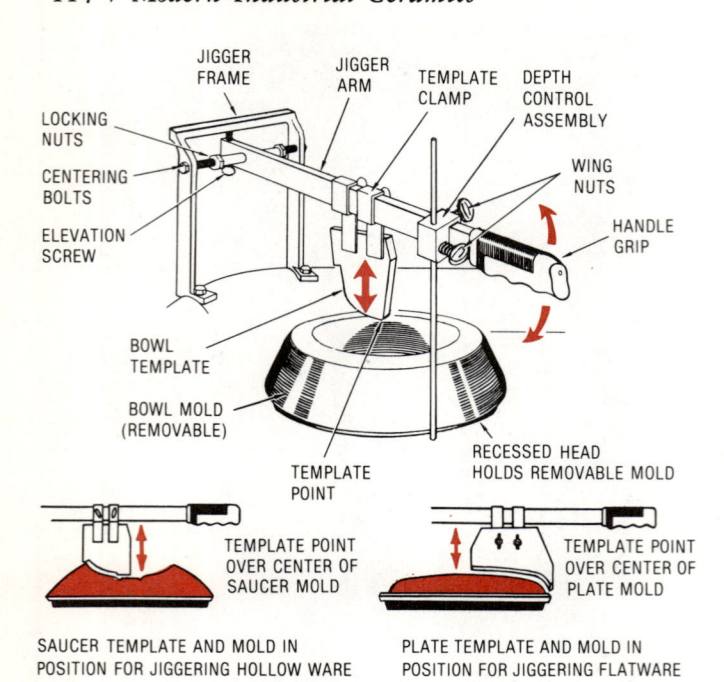

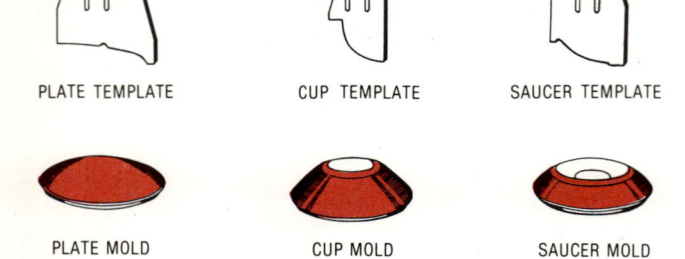

Fig. 33-4. A manually controlled jiggering machine with typical molds and templates. This type of jiggering machine is frequently used by small industrial firms and educational institutions. (Courtesy American Art Clay Co., Inc.)

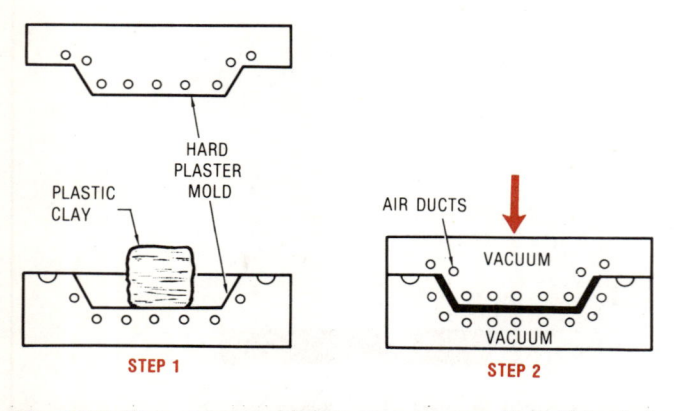

Designing Molds

There is one major rule for designing molds. The molds should be 1 in. (2.5 cm) larger than the model in all directions (see Fig. 33-2). Actually, this means the height will be 1 in. (2.5 cm) larger; and the width and length will be 2 in. (5.1 cm) larger.

For example, the model for a square tile is ¼ × 4 × 4 in. (0.64 × 10.2 × 10.2 cm). The mold will be 1¼ × 6 × 6 in. (3.2 × 15.2 × 15.2 cm).

This will provide enough plaster for adequate absorption and strength. Using more plaster is wasteful. More plaster also makes the mold unnecessarily heavy.

When the dimension of any model exceeds 12 in. (30.5 cm), the thickness of the mold wall may be increased. However, this should be increased proportionately and not exceed 2 in. (5.1 cm).

Mold Making Terms

There are several terms unique to mold making that require defining. They are:

1. Casting box
2. Cottle
3. Fettle line
4. Joggle
5. Spare
6. Trimshelf

A *casting box* is a square or rectangular form. "Liquid" plaster (see Unit 31. *Batching of Plaster*), is poured over a model inside the casting box. The box shapes the exterior of the mold.

The casting boxes are usually made of wood or sheet metal. Asphalt roofing or shingles, glass, linoleum, plastics, and similar nonporous rigid materials can be used. Usually, the size of the casting box is adjustable

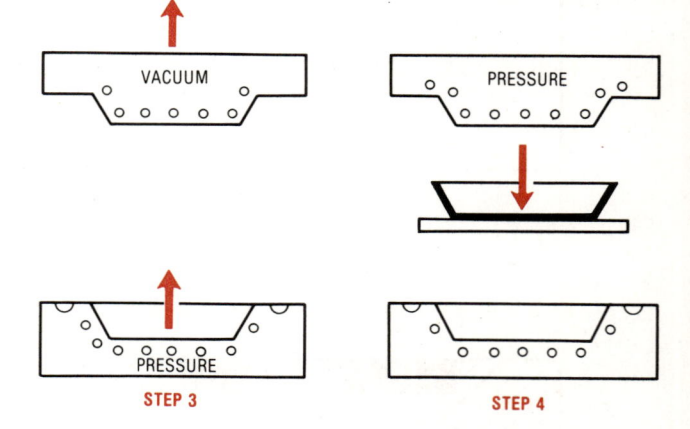

Fig. 33-5. Ram-press molds have two pieces with built-in air ducts. They form plastic clay by pressing when the top half of the mold is lowered into the bottom half. Hydraulic pressure is usually used to close the mold. Air releases the piece, as shown in Steps 3 and 4. The sequence is shown in Steps 1 through 4, as follows:

1. Place plastic clay in open mold
2. Press
3. Open and release ware from bottom half via air pressure
4. Release ware from top half of mold via air pressure

Mold Making / 115

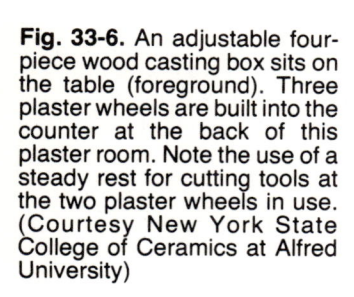

Fig. 33-6. An adjustable four-piece wood casting box sits on the table (foreground). Three plaster wheels are built into the counter at the back of this plaster room. Note the use of a steady rest for cutting tools at the two plaster wheels in use. (Courtesy New York State College of Ceramics at Alfred University)

(Fig. 33-6). Wedges, clamps, and plastic clay are used to hold and seal them. Also, they are usually easily disassembled.

Cottles function the same as the casting boxes. Their shape is different from the casting boxes. They are usually cylindrical or free form in shape. Thus, flexible sheet material, such as plastics, is used to make the cottles. Their size is adjustable, and they are easily disassembled. Also, they are secured and sealed, like the casting boxes.

A *fettle line* occurs on a clay piece where the pieces of a mold join (see Fig. 33-3). Actually, it is a parting line. The fettle lines occur when the clay is formed in molds of two or more pieces.

Joggles are alignment devices for two-piece press molds and for slip-casting molds of two or more pieces (see Fig. 33-3). Also, they prevent a mold from being assembled incorrectly. Alignment and correct assembly are insured by using two or more joggles of different sizes. The joggles are usually made of plaster, and they are part of the actual mold.

All correctly designed slip-casting molds have *spares*. A *spare* is an opening in the top of the mold, usually at the center (Fig. 33-7). The spares serve two functions: (1) they allow slip to be poured into the mold, and (2) they act as a reservoir for extra slip. The extra slip in the spare keeps the mold full when the water is absorbed. The clay in the spare is cut out before a piece is removed from the mold.

Spares can be made of several materials, including clay, plaster, and wood. Usually, the spare is attached to or is an integral part of the mold. Thus, when the "liquid" plaster is poured over the model, the spare is also formed.

A *trimshelf* is part of a spare. It is the flat shelf in the bottom of the spare (see Fig. 33-7). All correctly designed spares for slip-casting molds should have a trimshelf, if possible. Exceptions are the molds for objects with very small bases, such as figurines and salt and pepper shakers.

Plaster Mold Making Methods

The drape, one-piece press, and one-piece slip-casting molds are considered "simple" molds. The major methods for making these plaster molds are:

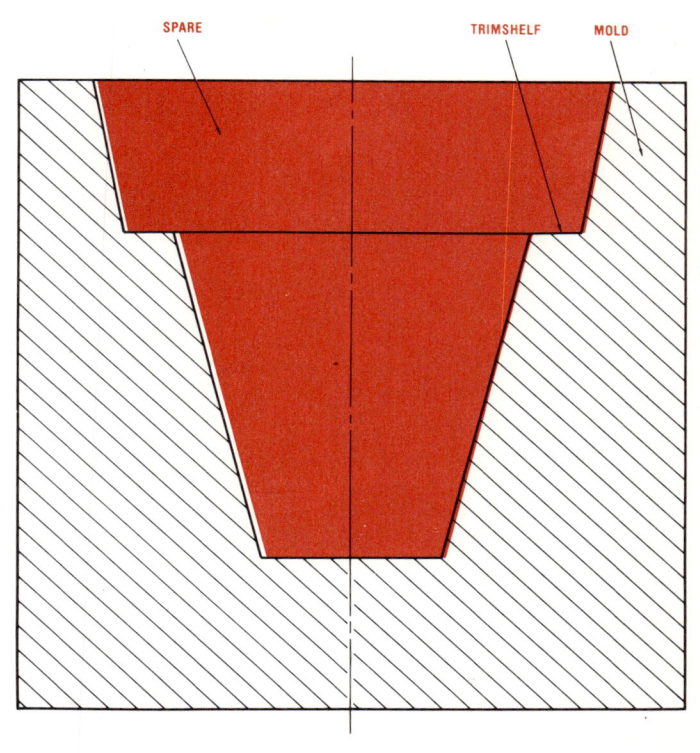

Fig. 33-7. A full-section drawing of a one-piece drain-type slip casting mold with a spare of adequate size and trimshelf.

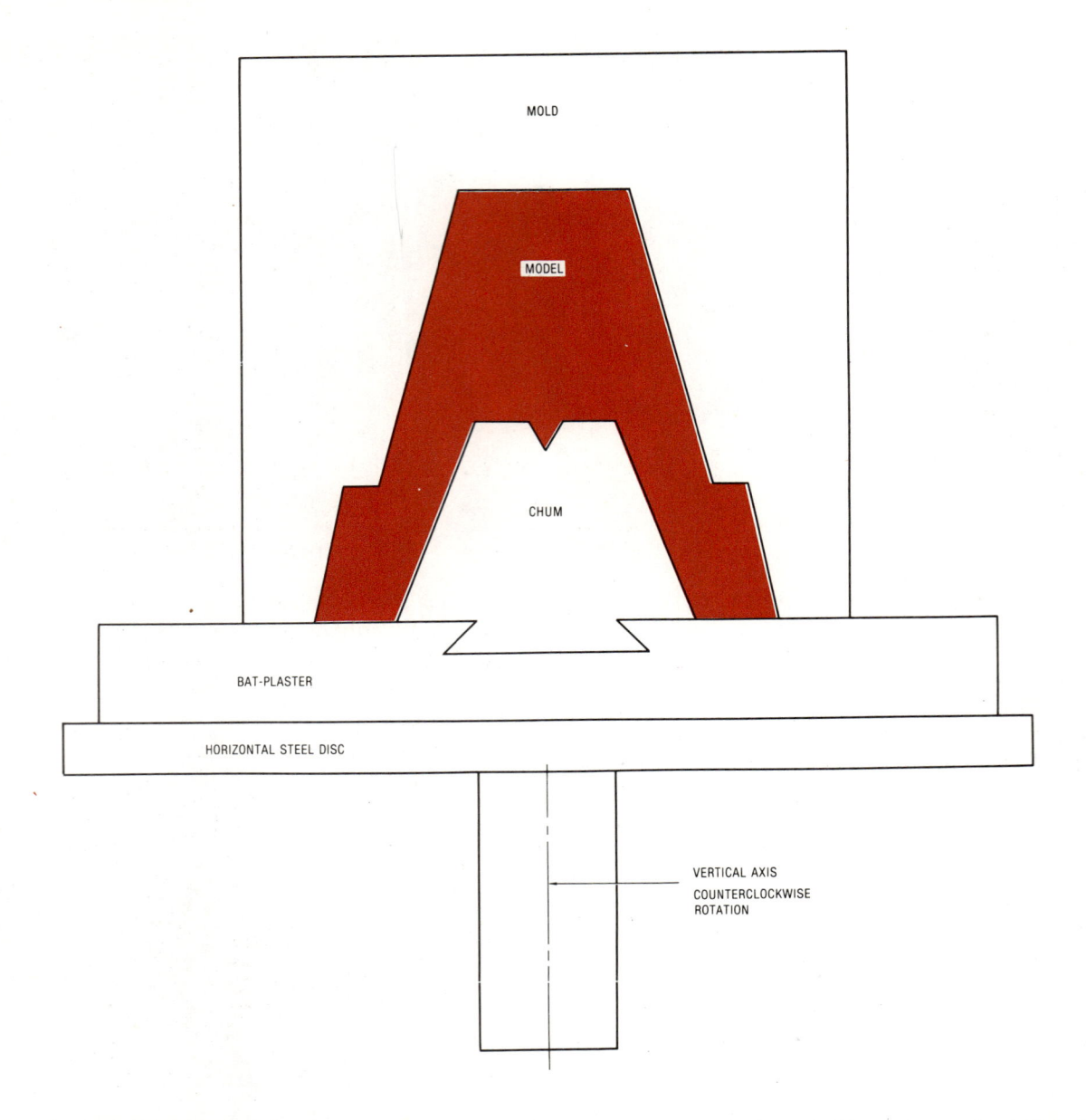

Fig. 33-8. A full-section drawing of a one-piece drain-type slip casting mold and model assembled on a plaster wheel. Note how the mold is made upside down.

1. Direct carving
2. Casting in box or cottle
3. Plaster wheel (see Figs. 33-6 and 33-8)

Drape molds are usually made by direct carving. They can be mass produced, using a case mold. A description of the case mold follows.

One-piece *press molds* and one-piece *slip-casting molds* are made with a casting box, cottle, or plaster wheel. Two-piece press and most slip-casting molds of two or more pieces are also made with a casting box or cottle.

Cylindrical molds are most easily made on a plaster wheel. These include drape, one-piece press, and slip-casting molds. However, the plaster wheel is most frequently used to make the one-piece slip-casting molds (see Fig. 33-8). Two- and three-piece slip-casting molds can also be made on a plaster wheel. Refer to the Unit 32. *Model Making* for discussion of the plaster wheel.

Block and Case Molds

The preceding methods are used only when one mold is to be made. When several molds are needed, a block and case mold is made (Fig 33-9).

Mold Making / 117

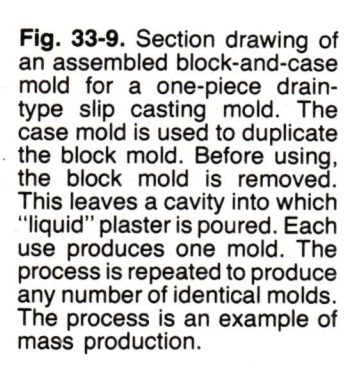

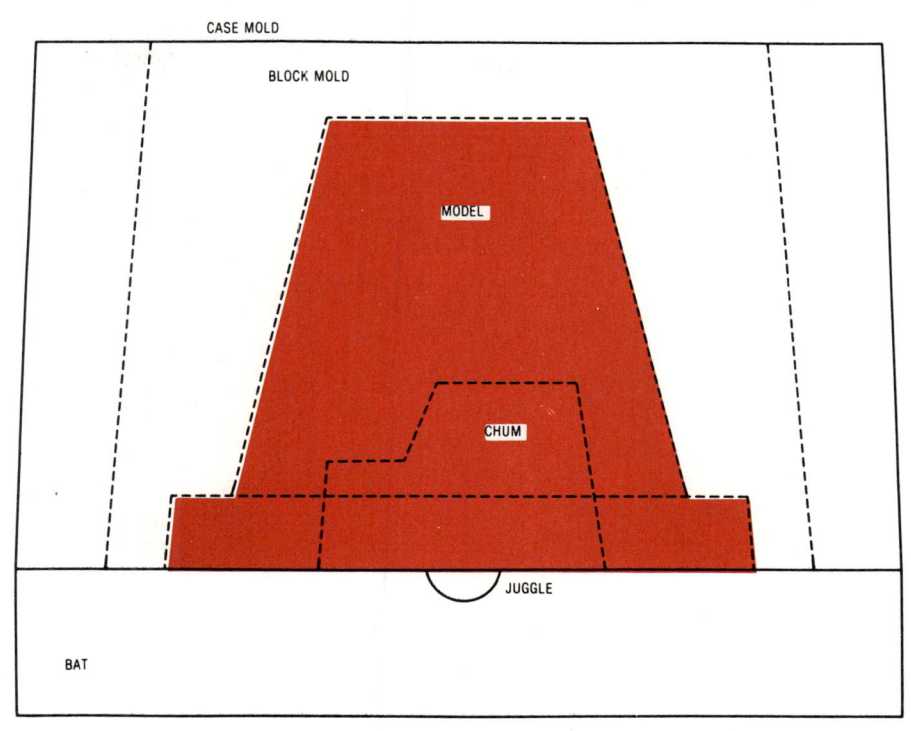

Fig. 33-9. Section drawing of an assembled block-and-case mold for a one-piece drain-type slip casting mold. The case mold is used to duplicate the block mold. Before using, the block mold is removed. This leaves a cavity into which "liquid" plaster is poured. Each use produces one mold. The process is repeated to produce any number of identical molds. The process is an example of mass production.

The original mold used to make a case mold is called a *block mold*. It has to be sealed with liquid soap. Therefore, since it cannot absorb water, its only use is to make more case molds.

A *case mold* is a special form that allows a mold to be duplicated. It may be made of or in combination with plaster, plastics, sheet metal and/or wood. "Liquid" plaster is poured into the case mold. When the plaster chemically sets, the new mold is removed, dried, and ready for sale or use.

The case mold can be reused to produce numerous identical molds. A case mold can make only one specific mold. Thus, a separate case mold is needed for each mold shape that is duplicated or mass produced.

A case mold usually has two or more pieces. They are easily assembled and disassembled.

Jiggering Molds

Jiggering molds are made directly on the machine or in a case mold. When they are made on the machine (Fig. 33-10), the procedure is similar to that used for the plaster wheel. One exception occurs in the turning operation. A template is used on the jiggering arm to cut the "soft" plaster into the shape of the mold (see Fig. 33-10).

The procedure above is used when only one mold is to be made. When several molds are needed, a block and case mold is made. The block or original mold is made on the jiggering machine as described above. Then, the case mold would be made. (Return to the previous discussion of the block and case mold in this unit.)

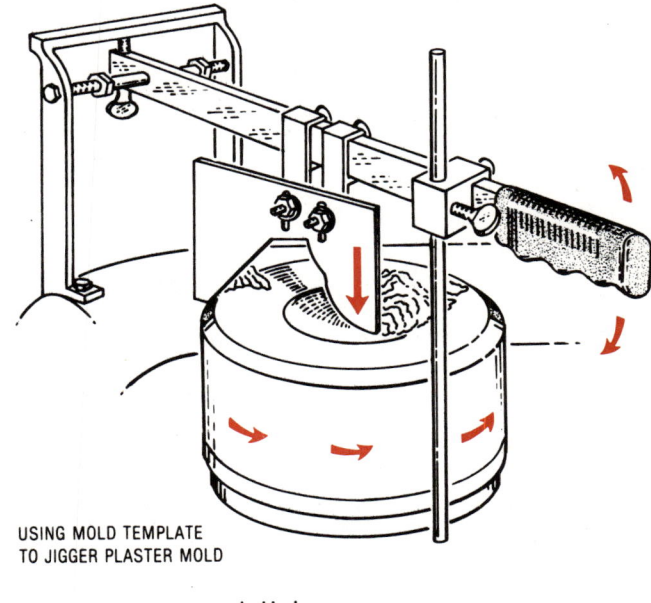

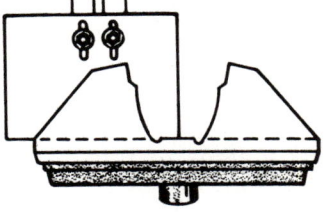

Fig. 33-10. Making a jiggering mold on a manually controlled machine. A case mold would be made and used when ten or more molds are needed. (Courtesy American Art Clay Co., Inc.)

Ram-Press Molds

Ram-press molds are made by using a casting box or cottle. These molds can be made in cubic, rectangular, or cylindrical shape. Industrial ram-press molds are usually cast in steel housings. Each section of the two-piece mold is cast in a separate housing. After casting, the mold remains in the housing which is mounted in a hydraulic press. Ram-press molds contain an imbedded air line and air passages. Thus, these molds have to be made individually. They cannot be made in a case mold.

Procedures

General procedures for making the plaster molds are limited to the "simple" molds. These include the drape (see Fig. 33-1), one-piece press (see Fig. 33-2), and one-piece slip-casting (see Fig. 33-7) molds.

The model making aspect is included, if applicable, in each procedure. This makes each procedure complete.

Drape Mold

The general procedure for making a plaster drape mold by the casting box or cottle method includes:

A. Designing/planning
  1. Full-scale working drawing(s)
  2. Make full-scale template(s) (for example, top, bottom, profile)
  3. Determine the form size
  4. Compute materials needed (see Unit 31. *Batching of Plaster*)
B. Manufacture mold
  1. Prepare form (casting box or cottle)
  2. Batch materials (see Unit 31. *Batching of Plaster*)
  3. Pour (see Unit 31. *Batching of Plaster*)
  4. Form removal (see Unit 31. *Batching of Plaster*)
  5. Direct carve the shape of the drape mold.
    a. Use hand tools and/or machines
    b. Use drawings and templates
    c. Use water to wet any air holes. Then fill them with freshly mixed "liquid" plaster.
  6. Finish mold exterior
  7. Dry mold (air or forced dry in a drying cabinet)

One-Piece Press Mold

The general procedure for making a one-piece plaster press mold by the casting box or cottle method includes:

A. Designing/planning
  1. Model
    a. Full-scale working drawing(s)
    b. Determine form size
    c. Compute materials needed (see Unit 31. *Batching of Plaster*)
  2. Model mount (recommended, but actually optional)
    a. Full-scale working drawing
    b. Determine form size
      (1) Thickness: not more than 1 in. (2.5 cm)
      (2) Width and length: equal dimensions of mold
    c. Compute materials needed (see Unit 31. *Batching of Plaster*)
  3. Mold
    a. Full-scale working drawing(s)
    b. Determine form size
    c. Compute materials needed (see Unit 31. *Batching of Plaster*)
B. Manufacture model
  1. Prepare form (casting box or cottle usually)
  2. Batch materials (see Unit 31. *Batching of Plaster*)
  3. Pour (see Unit 31. *Batching of Plaster*)
  4. Form removal (see Unit 31. *Batching of Plaster*)
  5. Make model, such as direct carving
C. Manufacture model mount
  1. Materials
    a. Plaster (recommended)
    b. Plastics
    c. Wood
  2. Manufacture: if made of plaster, repeat the previous steps in this procedure for B. 1–4 above.
  3. Lay out model location
  4. Adhere or attach model to mount
D. Manufacture mold
  1. Prepare model
    a. If any air holes, fill—usually, with synthetic clay
    b. Soap—3-6 coats of liquid mold soap
  2. Prepare form (casting box or cottle)
  3. Batch materials (see Unit 31. *Batching of Plaster*)
  4. Pour (see Unit 31. *Batching of Plaster*)
  5. Form removal (see Unit 31. *Batching of Plaster*)
  6. Remove model
  7. Finish mold exterior
  8. Dry mold (air dry or force dry in a drying cabinet)

One-piece slip-casting mold

The general procedure for making a one-piece slip-casting mold on a plaster wheel includes:

A. Designing/planning
  1. Model
    a. Full-scale working drawing(s) (Full section front view recommended)
    b. Determine cottle size
      (1) Add ¾ in. (19.0 mm) to diameter for truing
      (2) Add ¼ in. (6.4 mm) to height for truing
    c. Compute materials needed (see Unit 31. *Batching of Plaster*)
  2. Mold
    a. Full-scale working drawing(s) (Full section front view recommended)
    b. Determine cottle size

(1) Add ¾ in. (19.0 mm) to diameter for truing
(2) Add ¼ in. (6.4 mm) to height for truing
   c. Compute materials needed (see Unit 31. *Batching of Plaster*)
      (1) Compute
         (a) Volume of cottle for mold
         (b) Volume of finished model
           Note: Use smallest diameter and treat model as a simple cylinder, even though it is composed of two frustums of cones
      (2) Subtract
         (a) Mold cottle volume_____
         (b) Minus finished model volume_____
                              ans._____

B. Manufacture model
  1. Prepare plaster wheel
    a. Clean
    b. Soap (several light coats)
      (1) Chum (if used)
      (2) Bat
  2. Prepare cottle
  3. Batch materials (see Unit 31. *Batching of Plaster*)
  4. Pour (see Unit 31. *Batching Of Plaster*)
  5. Form removal (see Unit 31. *Batching of Plaster*)
  6. Turn model
    a. Make all cuts downward
    b. Use scrapers
    c. Wet sand
  7. If there are any air holes, fill—usually, with synthetic clay (model only)
  8. Dry (optional)
  9. Clean up wheel; then soap chum and bat

C. Manufacture mold
  1. Prepare model and plaster wheel
    a. Fill air holes in model only
    b. Soap
  2. Prepare cottle
  3. Batch materials (see Unit 31. *Batching of Plaster*)
  4. Pour (see Unit 31. *Batching of Plaster*)
  5. Cottle removal (see Unit 31. *Batching of Plaster*)
  6. Turn mold
    a. Make all cuts downward
    b. Use scrapers
    c. Wet sand
  7. Finish mold exterior
    a. Grips (optional)
    b. Round edges
  8. Dry mold (air or force dry in a drying cabinet)
  9. Clean up wheel, then soap chum and bat

Procedures for Using Plaster Molds

The general procedures for using plaster molds to form clay are included later in Section XIV. *Whiteware Industry*. Please refer to that section for discussions, procedures, and illustrations.

SECTION IX

Lime Industry

Unit 34: Material Science

Three major hydrosetting materials are used. They are plaster, lime, and Portland cement. Plaster is manufactured by the gypsum industry which was discussed in Section VII. This section is concerned with the lime industry. Section X deals with the Portland cement industry.

The manufacture of these hydrosetting materials can be expressed by a simple word equation:

$$\text{raw material(s)} \xrightarrow{\Delta} \text{primary product}$$

Hydrosetting means that a powdered material will chemically set or change into a homogeneous solid when water is added. Adding water to the above word equation, it becomes:

$$\text{raw material(s)} \xrightarrow{\Delta} \text{primary product} + \text{water} \longrightarrow \text{homogeneous solid}$$

We can now see how lime relates to the previous and following sections.

Definition and Properties

Lime is a calcium compound. It is a primary product that is usually sold in a powdered form. There are three types:

1. Agricultural lime
2. Quicklime
3. Slaked lime

The chemical composition of these types of lime are:

1. Agricultural—calcium carbonate ($CaCO_3$)
2. Quicklime—calcium oxide (CaO)
3. Slaked—calcium hydroxide ($Ca(OH)_2$)

Physically, lime is usually *white*. Impurities may cause an off-white color. Only quicklime is caustic. It has an affinity for water.

Raw Materials

Natural raw materials that are sources of calcium carbonate ($CaCO_3$) used by the lime industry include:

1. Limestone
2. Dolomite
3. Calcite
4. Marble
5. Chalk
6. Aragonite
7. Coral
8. Calcareous marl
9. Stalactites
10. Stalagmites

Limestone is the major raw material used by this industry. It is a sedimentary rock, usually gray in color. Chemically, it is calcium carbonate ($CaCO_3$).

Dolomite is also a sedimentary rock, usually dark gray in color. Chemically, it is a calcium and magnesium carbonate ($CaCO_3 \cdot MgCO_3$). Lime made from this rock is called dolomitic lime.

Calcite is a mineral. It is the crystal form of calcium carbonate (CaCO$_3$).

Marble is a metamorphic rock. It is either metamorphic limestone or dolomite.

Natural *chalk* is a soft white form of calcium carbonate. *Aragonite* includes shells, such as oyster shells. *Coral* is another form of aragonite. *Calcareous marl* is a limestone and clay mixture. Calcareous means that it is largely composed of calcium carbonate. *Stalactites* and *stalagmites* are formed in caves by dripping water that contains calcium carbonate. All these forms of calcium carbonate are found throughout the earth.

Unit 35: Procurement

Man procures materials either by harvesting or by extracting. The ceramic industries, including the lime industry, extract raw materials from the earth's crust.

Prospecting and Developing

A flow sheet for procuring raw materials for the lime industry is diagrammed in Fig. 35-1. Prospecting processes previously discussed are used by this industry. The task is not exceedingly difficult, since limestone is a universal material. The large number of raw materials listed in Unit 34 also makes prospecting an easier job.

The developing processes discussed previously are also used by the lime industry.

Winning

The three types of winning are *superficial, subterranean, and subaqueous.* All three types of winning are used by the lime industry. The location of the raw material determines the type of winning.

Superficial or open-pit winning is the method most frequently used by this industry (Fig. 35-2). This is due to the fact that limestone, the major raw material, is usually present on or near the earth's surface.

Superficial (Fig. 35-3) and subterranean processes are used to win limestone, dolomite, calcite, marble, chalk, and calcareous marl. Stalactites and stalagmites are won only by subterranean processes.

Subaqueous processes are used to win aragonite, coral, and calcareous marl.

Most of the material is placed in stockpiles after it has been won. Stockpiling insures continuous transformation plant operation (Fig. 35-4). Some companies transport the raw material directly to the plant. Then, storage bins are used to insure continuous operations. Very little rock is distributed directly from winning. Rock that is distributed would most likely be used as broken stone, not as a lime product.

Fig. 35-2. Superficial winning of limestone in Kentucky. The size of this open pit is indicated by the trucks in the background. Note the drilling machine and boxes of explosives in the foreground. (Courtesy National Limestone Institute, Inc.)

Fig. 35-1. A typical vertical flow sheet for procurement of raw materials used by the lime industry.

Fig. 35-3. A power shovel loading limestone in the open pit. Note the heavy-duty truck and drilling machine. (Courtesy The National Lime and Stone Company)

123

124 / *Modern Industrial Ceramics*

Fig. 35-4. Stockpiling (right foreground) insures continuous transformation plant operation. Note the irregular shapes and different sizes of the rock in the stockpile. These are the results of the explosives used in winning. A portable transformation—for example, crushed stone, plant is in the background. (Courtesy National Limestone Institute, Inc.)

Unit 36: Transformation

Transforming a raw material into lime is diagrammed in Fig. 36-1. It is a typical flow sheet for all three types of lime—agricultural lime, quicklime, and slaked lime. Transformation plants are often located adjacent to procurement (Fig. 36-2).

Beneficiation

Beneficiation refines the raw material into a usable form. Both comminution and the classifying processes are used.

Agricultural lime requires only beneficiation (Fig. 36-3). Primary and secondary crushing are usually used to produce this powdered lime. For example, the machines used may be a jaw crusher and a hammermill. Sieves are used to maintain a uniform fine-powder product. Most of the agricultural lime is sold in bulk.

Raw material to be calcined requires only primary crushing (Fig. 36-4). Sieves remove the oversize pieces, and they are returned to the primary crusher.

Calcining

Calcining was previously defined as the use of heat in a kiln, but not to the point of fusion, to cause a chemical and/or physical change. This heat-treating process is usually done in a vertical kiln or in a rotary kiln. Both are continuous kilns.

Vertical Kiln

The vertical or shaft kiln resembles a tall cylinder. Usually, these kilns are built with a steel exterior, refractory lining, and cylindrical shape (Fig. 36-5). This kiln was discussed previously. Refer to Section VI—*Kiln Industry* to review how it operates.

Crushed raw material is fed into the top of the kiln. It is preheated in the upper portion. Then, as it moves

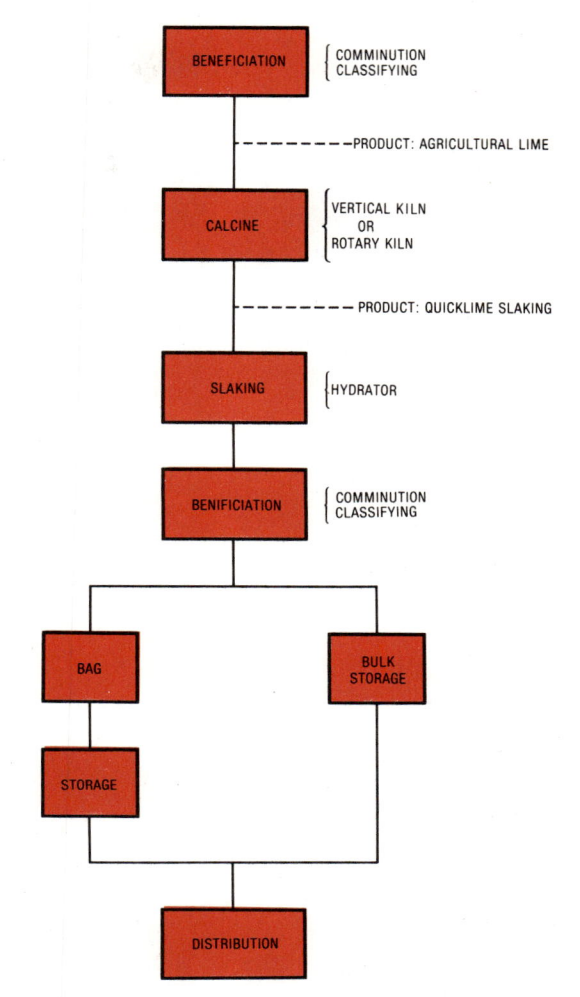

Fig. 36-1. A typical vertical flow sheet for the transformation of a raw material, such as limestone, into lime. Lime is usually sold in a powder form.

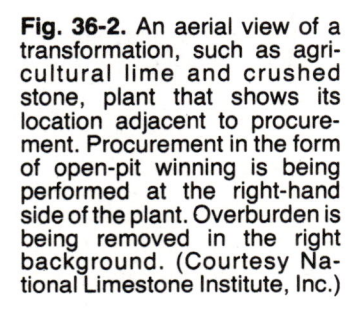

Fig. 36-2. An aerial view of a transformation, such as agricultural lime and crushed stone, plant that shows its location adjacent to procurement. Procurement in the form of open-pit winning is being performed at the right-hand side of the plant. Overburden is being removed in the right background. (Courtesy National Limestone Institute, Inc.)

126 / Modern Industrial Ceramics

Fig. 36-3. A stationary agricultural lime plant in Iowa. The primary crusher is located at the right, where the truck is unloading. Note the use of continuous-belt conveyors, hoppers, and stockpiles. (Courtesy National Limestone Institute, Inc.)

Fig. 36-4. A truckload of rock being dumped into a primary crusher. The continuous-belt conveyor moves the crushed material to the next machine process. (Courtesy National Limestone Institute, Inc.)

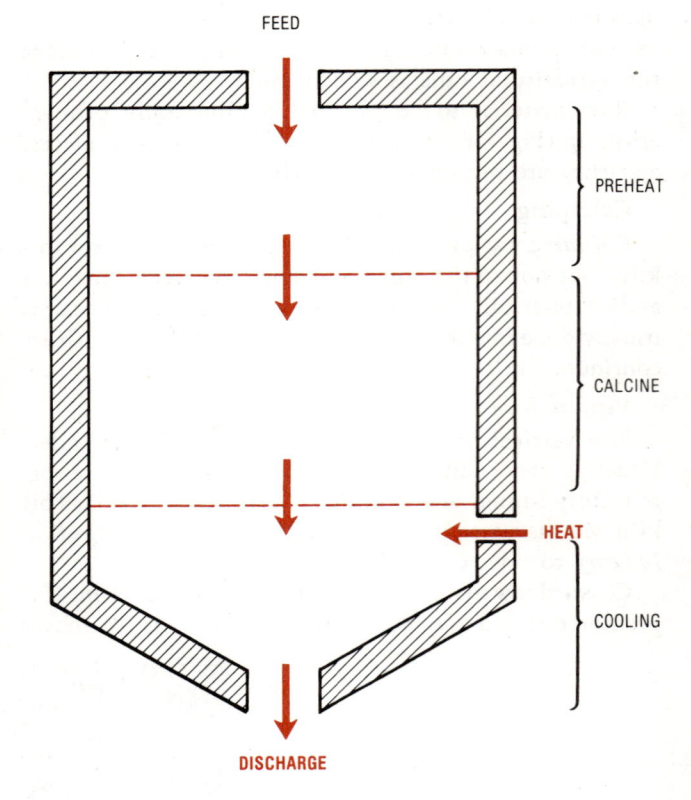

Fig. 36-5. Simplified full-section drawing of a typical vertical or shaft kiln. This kiln usually has a steel exterior and refractory lining. Each of the three zones is full of material that slowly moves downward.

downward inside the kiln, it is calcined. Cooling takes place at the bottom of the kiln.

Rotary Kiln

The rotary kiln is a steel cylinder with a refractory lining that is inclined to the horizontal, usually ½ in. (12.7 mm) per ft (30.5 cm). Large gears slowly rotate (usually 1 rpm) the entire kiln, when operating (Fig. 36-6). This kiln was also discussed in Section VI—*Kiln Industry*. Refer to that section for a review of how it operates.

Raw material is fed into the higher end of the kiln. The inclination and rotation cause the feed to move slowly downward through the kiln. The burner is at the lower end of the kiln. Thus, the heat rises and travels upward through the length of the kiln.

Preheating occurs in the higher portion of the kiln (see Fig. 36-6). Calcining occurs toward the lower end, where the temperature is highest. Cooling takes place after discharge from the kiln.

Process

The calcining process requires about 2500°F (1371°C). The heat drives off carbon dioxide (CO_2) from the calcium carbonate ($CaCO_3$), as shown by the chemical equation:

$$CaCO_3 \xrightarrow{\Delta} CaO + CO_2 \uparrow$$

Calcium oxide (CaO) is the result of calcining. It is known by several trade names:

1. Quicklime
2. Burnt lime
3. Caustic lime

This lime must be treated with *caution*. It readily combines with water (H_2O) and gives off heat while doing so. Quicklime does not require further processing.

Slaking

The term *slake* means to hydrate or to allow a material to take on water at its own rate. A hydrator or slaker is the machine used for slaking (Fig. 36-7). Both batch or periodic and continuous hydrators are used. The machine mechanically mixes the quicklime and water. The chemical reaction is:

$$CaO + H_2O \longrightarrow Ca(OH)_2 + heat \uparrow$$

This chemical reaction produces heat. Technically, it is called the heat of hydration.

Calcium hydroxide, $Ca(OH)_2$, is the result of slaking. It is known by several trade names as:

1. Slaked lime
2. Builder's lime
3. Hydrated lime

Storage

Bagging and bulk storage complete the transformation process diagrammed in Fig. 36-1. Lime is sold in paper bags, such as 50 and 100 pounds (23 and 45 kilograms), and in bulk. The bagging process is shown in Fig. 36-8. Bulk storage and bulk shipping by rail are shown in Fig. 36-9.

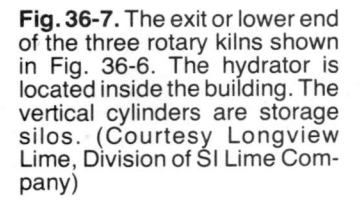

Fig. 36-6. Two views of rotary kilns. Compare the length of the three rotary kilns, the long slightly inclined to the horizontal cylinders, with the rail hopper cars (top). Limestone is fed into the higher or left-hand end of the kilns, and the calcium oxide exits at the lower or right-hand end of the kiln.

A closeup of the large gear and bearing that rotates this kiln is shown (bottom). Rotation is usually 1 rpm. The structure is a preheater, and the raw materials feed at the higher or left-hand end. (Courtesy Longivew Lime, Division of SI Lime Company)

Fig. 36-7. The exit or lower end of the three rotary kilns shown in Fig. 36-6. The hydrator is located inside the building. The vertical cylinders are storage silos. (Courtesy Longview Lime, Division of SI Lime Company)

128 / *Modern Industrial Ceramics*

Fig. 36-8. Packaging hydrated lime in 50-lb (23 kg) paper bags. Three bags are filled at the same time by the semiautomatic machine. The operator positions an empty bag and oversees the machine. Note the clean-air system the operator is wearing. (Courtesy Longview Lime, Division of SI Lime Company)

Fig. 36-9. Bulk-lime storage silos that permit direct loading of railroad cars for shipping. Similar bulk-storage silos are used to load trucks, barges, and ships. (Courtesy Longview Lime, Division of SI Lime Company)

Unit 37: Utilization

Products of the lime industry are used in manufacturing and construction. The major consumers of lime are the agricultural, chemical, construction, metals, and refractory industries.

Manufacturing

There are three groups of manufactured lime products. They are identified by their chemical composition:

1. Calcium carbonate, or $CaCO_3$
2. Calcium oxide, or CaO
3. Calcium hydroxide, or $Ca(OH)_2$

Calcium Carbonate

Agricultural lime is usually powdered limestone ($CaCO_3$) or dolomite ($CaCO_3 \cdot MgCO_3$). It is used for soil improvement (Fig. 37-1). This lime "sweetens" an acid soil.

Fig. 37-1. Broadcasting agricultural lime on farmland. The same truck is used for transporting and spreading the lime. Most agricultural lime is sold in bulk. (Courtesy National Limestone Institute, Inc.)

Powdered calcium carbonate may also be used in fertilizer. Fertilizer is used to enrich lawns, gardens, and soil in general.

Trapshooters and skeet shooters use clay pigeons for targets. The targets are made of calcium carbonate and pitch.

The ceramic industries call calcium carbonate *whiting*. Whiting is used primarily as a flux in clay bodies, glaze, porcelain enamels, and glass. A ceramic flux lowers the melting temperature.

A major use of calcium carbonate is to manufacture calcium oxide. The manufacturing process was discussed in Unit 36.

Calcium Oxide

The metals industry is a major consumer of calcium oxide. The iron and steel industry uses it as a flux. As a flux, it purifies iron or steel when it is melted in a furnace. Calcium oxide also serves a similar purifying role in the manufacture of other metals, such as magnesium and copper.

Other uses include making lye for soap, calcium carbide for acetylene, and in waste or sewage treatment. The calcium oxide neutralizes the acids and removes other impurities from waste and sewage.

A major use of calcium oxide is in manufacturing calcium hydroxide. The manufacturing process was discussed in Unit 36.

Calcium Hydroxide

Calcium hydroxide is used in water treatment for softening and purifying water (Fig. 37-2). Water treatment includes drinking water and the waste-water plants.

Today, the chalk for blackboards is a synthetic product. It is usually made of calcium hydroxide and plaster.

Other uses of calcium hydroxide include bleaching agents, the lines on athletic fields, and masonry cement.

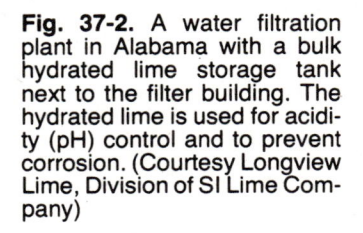

Fig. 37-2. A water filtration plant in Alabama with a bulk hydrated lime storage tank next to the filter building. The hydrated lime is used for acidity (pH) control and to prevent corrosion. (Courtesy Longview Lime, Division of SI Lime Company)

130 / Modern Industrial Ceramics

This cement is used for mortar, a basic construction material.

Construction

Calcium hydroxide or slaked lime is the major lime industry product used in construction. The major uses are in mortar, stucco, finish coat of wet-wall plaster, whitewash, and soil stabilization.

Mortar

Mortar is a bonding agent for stone, bricks, and concrete blocks (Fig. 37-3). The two types of mortar are lime mortar and cement mortar. Lime mortar is seldom used today. It was made of slaked lime, masonry sand, and water. This mortar is slow in setting and is not a strong bonding agent, as is the cement mortar. Incidentally, masonry sand contains particles only ⅛ in. (3.18 mm) and finer.

Cement mortar is made of Portland cement, lime, masonry sand, and water. A typical mix is: 1 part (by volume), Portland cement; 2 parts, slaked lime; 3 parts, masonry sand; and enough water to make a plastic or trowelable mass. Lime is not required when masonry cement is used. The mix would be: 1 part, masonry cement; and 3 parts, masonry sand.

Other Uses

Stucco is a continuous coating applied wet to exterior walls. Usually, its composition and ratios are the same as for cement mortar. Stucco is usually applied in two coats, like wet-wall plaster.

Plaster is a continuous coating applied wet to interior walls (see Fig. 37-3). This was discussed previously in Section VII—*Gypsum Industry*. Slaked lime is usually used in the second or finish coat of plaster. The lime helps in making the plaster harder and adds to its whiteness.

Whitewash is a coating similar to paint. Basically, it is a mixture of slaked lime and water. It is applied like paint to the interior walls of barns and poultry buildings.

The soil is often stabilized by adding slaked lime. Basically, the process involves thorough mixing, dampening with water, and compressing by rolling or other means. Roads, airport runways, earthen dams, and building blocks are some examples of places where soil stabilization has been used.

Fig. 37-3. A cross-section drawing of a brick veneer house showing the location and use of cement mortar and plastered interior wall. There are at least five ceramic products used in this house. How many of these are used in the construction of your home?

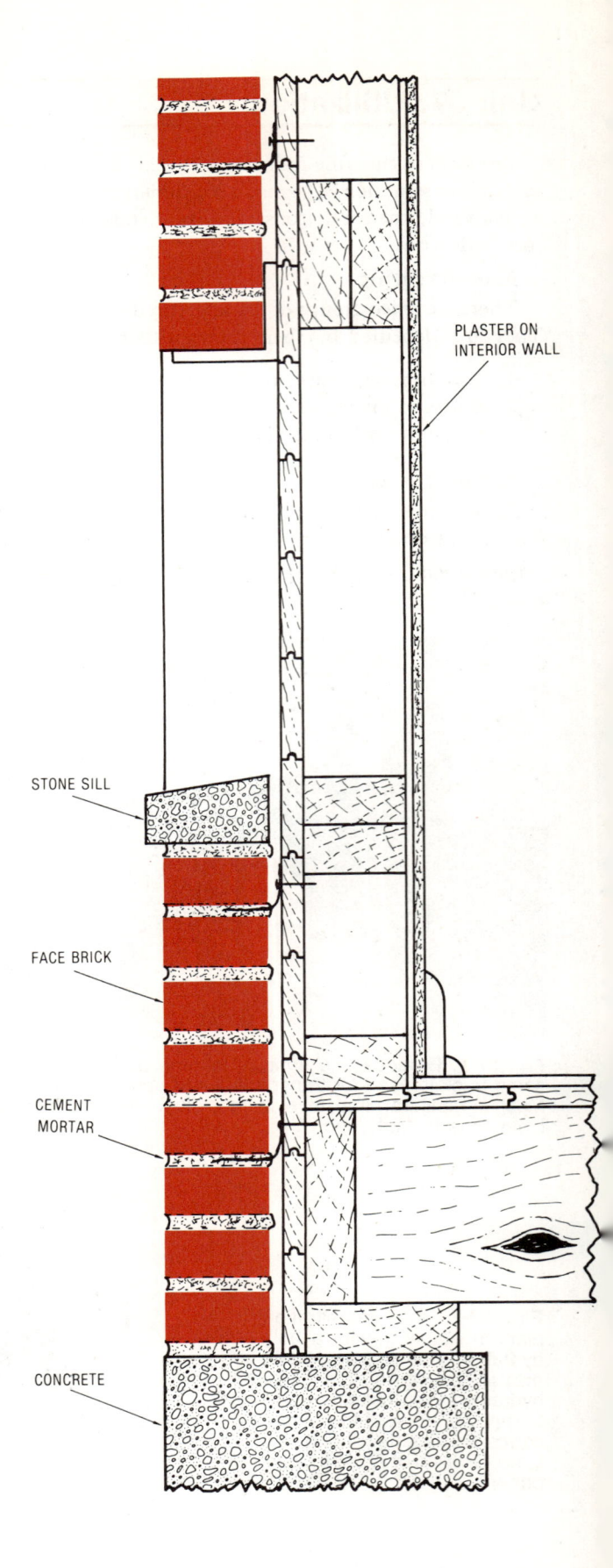

Unit 38: Disposition

A product is disposed of either by reclaiming or by discarding. One type of reclaiming is the removal of the fine particles from the kiln exhaust at lime plants (Fig. 38-1). These particles can be recycled as calcium oxide (CaO).

Reclaiming of lime products is usually impossible. Also, it is often undesirable. For example, agricultural lime becomes a part of the soil. Farm crops will consume the lime and eventually remove it from the soil.

Often, there is little or nothing to reclaim. The lime has become chemically combined with another product, such as glass or mortar. Then, it is the disposition problem of another industry, such as the glass and construction industries.

The disposition of glass will be discussed later in Section XVII—*Glass Industry.* Mortar can only be discarded after it has been used. Usually, it is discarded as hard fill and, therefore, is beneficial.

It is rarely necessary to discard any type of lime. If any types of lime are discarded, they do not usually create an ecological problem. For example, calcium carbonate ($CaCO_3$) is a natural material, calcium oxide (CaO) would combine with water that would change it into calcium hydroxide ($Ca(OH)_2$, and calcium hydroxide would slowly combine with carbon dioxide (CO_2). This would recarbonate the lime or return it to calcium carbonate.

Fig. 38-1. Kiln exhaust stacks with air cleaning equipment at a lime plant. (Courtesy The National Lime and Stone Company)

SECTION X

Portland Cement Industry

Unit 39: Material Science

The third major hydrosetting material introduced here is Portland cement. The other two hydrosetting materials, plaster and lime, were introduced in Section VII—*Gypsum Industry* and Section IX—*Lime Industry*.

The manufacture of hydrosetting materials has been previously expressed by a simple word equation:

$$\text{raw material(s)} \xrightarrow{\Delta} \text{primary product}$$

This equation also describes the manufacture of Portland cement. However, there are four major differences that distinguish Portland cement from plaster and lime. Briefly, these differences are:

1. *Raw material(s)*. Plaster and lime each require only one raw material. Portland cement usually requires at least three different raw materials.
2. *Heat treatment*. Plaster and lime each require calcining, but at different temperatures. Plaster requires a low temperature (340°F or 171°C), while lime requires a high temperature (2500°F or 1371°C). Calcining requires heating in a kiln, but not to the point of fusion. Portland cement also requires a high temperature (up to 3000°F or 1649°C) for heat treating. However, the heat-treating process, burning, causes partial fusion.
3. *Primary product*. Plaster and lime are primary products that are composed of a single chemical compound, such as $2CaSO_4 \cdot H_2O$ and CaO. Portland cement is composed of four major chemically complex compounds—for example, tetracalcium aluminoferrite ($4CaO \cdot Al_2O_3 \cdot Fe_2O_3$).
4. *Addition*. Only to the Portland cement is a material added after heat treating. The addition is raw rock gypsum.

These differences set the stage for our discussion of the Portland cement industry. They also indicate that Portland cement and its manufacture are more complex than plaster and lime.

Definitions

Portland cement is a hydrosetting powder. It is gray or white in color and composed of very fine particles (Fig. 39-1). The powder will pass a 200-mesh sieve.

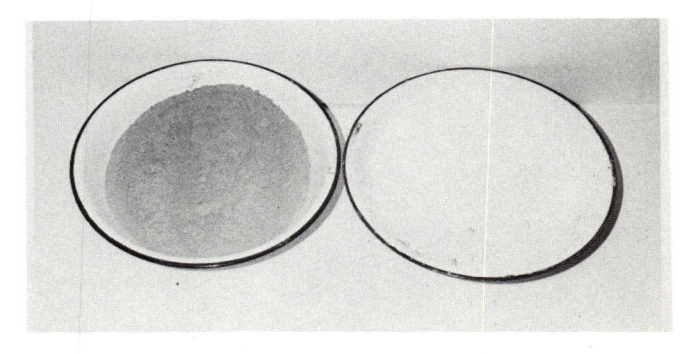

Fig. 39-1. Portland cement is a fine powder. It is made in two colors—gray (left) and white (right).

Concrete is a homogeneous solid, usually made from Portland cement, fine aggregates, coarse aggregates, and water (Figs. 39-2 and 39-3). It is incorrect to call concrete "cement," as is done all too frequently! Note the differences: (1) Portland cement is a fine powder (Fig. 39-1), but concrete is a homogeneous hard dense solid (Fig. 39-4); and (2) Portland cement hydrosets, but concrete does not. Concrete is dealt with in Section XI—*Concrete Industry*.

Mortar is technically a special type of concrete. It is used to bond or adhere bricks, concrete blocks (Fig. 39-5), or similar materials together, such as a brick wall. The usual composition of mortar is Portland cement, lime, masonry sand, and water.

Properties

Portland cement is composed of four major chemical compounds, as shown in Table 39-1.

Chemical

Also, Portland cement is a hydrosetting material. This means that Portland cement will chemically set or change into a homogeneous solid when water is added.

Physical

Portland cement is made in two colors—gray and white (Fig. 39-1). The Portland cement most frequently used is

Fig. 39-2. Fine and coarse aggregates for concrete. The three sections (bottom right) contain coarse aggregates. The other six sections contain fine aggregates. (Courtesy Portland Cement Association)

Fig. 39-3. The concrete used for this patio was made of Portland cement, fine and coarse aggregates, and water. The concrete becomes a homogeneous solid when the Portland cement hydrosets.

Fig. 39-4. Concrete is a homogeneous hard dense solid. A concrete brick, a patio block, and two 8-in. blocks are shown (top) and stockpiles of several different precast concrete products are shown in the photo (bottom).

gray in color. It is a very fine powder that passes a 200-mesh sieve (Fig. 39-6). A 200-mesh sieve has 40,000 openings per square inch.

Raw Materials

The four basic components needed to make Portland cement are:

1. Lime (CaO)
2. Silica (SiO$_2$)
3. Alumina (AL$_2$O$_3$)
4. Iron (Fe$_2$O$_3$)

Fig. 39-5. A wall made of concrete blocks. Mortar was used to bond the blocks together.

Table 39-1. Major Chemical Compounds in Portland Cement

Compound	Formula	Abbreviation*
Tricalcium silicate	3CaO·SiO$_2$	C$_3$S
Dicalcium silicate	2CaO·SiO$_2$	C$_2$S
Tricalcium aluminate	3CaO·Al$_2$O$_3$	C$_3$A
Tetracalcium aluminoferrite	4CaO·Al$_2$O$_3$·Fe$_2$O$_3$	C$_4$AF

* Simplified chemical abbreviations developed and used by only the Portland cement industry.
(Courtesy Portland Cement Association)

The technical names for these four components are:

1. Calcareous (CaO)—lime
2. Siliceous (SiO$_2$)—silica
3. Argillaceous (Al$_2$O$_3$)—alumina
4. Ferriferous (Fe$_2$O$_3$)—iron

The materials that provide these components are listed in Table 39-2 and can be seen in Fig. 39-7. Sometimes,

Fig. 39-6. Portland cement is a fine powder that passes the 200-mesh sieve shown above. A 200-mesh sieve has 200 openings per linear inch or 40,000 openings per square inch.

Table 39-2. Typical Sources of Materials Used in Manufacture of Portland Cement

Components				
Lime	Silica	Alumina	Iron
Cement rock	Sand	Clay	Iron ore
Limestone	Traprock	Shale	Iron calcine*
Marl	Calcium silicate	Slag	Iron dust
Alkali waste*	Quartzite	Fly ash*	Iron pyrite
Oyster shell	Fuller's earth	Copper slag*	Iron sinters*
Coquina shell		Aluminum ore refuse*	Iron oxide
Chalk		Staurolite	Blast furnace flue dust*
Marble		Diaspore clay	
		Granodiorite	
		Kaolin	

* Manufactured materials. All others are natural materials.
(Courtesy Portland Cement Association)

136 / *Modern Industrial Ceramics*

Fig. 39-7. Typical raw materials used to make Portland cement. The lime components (left), plus silica, alumina, and iron components (center) yield clinker (after burning), plus gypsum yields Portland cement. (Courtesy Portland Cement Association)

Fig. 39-8. Joseph Aspdin named and patented Portland cement in 1824. He was an English stonemason. (Portland Cement Association)

a single raw material can provide two or three components. For example, clay and shale are hydrous alumina silicate ($Al_2O_3 \cdot 2SiO_2 \cdot 2H_2O$). They provide two components, silica and alumina. A red clay or shale could provide three of the components—silica, alumina, and iron.

However, all this can be simplified. The typical or basic raw materials (usually three) used to make Portland cement are usually limestone, sand, and red clay or shale.

After the above materials are burned, another material is added. The addition is raw rock gypsum. It functions as a retarder to prevent a "flash" set when water is added to the mixture.

Background

Joseph Aspdin invented Portland cement and patented it in 1824 (Fig. 39-8). He was an English mason. Aspdin named his product Portland cement, since it resembled the rock on the Isle of Portland, England.

Portland cement was first brought to the United States as ballast in ships. This was about 1870. The first Portland cement made in the United States was produced at Coploy, Pennsylvania in 1872.

Since 1872, many changes and improvements have been made in the manufacture of Portland cement. Modern procurement and transformation processes used to make Portland cement are discussed in Units 40 and 41.

Unit 40: Procurement

Procuring the raw materials for the Portland cement industry is diagrammed in Fig. 40-1. The prospecting and developing processes discussed previously are used by this industry.

Three factors which make material procurement for this industry different from those previously discussed are:

1. Both raw and manufactured materials may be used to make Portland cement, as shown in Table 39-2.
2. Several materials are needed to provide the four chemical components necessary. For example, at least three materials are usually needed, such as limestone, sand, and red shale.
3. Locating material suppliers is necessary. Purchase of manufactured material and one or more of the raw materials may be required. Some Portland cement plants purchase all their materials. The purchase of material depends on the location of the plant, number of materials available, and the quantity of materials available.

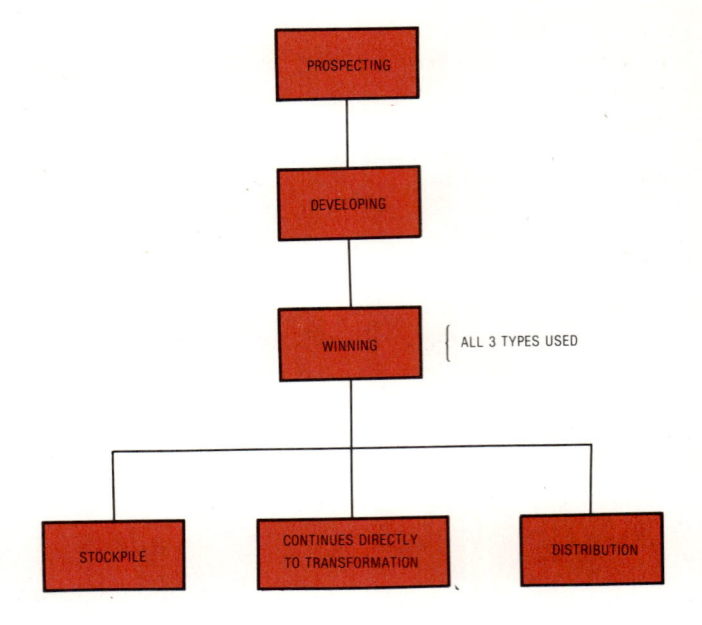

Fig. 40-1. A typical vertical flow sheet for procurement of raw materials used by the Portland cement industry.

Fig. 40-2. Aerial view of a modern cement plant. Deposits of raw materials are located at the left foreground. (Courtesy Portland Cement Association)

138

Plant Location

Portland cement plants are usually located near the deposits (Fig. 40-2) and/or suppliers. Also, they may be located near steel mills, which are a source of slag. Or they may be located near bodies of water where materials can be received and Portland cement transported by barges and ships.

Winning

All three types of winning or mining are used by this industry. The three types are:

1. Superficial
2. Subterranean
3. Subaqueous

The location of the raw materials determines the type of winning used.

Raw materials deposits on or near the earth's crust would be won by *superficial* processes (Fig. 40-3). This type of winning is most often used by the Portland cement industry and its suppliers. Many of the raw materials, such as limestone, sand, clay, and shale, are available on or near the earth's surface.

Subterranean processes (Fig. 40-4) are used to obtain those materials that are too deep for superficial winning. Marble, quartzite, and granodiorite are typical examples of these raw materials.

Subaqueous processes are used to obtain materials that are under water. These include marl, shells, and clay.

Helping Ecology

Raw material that is unsuitable for its typical use can be used to make Portland cement. For example, the marble and granodiorite that the dimension stone industry rejects, due to cracks or other flaws, can be used in making Portland cement.

The use of ash and slags helps reduce the disposition problems of other industries. This is especially helpful to the iron and steel industry.

Stockpile

Materials for making Portland cement are usually stockpiled at the plant (Fig. 40-5). This insures continuous plant operation. Stockpiling is a form of insurance that eliminates many problems caused by the weather, labor, or mechanical breakdown.

Fig. 40-3. Superficial winning of raw material for a Portland cement plant. A power shovel loads a truck that will transport the rock to the plant. (Courtesy Portland Cement Association)

140 / *Modern Industrial Ceramics*

Fig. 40-4. A drilled hole is being filled with explosives. Blasting is used in both superficial and subterranean winning. (Courtesy Portland Cement Association)

Fig. 40-5. Stockpiling raw materials insures continuous plant operation. Stockpiles help eliminate problems caused by weather, labor, or mechanical breakdown. (Courtesy Portland Cement Association)

Unit 41: Transformation

Transformation of materials into Portland cement is diagrammed in Fig. 41-1. A pictorial flow sheet for both procurement and transformation is provided in Fig. 41-2. Together, the two flow sheets provide a comprehensive picture of all the Portland cement manufacturing processes. The sequence of the processes is clearly shown in each flow sheet.

Beneficiation

The type of crusher used in a plant depends on the type and amount of raw or manufactured material that must be reduced in size. Rock, such as limestone, is usually reduced by two crushers—primary and secondary. The primary crusher, for example, a gyratory (Fig. 41-3) or jaw crusher, reduces the rock to 5 in. (12.7 cm). The secondary crusher, such as a hammermill, reduces the rock to ¾ in. (1.9 cm), or less. Oversize particles are recycled through the crushers. Each material is stored separately after crushing.

Wet and Dry Process

The process of fine grinding may be done either wet or dry. A plant will use only one of the two processes.

Wet Process

About 60 percent of the industry uses the wet process. It is the older of the two processes. It was first used in Europe before 1872. The wet process is used when some of the raw materials are moist, such as clay and marl.

The advantages of the wet process are: (1) reduces dust, and (2) the liquid called slurry is easy to move. Disadvantages are: (1) the slurry must be agitated during storage, otherwise particles settle out of the water; and (2) the extra water, about one-third of the slurry, has to be removed at the beginning of the burning process. This consumes more heat and thus more fuel than the dry process.

The wet process consists of seven steps: (1) Each material is stored separately after crushing, as previously mentioned, (2) A chemical analysis is done on each material (Fig. 41-4). This determines the amount of each material to be used. (3) Materials are physically proportioned when they are conveyed to fine grinding. (4) Water is added as the materials are fed into fine grinding. Very large ball mills are used for fine grinding (Fig. 41-5). About 85 percent of the particles will pass a 200-mesh sieve after this process. (5) Classifying causes the oversize particles to be recycled through the mill. (6) Further chemical analysis may require blending of slurry from different storage tanks to achieve precise proportions prior to burning. (7) Slurry is agitated by air pressure and/or mixers while in the storage tanks (Fig. 41-6). The tanks insure continuous kiln operation.

Dry Process

About 40 percent of the industry uses the dry process. This process is used when the raw materials are dry.

The advantages of the dry process are: (1) no water is required, and (2) the burning process is faster, since there is no excess water to be removed. Disadvantages of the dry process are: (1) dust collectors are necessary, due to the fine particles and (2) sometimes, heating is required to remove surface moisture. Drying prevents compacting in the mill and clogging of the sieves.

This process consists of seven steps. The steps are identical to those in the wet process, except: (1) no water

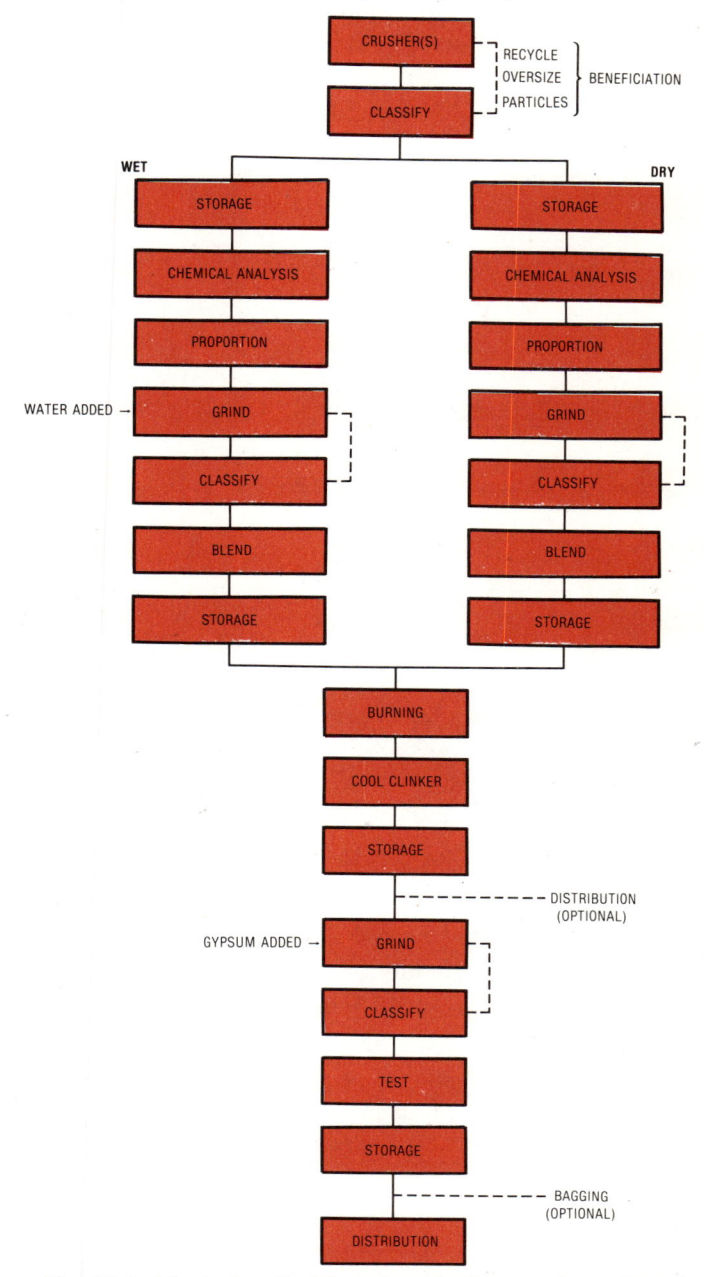

Fig. 41-1. A typical vertical flow sheet for the transformation of materials into Portland cement. The final product, Portland cement, is in the form of a powder.

142 / *Modern Industrial Ceramics*

1 STONE IS FIRST REDUCED TO 5-IN. SIZE, THEN 3/4 IN., AND STORED

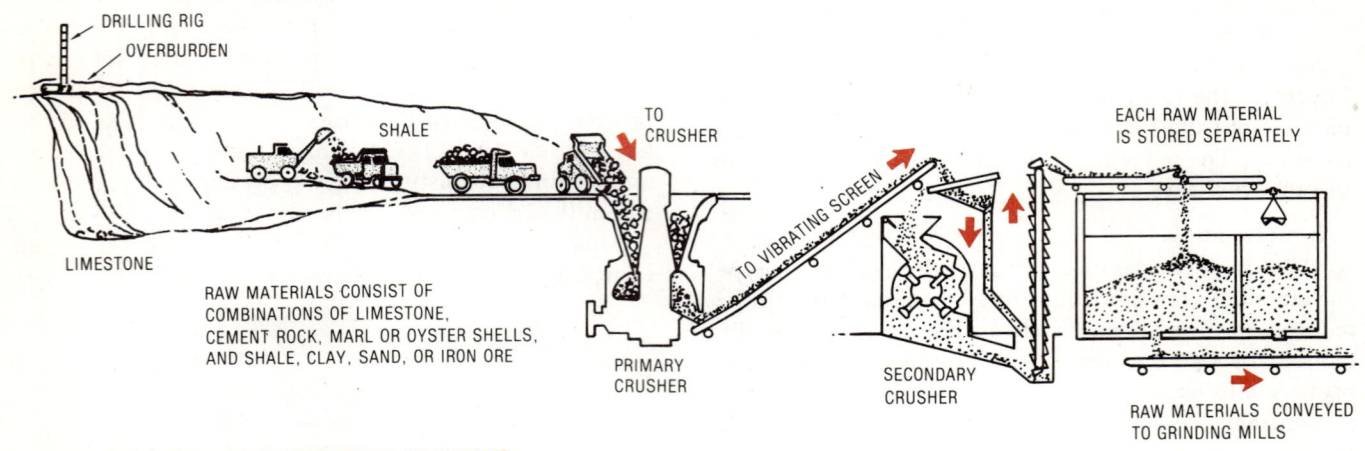

2 RAW MATERIALS ARE GROUND TO POWDER AND BLENDED

RAW MATERIALS ARE GROUND, MIXED WITH WATER TO FORM SLURRY, AND BLENDED

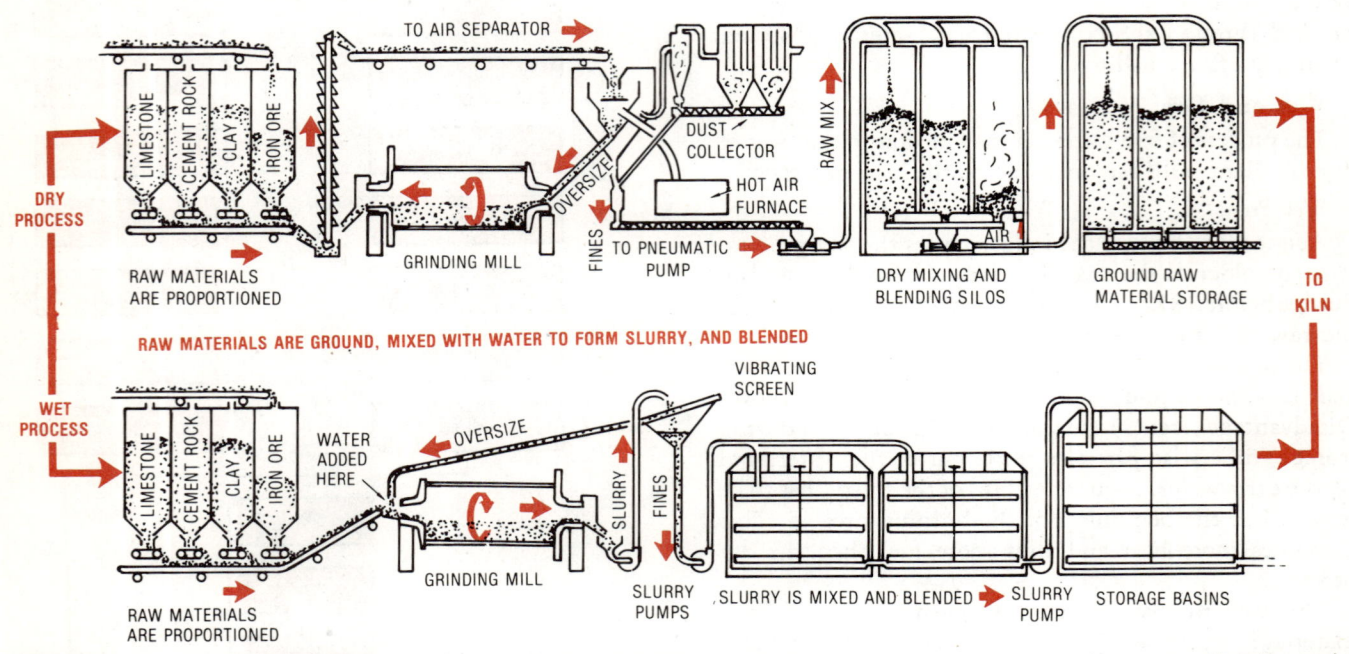

Fig. 41-2. A pictorial flow sheet for procurement of raw materials and their transformation into Portland cement. Note the use of

is added for fine grinding, and (2) storage silos or bins, instead of tanks, are used (Fig. 41-7).

Burning

This is the key process in making Portland cement. *Burning* is the use of heat in a kiln to the point of partial fusion, thereby causing a chemical and physical change.

Rotary Kiln

The rotary kiln is used for burning (Fig. 41-8). It is the largest piece of *moving* machinery used in any industry. This kiln is a refractory-lined steel cylinder. A layer of clinker is allowed to build up on top of the refractory to reduce heat loss. The dimensions of the average kiln are 10 feet (3 meters) or more in diameter and 300 feet (91.4 meters) or more in length. Compare this length with that of a football field, which is 100 yards, or 300 feet (91.4 meters), in length.

The rotary kiln is inclined toward the horizontal, usually ½ in. (12.7 mm) per ft (30.5 cm). The entire kiln slowly rotates (usually 1 rpm) on large gears (see Fig. 41-8).

Ground materials, wet or dry, are fed into the higher end of the cylinder. The inclination and rotation cause the feed to slowly move downward through the kiln. The burner is at the lower end of the kiln. Thus, the heat rises and travels upward through the length of the kiln, as

Transformation / 143

3 BURNING CHANGES RAW MIX CHEMICALLY INTO CEMENT CLINKER

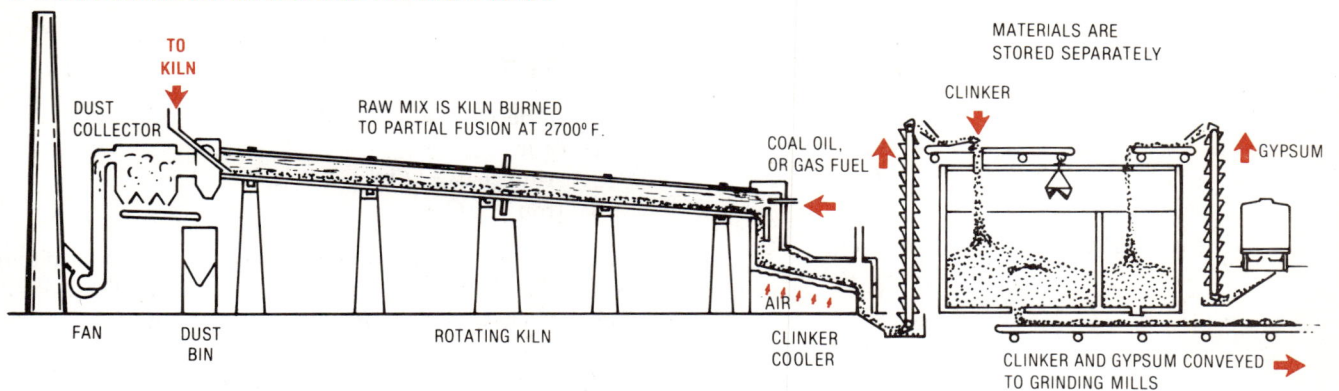

4 CLINKER WITH GYPSUM ADDED IS GROUND INTO PORTLAND CEMENT AND SHIPPED

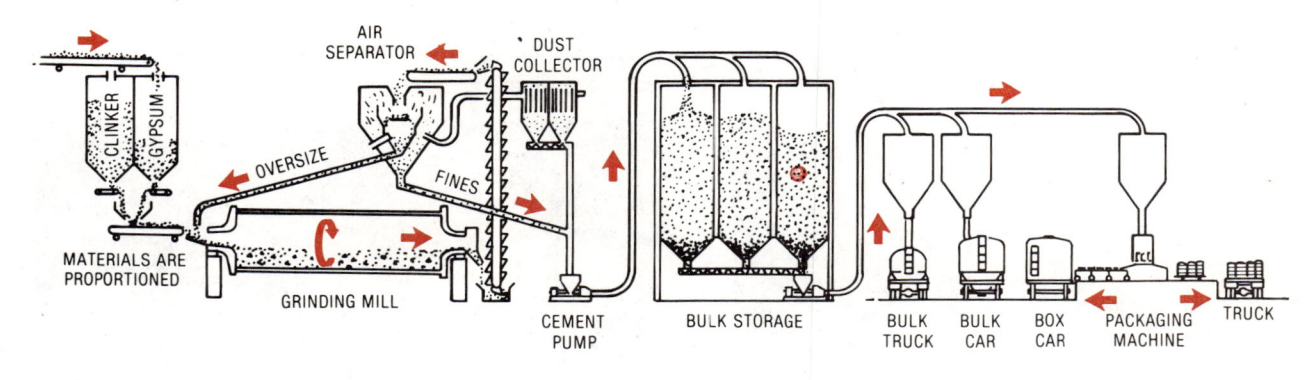

dust collectors. (Courtesy Portland Cement Association).

shown in Fig. 41-9. Combustible fuels, such as coal, oil, or natural gas, provide the heat.

Burning Process

Temperatures of 2600–3000°F (1427–1649°C) are produced by the above fuels. The 4-hr burning process causes physical and chemical changes.

During the burning process the materials are first dried, then heated, and finally burned at the lower end—where the temperature is highest. Each zone occurs in approximately one-third the length of the kiln, as shown in Fig. 41-9. Cooling occurs after the clinker emerges from the kiln. A rotary cooler is frequently used to recover heated air. The preheated air is returned to the burner. This improves burning efficiency.

Physical Changes

The physical change is easily seen. Powdered materials passing a 200-mesh sieve are fed into the kiln. Clinker exits the kiln. The clinkers are usually black in color. They can be described as round, pea to marble in size, and glass-hard balls. The clinker is harder than the rock from which it was produced.

Fig. 41-3. Rock is dumped into a gyratory crusher. This crusher has a steel cone in the center that presses other rock against the outside steel wall to reduce its size. (Courtesy Portland Cement Association)

Chemical Changes

The chemical changes cannot be seen visually. They are complex and important. Typical raw materials in the feed are:

1. Limestone ($CaCO_3$)
2. Sand (SiO_2)
3. Red shale ($Al_2O_3 \cdot 2SiO_2 \cdot 2H_2O + Fe_2O_3$)

These materials provide the four components or oxides needed to make Portland cement:

1. Lime (CaO)
2. Silica (SiO_2)
3. Alumina (Al_2O_3)
4. Iron (Fe_2O_3)

These four oxides combine during burning to form four new major chemical compounds, as shown in Table 41-1. The percentage of each oxide present in the finished product is listed in Table 41-2.

Burning produces two by-products—water (H_2O) and carbon dioxide (CO_2). Surface water, such as the excess water from the wet process slurry, if used, and chemically combined water, such as that in clay or shale, are driven off in the drying and heating zones (see Fig. 41-9). Carbon dioxide, such as from limestone, is driven off in the heating and burning zones.

Fig. 41-4. Each material used to make Portland cement is chemically analyzed. This determines the amount of each material to be used. (Courtesy Portland Cement Association)

Fig. 41-5. Large ball mills used for fine grinding materials. Many steel balls of various sizes are used in each mill. (Courtesy Portland Cement Association)

Fig. 41-6. Slurry for the wet process is mixed and stored in these huge tanks prior to burning. (Courtesy Portland Cement Association)

Fig. 41-7. Materials for the dry process are stored in bins. A traveling crane with a clamshell fills and removes the material from the bins. (Courtesy Portland Cement Association)

Transformation / 145

Fig. 41-8. Three rotary kilns are shown above. They are 10 ft (3 meters) or more in diameter and 300 ft (91.4 meters) or more in length. Compare the length to that of a football field. This kiln is usually inclined ½ in. per foot. (Courtesy Portland Cement Association)

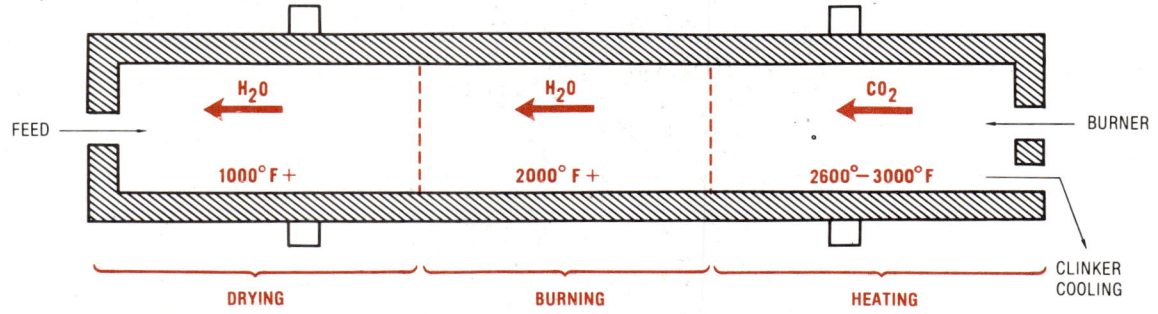

Fig. 41-9. Full-section drawing of a rotary kiln, showing the burning process. Note: (1) the three zones (drying, heating, and burning), plus clinker cooling outside the kiln; (2) the approximate temperatures in each zone; and (3) the by-products of each zone.

Table 41-1. Major Chemical Compounds in Portland Cement

Compound	Formula	Abbreviation*
Tricalcium silicate	$3CaO \cdot SiO_2$	C_3S
Dicalcium silicate	$2CaO \cdot SiO_2$	C_2S
Tricalcium aluminate	$3CaO \cdot Al_2O_3$	C_3A
Tetracalcium aluminoferrite	$4CaO \cdot Al_2O_3 \cdot Fe_2O_3$	C_4AF

* Simplified chemical abbreviations developed and used by only the Portland cement industry.

(Courtesy Portland Cement Association)

Table 41-2. Oxide Composition of Type 1 or Normal Portland Cement

Oxide	Range (percent)
Lime (CaO)	60-66
Silica (SiO_2)	19-25
Alumina (Al_2O_3)	3-8
Iron (Fe_2O_3)	1-5
Magnesia (MgO)	0-5
Sulfur trioxide (SO_3)	1-3

(Courtesy Portland Cement Association)

Clinker

The clinker has been physically and chemically described above. It can be stored a long period of time, if necessary, since it does not deteriorate. Or it can be shipped to other plants for the final grinding process.

Final Grinding

Final grinding again reduces the particles until they pass a 200-mesh sieve (Fig. 41-10). Ball mills are frequently used for this process. Classifying recycles the oversize particles to the mill. A small amount of gypsum ($CaSO_4 \cdot 2H_2O$) is added during this final *dry* grinding process. Gypsum, usually not over 3 percent by weight, is added to regulate the setting time of the Portland cement. Actually, the gypsum is a retarder to prevent a "flash set," when water is added. A "flash set" occurs when water is added to the powdered Portland cement, and it immediately chemically sets or forms a homogeneous solid.

Final grinding completes the three important processes that made Portland cement successful for Joseph Aspdin in 1824. The three processes were: fine grinding, before burning; burning, or partial fusion; and fine grinding, after burning.

Completion

Quality control testing is done after final grinding to maintain *American Society for Testing Materials (ASTM)* and other standards. Bulk cement is stored in weathertight silos (Fig. 41-11). Then, it is either bagged or shipped in bulk.

Paper bags with a small one-way paper valve are used. Portland cement enters the bag through the one-way valve when it is on a packaging machine (Fig. 41-12). The bags are filled by weight to 94 pounds (42.6 kilograms). The pressure of the cement inside the bag closes the valve through which it was filled.

Bagged Portland cement is most frequently shipped by truck and by boxcar. Bulk Portland cement is carried in ships, barges, bulk or tanker trucks, and bulk or hopper-bottom railroad cars (see Fig. 41-2).

Fig. 41-11. Bulk Portland cement is stored in these silos. Then, it is either bagged or shipped, as in the bulk trucks shown above. (Courtesy Portland Cement Association)

Fig. 41-10. Ball mills are used for final grinding. This reduces the clinker and gypsum until they will pass a 200-mesh sieve. (Courtesy Portland Cement Association)

Fig. 41-12. A packaging machine fills paper bags of Portland cement via a one-way valve. The bags are filled by weight (94 pounds). (Courtesy Portland Cement Association)

Unit 42: Utilization

Portland cement is a primary product used in manufacturing and construction. First, as potential consumers of Portland cement, there are several facts we should know.

Consumer Information

The color of Portland cement is either gray or white (Fig. 42-1). Its usual color is gray, due to the iron oxide in the materials used. Rust on the various machine contact surfaces also contributes to the gray color. White Portland cement is made from materials free from iron oxide. All the contact surfaces of the machine are also iron free. They may be covered with ceramic plates, such as alumina plates. The special materials and processing have two effects: (1) few companies manufacture it, since a complete separate plant is required; and (2) the cost is higher.

Fig. 42-1. Portland cement is either gray (left) or white (right). Its usual color is gray.

Fig. 42-2. A 200-mesh sieve has 200 openings per linear inch or 40,000 openings per square inch.

Particle Size

The particle size of the powdered Portland cement is very small. The particles are measured in microns or micrometers ($1/25,000$ inch). Nearly all the powder will pass a 200-mesh sieve, which has 40,000 openings per square inch (Fig. 42-2).

Weight

The traditional unit of measure for this industry in the United States is a *barrel,* which weighs 376 pounds (170.5 kilograms). This dates back to the period when wooden barrels were used for shipping. Today, Portland cement is sold in paper bags weighing 94 pounds or 42.6 kilograms (Fig. 42-3). One bag is equal to 1 cubic foot (28.3 cu decimeters). Four bags are equal to 1 barrel (170.5 kilograms). The manufacturing and construction companies that consume large quantities of cement usually purchase Portland cement in bulk.

Types

The various types of Portland cement and the special Portland cements available today in the United States are listed in Tables 42-1 and 42-2. All of those listed in Table 42-1 are gray in color, except for white Portland cement. The different types of Portland cement result from varying or adding to their composition. For example, the major chemical compounds are varied in Types I-V, as shown in Table 42-3.

Hydration

Hydration means to combine with water. When water is added to the powdered Portland cement, it hydrates and chemically sets or changes into a homogeneous solid. The Portland Cement Association indicates that hydration causes the following:

1. Tricalcium silicate hardens rapidly and is largely responsible for initial set and early strength. In

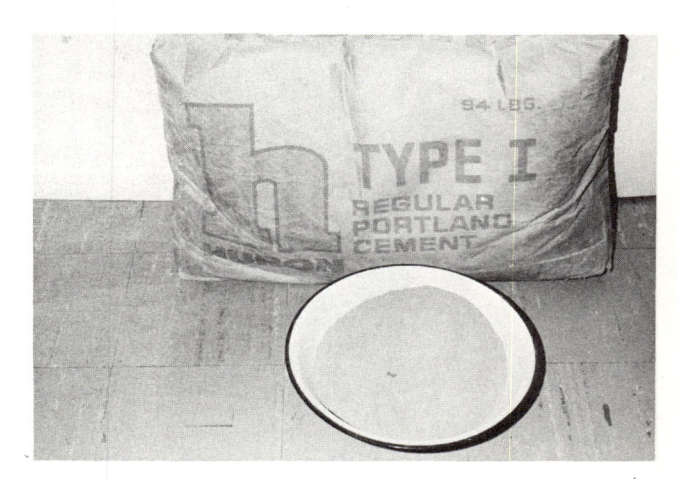

Fig. 42-3. A paper bag of Portland cement weights 94 pounds (42.3 kilograms). This is equal to 1 cubic foot (23.3 cubic decimeters).

148 / Modern Industrial Ceramics

Table 42-1. Types of Portland Cement

ASTM	Designation	Name and/or Use
C150	Type I	Normal or regular; general-purpose cement
C150	Type II	Moderate sulfate
C150	Type III	High early strength
C150	Type IV	Low heat of hydration; used in dams
C150	Type V	Sulfate resisting
C150	Type IA	
C150	Type IIA	Air-entraining
C150	Type IIIA	
C150	White	
C595	Type IS	Portland blast furnace slag cement
C595	Type IS-A	Same with air-entraining
C595	Type IP	
C595	Type IP-A	Portland-pozzolan cement
C595	Type P	(A = air-entraining)
C595	Type PA	
C91	Masonry	Portland cement, slaked lime, and possibly other materials

Table 42-2. Special Portland Cements*

Name	Details and/or Use
Oil-well	Seal oil wells
Waterproof	Available in either gray or white
Plastic	Mortar, plaster, stucco
Expansive	Expands as it sets
Regulated set	Controlled setting time

* Not covered by ASTM specifications

general, the early strength of Portland cement concrete is higher with increased percentages of C_3S.

2. Dicalcium silicate hardens slowly and contributes largely to strength increase at ages beyond 1 week.
3. Tricalcium aluminate liberates a large amount of heat during the first few days of hardening. It also contributes slightly to early-strength development. Cements with low percentages of this compound are especially resistant to soils and waters containing sulfates.
4. Tetracalcium aluminoferrite formation reduces the clinkering temperature, thereby assisting in the manufacture of cement. It hydrates rather rapidly but contributes very little to strength.

Manufacturing

Portland cements (see Table 42-1) are used to make a large number of secondary products. This number can be reduced to several major product groups:

Table 42-3. Typical Calculated Compound Composition and Fineness of Portland Cements

Type of portland cement		Compound composition, percent*				Fineness, sq.cm. per g.**
ASTM	CSA	C_3S	C_2S	C_3A	C_4AF	
I	Normal	50	24	11	8	1,800
II		42	33	5	13	1,800
III	High Early Strength	60	13	9	8	2,600
IV		26	50	5	12	1,900
V	Sulfate Resisting	40	40	4	9	1,900

*The compound compositions shown are typical. Deviations from these values do not indicate unsatisfactory performance. For specification limits see ASTM C150 or CSA A5.

**Fineness as determined by Wagner turbidimeter test.

CSA = Canadian Standards Association
(Courtesy Portland Cement Association)

Fig. 42-4. Precast concrete blocks: The standard 8-in. (width) block (left), and loading a delivery truck with 4-in. (width) blocks (right).

1. Precast concrete (Fig. 42-4)
2. Cement paint
3. Special Portland cements
4. Bagged or packaged dry, ready mixed
5. Ready-to-use, such as for patching

Precast concrete may be plain, reinforced, or prestressed. These products include concrete bricks, blocks, steps, and girders (Fig. 42-5). This is discussed in Section XI—*Concrete Industry*.

Cement paint is used on exterior walls and interior walls, such as basement walls. Some special cements, such as waterproof and plastic cements, are made by adding materials to Portland cement.

Fig. 42-6. Pouring concrete in place for a monostructure. This concrete slab contains reinforcing wire (background).

Bagged or packaged dry ready mixed mortar, concrete, and patching mixes are available. All you need to add is the water. Ready-to-use preparations, such as for patching and caulking, are also available.

Construction

Portland cements are extensively used to make concrete for construction. Portland cement is combined with aggregates and water to make concrete. Concrete poured in place is called a monostructure (Fig. 42-6). Typical examples are sidewalks, streets, and highways. This is discussed further in Section XI—*Concrete Industry*.

Summary

Most of the Portland cement manufactured each year is used in concrete. This includes both the precast and the poured-in-place concrete. Thus, we need to investigate concrete further.

Fig. 42-5. Precast concrete products include the three block shapes and the one brick.

Unit 43: Disposition

Portland cements (see Table 42-1) are seldom used alone. Therefore, technically, this industry does not have any disposition problems.

When Portland cement hydrates, it changes chemically and physically into some form of concrete. Therefore, disposition becomes the problem of the concrete industry (Fig. 43-1).

Reclaiming

Concrete can be reclaimed. Today, it is reused for hard land fill (Fig. 43-2), *riprap* (Fig. 43-3) and for similar places where large irregular pieces are needed. It is also recycled by crushing. Crushed concrete is used in place of crushed stone, as in a road base, base for concrete slabs, and coarse aggregate for concrete.

Discarding

Some discarding of used concrete still occurs, such as in dumps (Fig. 43-4). This can have a positive effect, if the dump operator uses it as a hard fill or cover layer.

Fig. 43-1. This problem area is adjacent to a ready mixed concrete plant. It has been used as a dump for extra concrete that is left in the trucks after delivery.

Fig. 43-2. Large concrete pieces from nearby highways have been reused as hard land fill.

Disposition / 151

Fig. 43-3. Pieces of concrete and old concrete block have been reused as riprap on a river bank.

Fig. 43-4. Discarding of used concrete still occurs. This dumping was unfortunately done in an open field. It should have been dumped in a landfill area.

SECTION XI

Concrete Industry

Unit 44: Material Science

Earlier, we stated that most of the Portland cement manufactured each year is used in concrete. A word equation shows how Portland cement relates to concrete.

raw materials $\xrightarrow{\Delta}$

Portland cement + aggregates + water ⟶ concrete

Definition

Concrete is a manufactured material that resembles rock. It may be defined by describing its composition. It is a homogeneous solid that is usually made from Portland cement, fine aggregates, coarse aggregates, and water.

Properties

Chemical properties—concrete is inert to almost everything except salt and some sewage.

Physical properties—concrete has a very high compressive strength or resistance to crushing. It has a low tensile strength or resistance to being pulled apart.

Materials

Both natural raw materials and manufactured or synthetic materials are used to make concrete. A typical concrete, and most of it, is made of four materials:

1. Portland cement
2. Fine aggregates
3. Coarse aggregates
4. Water

Two optional materials may also be included:

1. Colorant
2. Admixture(s)

The special concretes use different materials. These will be discussed later.

Portland Cement

Sixteen types of Portland cement were listed in Table 42-1. However, for most school projects and for work around the home, the Type I Portland cement is used (Fig. 44-1). This is gray in color. It is the normal, regular, or general-purpose Portland cement.

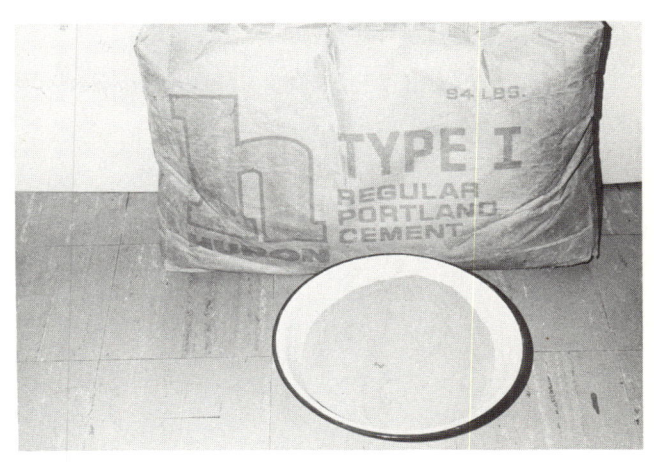

Fig. 44-1. Type I Portland cement is gray in color. It is frequently used for school and home projects.

Fine aggregates

Fine aggregates are materials that will pass a ¼-inch sieve. Grit, a concrete sand containing stone upward to ¼ inch is the usual material used for regular concrete (Fig. 44-2). Grit is a natural raw material. Manufactured materials, such as perlite and vermiculite, are used in place of grit to make lightweight and insulating concretes in which both weight and compressive strength will be reduced.

Fig. 44-2. Fine and coarse aggregate are separated into nine different size groups. Fine aggregates will pass a ¼-in. sieve. Coarse aggregates (the three groups at the lower right) do not pass a ¼-in. sieve. (Courtesy Portland Cement Association)

Coarse Aggregates

The coarse aggregates are materials that do not pass a ¼-inch sieve (Fig. 44-2). Crushed stone and/or gravel are used for regular concrete. Both are natural raw materials.

The following materials are typical examples of those used for lightweight and insulating concrete:

1. Pumice—natural material
2. Cinders—manufactured material
3. Perlite—manufactured material
4. Vermiculite—manufactured material

These materials will reduce the weight and compressive strength of the concrete.

Water

Only clean drinking water (H_2O) should be used to make concrete.

Colorant

The normal color of concrete is gray. This can be changed by adding a powdered colorant when dry batching. Using a colorant is optional.

Cement colors are usually sold in 1-pound (454-grams) packages. Basically, they are metallic oxides, such as red iron oxide for red. There are three important items concerning colored concrete:

1. Use white Portland cement, white sand, and white marble for white concrete.
2. Use these materials for the purer colors.
3. Do not use more than 10% colorant by weight of the Portland cement.

Admixtures

Using admixtures is also optional. Admixtures are an addition to the usual concrete materials. They may be powder or liquid. They may be added during dry or wet batching.

The ten admixture groups are:

1. Accelerators, such as calcium chloride in solution, are frequently used
2. Retarders, such as sugar, are frequently used
3. Air entraining agents, such as soapy substances, are frequently used
4. Gas-forming agents
5. Cementitious materials
6. Pozzolans
7. Alkali aggregate expansion inhibitors
8. Damp-proofing and permeability reducing agents
9. Workability
10. Grouting agents

The accelerators, retarders, and air entraining are the most frequently used groups.

Unit 45: Batching Concrete

Five important items in batching concrete are discussed here. These are: (1) ratios; (2) amounts; (3) forms; (4) batching procedure; and (5) terms.

Ratios

Two ratios are important. They are: (1) dry materials ratio, and (2) cement–water ratio.

The dry materials ratio is a *volume* ratio. The standard volume is 1 cubic foot (28.3 cu decimeters). However, the volume could be smaller or larger, for example, a cubic yard (764.6 cu decimeters). Small projects at school may only require the use of a beaker, such as 100 milliliters. A shovel could be used for home projects. The same volume measure must be used for all three of the dry materials.

Two of the ratios frequently used are the 1:2:2 and the 1:2:3 ratios. The 1:2:3 ratio, for example, means that one (1) volume of Portland cement, two (2) volumes of fine aggregate, and three (3) volumes of coarse aggregate are used in the mixture.

The amount of water used is very important. Six or seven gallons of water per bag of Portland cement is the recommended amount. If the fine aggregate (specifically *grit*) is damp, wet, or very wet see Table 45-1. Damp grit falls apart after squeezing some of it together in your hand; wet grit forms a ball; and very wet grit sparkles and wets your hand.

Small projects less than 1 cu ft (28.3 cu decimeters) require using the cement-water ratio. This is a *weight* ratio. It is based on 6 gallons (22.7 liters) of water per bag of Portland cement:

$$\frac{1 \text{ bag}}{6 \text{ gal}} = \frac{94 \text{ lb}}{6 \text{ gal}} = \frac{94 \text{ lb}}{48 \text{ lb}} = \frac{2 \text{ parts cement}}{1 \text{ part water}}$$

This ratio means that you use twice as much Portland cement as water by *weight*. Therefore, weigh the volume(s) of Portland cement and divide by *two* for the weight of water to be used. Use the metric scale to save time. For example, if the volume of Portland cement weighed 100 grams, then 50 grams or 50 milliliters of water would be used. One milliliter of water weighs 1 gram.

Amounts

The quantity of materials to use for a large concrete project, such as a patio, is computed in cubic feet or in cubic yards. The industry usually computes by the cubic yard of concrete. Cubic feet or cubic yards are easy amounts to work with, since a bag of Portland cement is equal to 1 cubic foot (28.3 cu decimeters).

The amounts of materials for the smaller concrete projects are determined with the following steps or procedure:

1. Plan project, such as a paperweight. Include the form, dry materials ratio (for example, 1:2:2), and cement-water ratio (2:1).
2. Select, assemble, or build the appropriate form—for example, a 9 oz (255 gm) clear plastic cup (disposable).
3. Select a volume measure in scale with the form—for example, a 100 ml beaker.
4. Fill the volume measure with coarse aggregate—for example, crushed stone.
5. Pour coarse aggregate into the form and repeat Step 4, until the form is full. Keep a tally of the number of volumes used—for example, 2.
6. Now, the number of volumes of coarse aggregate are known—for example, 2. Thus, the ratio selected, 1:2:2 in Step 1 will be used without adjustment. If four volumes of coarse aggregate were needed, the dry materials ratio (1:2:2) would be multiplied by two and the ratio 2:4:4 used.
7. Weigh 1 volume of Portland cement, using a metric scale—for example, 144 gm.
8. Divide the weight of the Portland cement, (Step 7) by two for the amount of water—for example, 144 ÷ 2 = 72 gm.
9. Measure the water with a metric graduate—1 gm = 1 ml.
10. Oil the form.
11. Batch the concrete.

Forms

There are three basic criteria for forms (Fig. 45-1). They should be rigid, removable, and oiled. School activities often do not require elaborate forms. Many sizes of paper and plastic cups are available which can save your time. You can recycle used coffee and soft drink cups to serve as concrete forms. Wash all used cups when recycling.

First select, assemble, or build the appropriate form. Oil *before* batching. The oil serves as a sealer and a separator. Used crankcase oil, or a form oil, is often used. Only a light coating of oil is necessary on plastic and metal forms.

Batching Procedure

The general procedure for batching of concrete includes:

Table 45-1. Water Content of Grit

Gallons of Water (per bag)	*When the grit is Damp	Wet	Very wet
6	5½	5	4¼
7	6¼	5½	4¾

* Reduce the gallons of water

155

156 / *Modern Industrial Ceramics*

1. Select, assemble, or build the form.
2. Select a volume measure in scale with the form.
3. Determine maximum size of coarse aggregate. *Note:* As a general rule, coarse aggregate should not exceed one-third (1/3) the thickness of the form—for example, a slab or wall; or if reinforcement is used one-third (1/3) the distance between it and the form.
4. Determine the amounts of materials—Portland cement, fine and coarse aggregate, and water. (See preceding procedure.)
5. Oil the form.
6. Dry batch (usually in a bowl or pan, if a small batch) until a uniform color is achieved.
7. Wet batch
    a. Add half (1/2) the water and mix
    b. Slowly add small amounts of the remaining water and mix.
        Note: More or less water may be required. This depends on the water content of the aggregates.
    c. A consistency that can be troweled, bringing a water sheen to the surface, is desired. This will be a relatively stiff consistency that does not segregate.
8. Place in the form. Mix is too stiff to pour (see Fig. 45-1).
9. Rod or vibrate to remove air. This makes a dense concrete.
10. Screed the top surface level with the form (see Fig. 45-1).
11. Smooth the surface to remove the rigids from the screeding.

Fig. 45-2. A cement mason floating and steel-troweling new concrete to produce a smooth hard surface. The mason has the wood float in his right hand and the steel trowel in his left hand. He is using knee boards to distribute his weight. This prevents indentations in the concrete. (Courtesy Portland Cement Association)

12. Optional finishing—when the water sheen has left the surface:
    a. Edge and groove (if a slab)
    b. Float (Fig. 45-2).
    c. Steel-trowel (Figs. 45-2 and 45-3).
    d. Other finishing.
13. Allow to set.
14. Remove form. ⎫
15. Cure.   ⎬ (Can be reversed)

Ready Mixed Concrete

Ready mixed concrete can be purchased for the larger projects, such as patios, driveways, and buildings. The minimum quantity that will be delivered is usually 1 cubic yard (0.76 cu meters). The ready mixed concrete

Fig. 45-1. Forms for this patio were made of 2 × 4's. The man using the shovel is placing the concrete. The other man is screeding the concrete level with the form. (Courtesy Portland Cement Association)

Fig. 45-3. A power trowel (foreground) is idle while the two cement masons hand-float the new concrete slab. The power trowel has three or four rotating, steel trowel blades. (Courtesy Portland Cement Association)

Fig. 45-4. A ready mixed concrete plant. Portland cement is stored in the silo (left), and the aggregates are in the hoppers above the truck being loaded. The hoppers are refilled by the conveyor (right).

Fig. 45-5. Ready mixed concrete trucks. The trucks are used for batching the delivery. These trucks can carry 7 cubic yards (5.3 cubic meters) of concrete.

Fig. 45-6. An unusually large ready mixed concrete truck discharging its load at the job site. The concrete is discharged via the chute at the rear of the truck. The moveable chute makes placing the concrete easier. (Courtesy National Ready Mixed Concrete Association)

firm does the batching for you. Thus, you would use only Steps 1, 4 (determine only volume), and 5; and Steps 8 through 15 in the preceding procedure. (See Figs. 45-4 through 45-6.)

Dry ready mixed concrete can be purchased in paper bags of various sizes. These sizes are adequate for small jobs—for example, a sidewalk block. You add the water and wet batch. Thus, you would use only Steps 1, 4 (determine only volume), and 5; and Steps 7 through 15 in the preceding procedure.

Terms

Three terms used in concrete work need defining—set, hydration, and curing. They are the results of the preceding batching procedure.

The *set* is a state of rigidity, after which you cannot work the new concrete. Water causes the chemical set to take place. This is known as *hydration*.

Curing involves controlling hydration. The two types of curing are air and water. *Air curing* means to leave the new concrete exposed to the air and allow evaporation to remove the water. *Water curing* means to retain the water in the new concrete. This allows hydration to continue and produce high compressive strength concrete. Several water curing methods are used:

1. Water—spray or flood
2. Mechanical barriers—cover with paper or plastic
3. Chemical membranes—spray-on coating

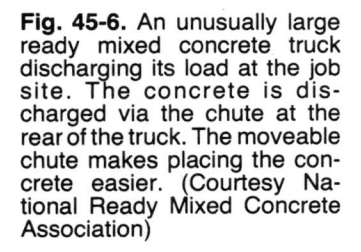

Unit 46: Manufacturing Concrete Products

Concrete product manufacturers usually purchase most, if not all, the materials they use. Their basic materials are Portland cement, fine aggregate, coarse aggregate, and water. The purchased materials are the primary products. Therefore, the manufactured concrete products are the secondary products.

Flow Sheet

Manufactured concrete products are made in a plant or place other than their final position of use. They are movable products. Their manufacture is diagrammed in Fig. 46-1.

Precast Concrete Products

The *precast* concrete products are manufactured. They are made in a plant or place other than their final position of use. Three major types of precast concrete products are:

1. Regular or plain
2. Reinforced
3. Prestressed

Regular

Regular or plain precast concrete products can be divided into three groups:

1. Blocks
2. Pipe or hollow tile
3. Other

The *blocks* are the most often used precast concrete product. The variety of shapes, sizes, and the number of blocks needed to build a project or structure, such as a wall, makes the precast blocks a major product.

Precast blocks can be subdivided into concrete and decorative blocks. The standard concrete block shapes and sizes can be seen in Fig. 46-2. They are made with regular and lightweight aggregates. The blocks are usually called *cinder* blocks when they are made of cinders. Concrete blocks are primarily used for load bearing or walls which support the weight of a roof.

Decorative blocks are primarily used for nonload-bearing walls. Some of the decorative blocks are:

1. Split block (Fig. 46-3)
2. Slump block
3. Grille block (Fig. 46-4)
4. Screen block (Fig. 46-4)
5. Patterned block (Fig. 46-5)
6. Blocks with special finishes

Small-diameter pipe or hollow tile are made of regular or plain precast concrete. The concrete drain tile are a typical example. They are small in diameter—for example, 3 in. (7.6 cm) and 12 in. (30.5 cm) in length.

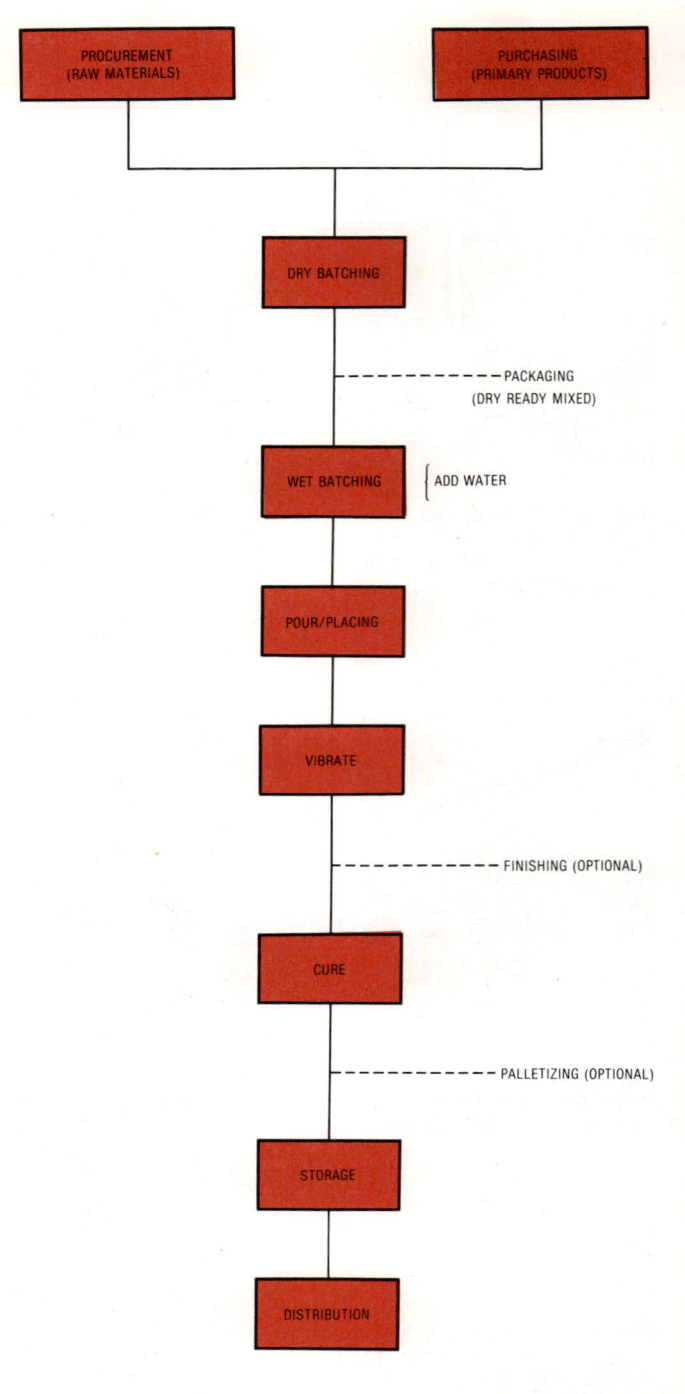

Fig. 46-1. A typical vertical flow sheet for the manufacture of precast concrete products.

Larger diameters and lengths usually require reinforcement.

Other regular or plain precast concrete products include functional and ornamental items. Park benches, picnic tables, and birdbaths are examples of functional products. Statues of animals, birds, and religious figures are examples of ornamental products. Many of these products may also be made with reinforcement.

Manufacturing Concrete Products / 159

Reinforced

Precast concrete is reinforced with steel to increase the tensile strength. Tensile strength is resistance to being pulled apart.

Two types of reinforcing steel are used. These are usually deformed bars or rods and welded wire screen (Fig. 46-6). The sizes of reinforcing rod are shown in Table 46-1.

Examples of reinforced precast concrete include:

1. Pipe
2. Park benches

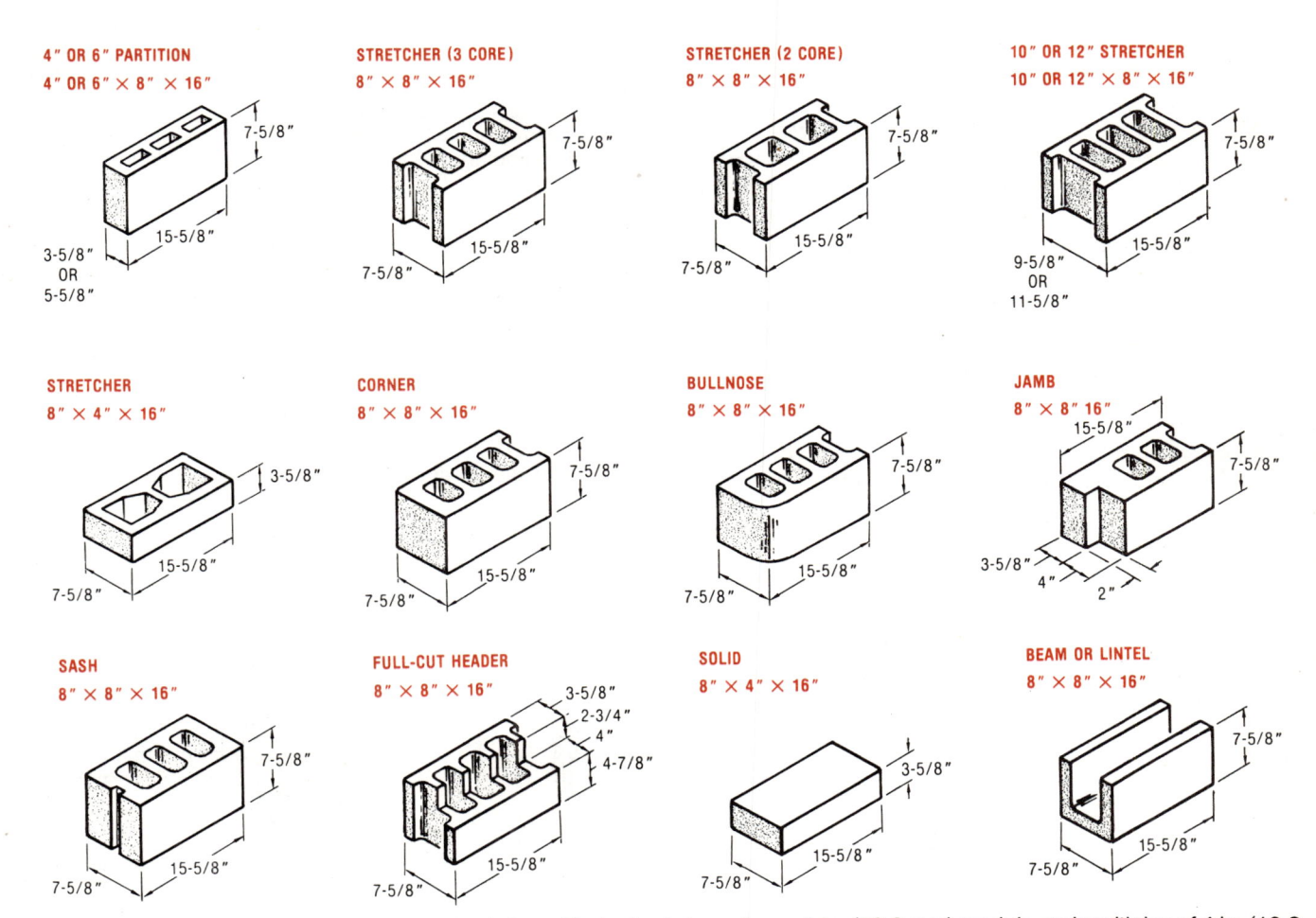

Fig. 46-2. Standard concrete block shapes and sizes. Their size is based on a 4-in. (10.2 mm) module and multiples of 4 in. (10.2 mm). The fractional dimensions shown allow for the mortar joint. Note the hollow core to reduce weight, to provide air space or space for insulation, and to make the blocks easier to pick up.

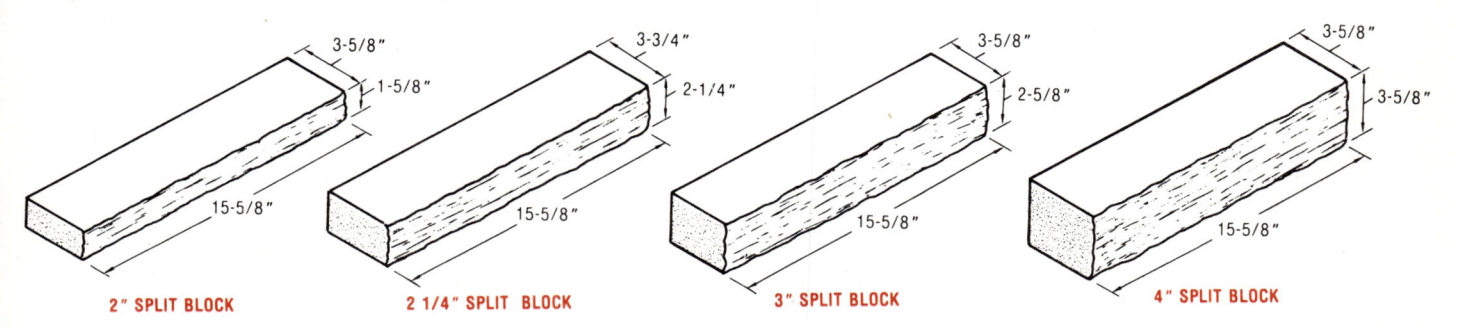

Fig. 46-3. The split block, a decorative concrete block, resembles natural stone. Four sizes are shown above. It is made by splitting a standard, 8-in. (20.3 cm) width, solid concrete block into two pieces. Splitting makes the rough textured stone-like surface. It is usually available in gray and other colors.

3. Picnic tables
4. Steps
5. Silo blocks
6. Beams
7. Highway guard rail posts

Prestressed

Precast concrete products are *prestressed* to make them stronger than reinforced products. Prestressed concrete has been precompressed before applying a load. Steel cable and rod are used with the concrete.

The two types of prestressing are *pretensioning* and *post tensioning*. Pretensioning means that tension is placed on the steel cable or rod before pouring. Post tensioning means that the tension is placed after the concrete cures.

Pretensioning is done by running a steel cable or rod the length of a form, as for a beam. Then, the steel cable or rod is stretched with a hydraulic jack, the concrete is poured, and it is cured. After thorough curing, the steel is released. This transfers the tension to the concrete. The bond between the steel and concrete holds the concrete under tension.

Post tensioning is done by pouring concrete into a form with the steel cable inside a pipe or treated rod, but not under tension. The tensioning steel usually runs the length of the form. After thorough curing and form removal, the concrete is stressed. The steel is free to stretch within the concrete, due to the pipe or treated rod that prevents bonding.

Examples of prestressed precast concrete include:

1. Beams
2. Roof girders
3. Bridge girders

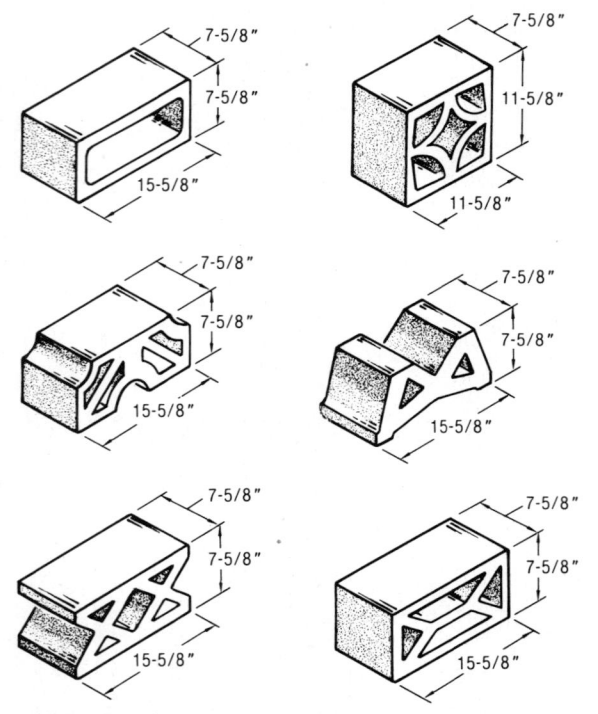

Fig. 46-4. Six examples of screen blocks (decorative concrete blocks). Screen blocks are lighter in weight and more open than grille blocks. Grille blocks are primarily used to reduce the effect of heavy winds without cutting off all air circulation. Screen blocks are primarily used to provide privacy and light. Grille blocks are usually 4 or 6 in. (10.2 or 15.2 cm) in width while screen blocks are either 4 or 8 in. (10.2 or 20.4 cm) in width.

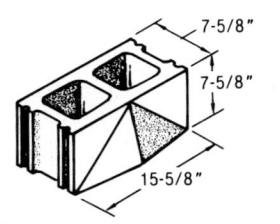

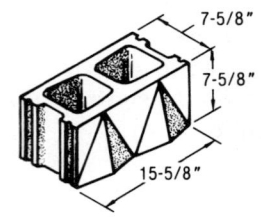

Fig. 46-5. Two examples of patterned blocks (decorative concrete blocks). These blocks have raised half pyramids on their face surface. Two blocks with the same surface pattern can be used to make a diamond—for example, place one block right side up and the other upside down. A pattern block can be combined with a regular block to create a wall dotted with diamond shapes. Pattern blocks are available in 4-, 8-, 10-, and 12-in. (10.2, 20.4, 25.4, and 30.5 cm) widths.

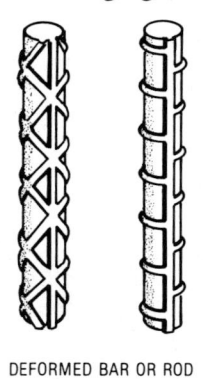

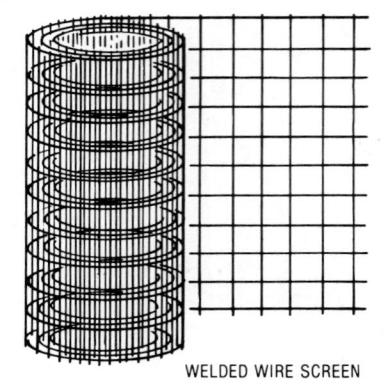

Fig. 46-6. Two types of reinforcing steel used in concrete to increase its tensile strength.

Table 46-1. Reinforcing Bar or Rod Numbers and Dimensions

Bar Number	Nominal Diameter (in.)	Weight (1lb per ft)	Cross-Section Area (sq in.)
2	¼	0.166	0.0500
3	⅜	0.376	0.1105
4	½	0.668	0.1963
5	⅝	1.043	0.3068
6	¾	1.502	0.4418
7	⅞	2.044	0.6013
8	1	2.670	0.7854
9	1/square	3.400	1.0000
10	1⅛	4.303	1.2656
11	1¼	5.313	1.5625

Unit 47: Concrete in Construction

Construction is building on a site in its final position of use. It involves the systems necessary to produce a structure that remains on the site on which it was built.

Concrete is a major construction material used today. The concrete is used in construction may be:

1. Monostructures
2. Special concretes
3. Assembled precast concrete products
4. Combination of precast and monostructures

Monostructures

Concrete *monostructures* are poured in place in the final site of their use. *Mono-* is a prefix meaning one (1). A monostructure is a single or complete structure.

Typical construction activities involved in concrete monostructures are diagrammed in Fig. 47-1. Three major types of concrete monostructures are:

1. Regular or plain
2. Reinforced
3. Prestressed

Regular

Regular or plain concrete monostructures are usually made of regular concrete. The regular concrete is made of Portland cement, grit, crushed stone and/or gravel, and water. It is usually subjected to only compression. Concrete has a high compressive strength. Examples include sidewalks, patios (Fig. 47-2), and footings.

Reinforced

Concrete monostructures are reinforced to increase their tensile strength. The same reinforcing materials that are used with precast concrete products are employed (Fig. 47-3). These were discussed and illustrated in Unit 46—*Concrete Products*. Examples include floors, walls, highways, and dams.

Prestressed

Prestressed concrete monostructures are precompressed

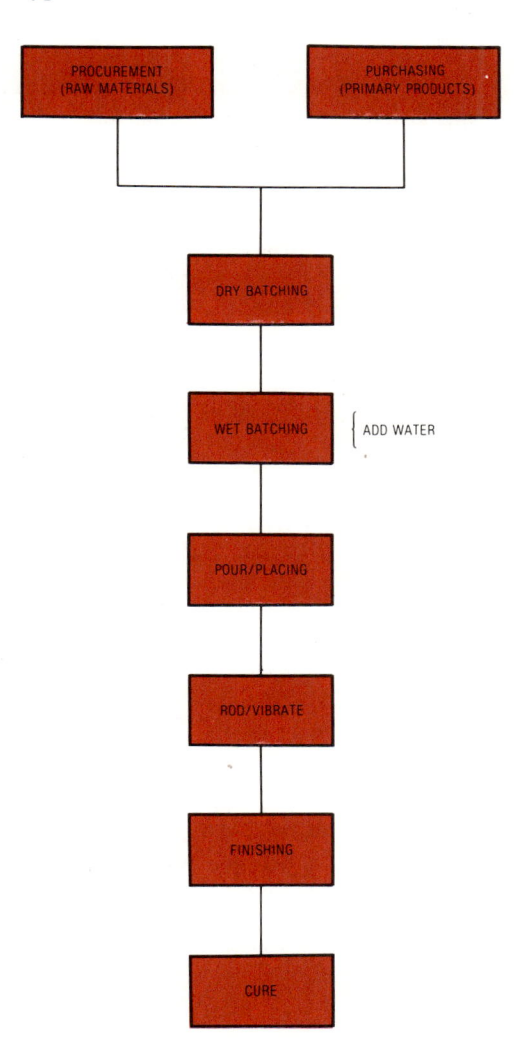

Fig. 47-1. A typical vertical flow sheet for the construction of concrete monostructures. Concrete monostructures are poured in place in their final position of use.

Fig. 47-2. This slab is regular or plain concrete. It is a monostructure that was poured in place. Note the roll of reinforcement wire.

Fig. 47-3. Steel reinforcing rod (foreground) is used to increase the tensile strength of concrete. Note how the concrete is being placed, using a bucket. The bucket is moved by a crane. (Courtesy Portland Cement Association).

162 / Modern Industrial Ceramics

before weight is applied in their final position of use. They may be either pretensioned or post tensioned. The same materials and procedures are used as with precast concrete products. These were discussed in Unit 46—*Concrete Products*. Examples include beams, girders, and bridges.

Special Concretes

Several of the special concretes are similar to the regular and reinforced concrete monostructures. However, there are enough differences to be considered a separate topic. Several special concretes are presented in Table 47-1.

Assembled Precast Concrete Products

Assembled precast concrete products are frequently referred to as masonry. The person who does the assembling is a mason. Mortar is the usual bonding agent for most of the assembled precast concrete products.

The act of assembling a concrete product such as a concrete block wall is a true and very typical construction activity. A discussion of the wide variety of skills and procedures required in construction would be too lengthy to be discussed here.

Examples of assembled precast concrete products include walls, patios, and silos. Reinforcement may also be added, as for a regular concrete block wall.

Combination

Precast concrete and concrete monostructures are frequently used together. Often, it is necessary to combine them. For example, the footing for a concrete block wall is a concrete monostructure. This is an example of a concrete monostructure that is hidden from view, especially when viewing the exterior. Precast concrete may also be hidden from your view, as in a foundation wall or when covered with stucco.

Job Site Tests

Job site tests are done at the construction location. Some of these tests may also be used when manufacturing precast concrete products. Then, the tests are usually conducted in a laboratory.

Aggregate Tests

Three job site *aggregate tests* are necessary when bank-run materials are being used. *Bank run* means that sand or gravel is used in its natural state as taken from the earth. The tests are *sieve analysis, silt test,* and *colorimetric test*.

The sieve analysis determines whether the aggregates are within a designated size range (Fig. 47-4). The analysis is also used to make sure the gravel does not contain fines.

The silt test determines the presence of extremely fine material in the grit. The colorimetric test checks for the presence and amounts of vegetable matter in the grit. Caution must be observed with this test, since a caustic material (sodium hydroxide) is used.

Batch Tests

Immediately after wet batching concrete, several tests may be used. Three of those frequently used are discussed briefly.

The *slump test* measures the consistency of the wet batch. This test indicates a too wet or too dry batch by the amount of slump after the cone is removed (Fig. 47-5).

Table 47-1. Special Concretes

Name	Differs from Regular Concrete	Usual Composition	Notes
Mortar	No coarse aggregate	Portland cement, lime	Bonding agent
Stucco	No coarse aggregate	Same as mortar	Continuous layer over exterior walls
Pneumatically applied	Usually no coarse aggregate	Portland cement, grit, water	Applied under air pressure, as in swimming pools
Terrazzo	No fine aggregate	White Portland cement, marble or granite, coarse aggregate, water	Ground and polished after curing to expose coarse aggregate, as in floors
Soil cement	No fine or coarse aggregate	Portland cement, dirt, water	90-95% dirt, 5-10% Portland cement, small amount of water. Must be compressed. Patios, walks, and blocks are some of its uses.

Concrete in Construction / 163

Fig. 47-5. The amount of water used affects the consistency of wet-batched concrete. This is indicated by the slump test. Examples above are too little slump (top); acceptable slump (center); too much slump (bottom). (Courtesy Portland Cement Association)

The *air content* test is being performed in Fig. 47-6. This test indicates the amount of entrained air in the batch.

The *compression test* determines whether the concrete has the specified compressive strength. Standard cylinders, procedures, and curing are used. The cylinders are 6 in.

Fig. 47-4. An aggregate sieve analysis: pouring a weighed sample into assembled sieves (top); operating mechanical sieve shaker (center); and results (bottom).

(15.3 cm) in diameter and 12 in. (30.5 cm) in height. Three cylinders are usually used for each age test (Fig. 47-7). If both job site and laboratory curing are used, six cylinders are used for each age test—usually 7 and 28 days.

Fig. 47-6. An air-content test being performed on wet batched concrete. (Courtesy National Ready Mixed Concrete Association)

Fig. 47-7. Six compression test cylinders, after filling at the job site. Three cylinders will be used for each age test— usually 7 and 28 days. (Courtesy Portland Cement Association)

SECTION XII

Clay Industries

Unit 48: Introduction

There are several similarities and differences between the *hydrosetting* materials industries (plaster, lime, and Portland cement) and the *clay* industries. The hydrosetting materials were expressed by a word equation:

$$\text{raw materials} \xrightarrow{\Delta} \text{primary products} + \text{water} \longrightarrow \text{homogeneous solid} + \text{heat} \uparrow$$

The clay industries can be expressed by a general word equation:

$$\text{raw material(s)} + \text{water} \xrightarrow{\Delta} \text{primary or secondary product} + \text{optional processing}$$

Similarities

Some *similarities* between the hydrosetting materials and the clay industries are:

1. A soft, squashy state occurs when water is added.
2. Both have product-forming—*after* heat treating for the hydrosetting materials and *before* heat treating for the clay.
3. Usually, high-temperature heat treatment for both.
4. Both have a hard, durable end product.

Differences

Some *differences* between the hydrosetting materials and the clay industries are:

1. When each are in a powdered form—after heat treating for the hydrosetting materials and as a raw material for the clay.
2. When water is added.
3. When heat-treated.
4. How a hard, durable end product is made—hydrosetting materials are chemically set, and clay is heat-treated (fired, for example).

Types of Products

The clay industries are a manufacturing industry. They manufacture both primary and secondary products. Most of these products are secondary products. Primary products, such as powdered raw materials, are purchased for the manufacture of secondary products.

Examples of primary clay products are brick and some art pottery. Wall tile and dinnerware are examples of secondary clay products.

Clay Products

The products of the clay industries can be divided into two major groups—*structural clay products* (SCP) and *whitewares*. Structural clay products can be subdivided into two groups—*bricks* (Fig. 48-1) and *tiles* (Fig. 48-2). Whitewares can be subdivided into *handmade ware* (Fig. 48-3) and *manufactured ware* (Fig. 48-4).

Definitions

Structural clay products (SCP) are manufactured clay products used primarily in construction. They include

166 / *Modern Industrial Ceramics*

Fig. 48-1. Bricks are typical and common structural clay products.

Fig. 48-2. Ceramic tile may be flat, as shown, or hollow. These glazed wall tile are 3 × 6 in. (8 × 15 cm). Drain tile and sewer pipe are examples of hollow tile. (Courtesy American Olean Tile Company)

bricks and tiles. The British refer to these as heavy clay products.

Whitewares are fired ware, glazed or unglazed, that usually have a fine textured white clay body. Products include handmade ware and manufactured ware.

These industries are discussed in Section XIII—*Structural Clay Products Industry* and Section XIV—*Whiteware Industry*.

Fig. 48-3. Typical handmade whiteware products made by art potters and hobby ceramists. The forming methods (left to right) are pinch, coil, slab, and thrown.

Fig. 48-4. Typical manufactured whiteware products are formed by machines.

Unit 49: Material Science of Clay

Frequently, the products of the clay industries are so commonplace that we take them for granted. Most of us have never stopped to think about their basic material—clay—what it is or how it originated. This discussion will answer these questions.

Formation of Clay

Igneous rocks and several minerals were formed when the earth's crust cooled; therefore, they were the primary materials. When they began to weather, the secondary materials were formed.

Clay is a secondary material in the earth's crust. It was formed by weathering of the mineral *feldspar*.

Weathering

Weathering decomposed and/or wore away the feldspar. Decomposition involves a breakdown or disintegration.

Water, temperature, and wind are the forms of weathering that have a major role in forming clay. Water is a major weathering force, since it can exist in all the three states of matter—solid, liquid, and gas. Rain, snow, and ice result in moving water. They cause chemical and physical changes in the feldspar.

Chemical Changes

The chemical change of feldspar into clay is explained by these equations:

Potash feldspar + water + carbon dioxide ⟶ kaolinite + quartz + potassium carbonate

or:

$K_2O \cdot Al_2O_3 \cdot 6SiO_2 + 2H_2O + CO_2 \longrightarrow Al_2O_3 \cdot 2SiO_2 \cdot 2H_2O + 4SiO_2 + K_2CO_3$

Potash feldspar ($K_2O \cdot Al_2O_3 \cdot 6siO_2$) is a typical feldspar. Water (H_2O) is available in one or more of the states of matter. Carbon dioxide (CO_2) is from the atmosphere. Kaolinite ($Al_2O_3 \cdot 2SiO_2 \cdot 2H_2O$) is the major clay mineral. Quartz (SiO_2) may be washed out of the clay by water. The potassium carbonate (K_2CO_3) is a soluble substance that is washed away by water.

Physical Changes

The physical changes are more obvious, since you can see the results. Feldspar and the rocks containing it are solids. They may occur as large massive deposits in the earth's crust. Weathering breaks down these massive deposits into clay. Clay is composed of very small flat plate-like crystals that are too small to be seen with the naked eye. Clay often contains excess water that causes it to be a soft, squashy material.

Clay formed from a pure feldspar will be white. However, impurities in the feldspar may change the color of the clay. Weathering forces may also contaminate the clay and change its color.

Time

No one really knows how long it takes to form clay. This process probably takes many years. Clay is now being formed and in larger quantities than we are using it. Clay appears to be a natural resource that will not be easily exhausted.

Clay Minerals

Kaolinite is the major clay mineral used for clay products. Chemically, it is hydrous silicate of alumina ($Al_2O_3 \cdot 2SiO_2 \cdot 2H_2O$). This chemical formula is the one ceramists use to represent all types of clay. Note that the water is chemically combined.

The clay minerals are listed in Table 49-1. Only kaolinite occurs abundantly. Thus, it is extensively used for clay products. *Vermiculite* expands when fired. Thus, it cannot be used for clay products.

Table 49-1. Clay Minerals

Mineral	Type of Crystal Structure
Kaolinite Nacrite Dickite Halloysite	2-layer sheet silicate
Illite Glauconite Chlorite	3-layer sheet silicate, not expandable
Montmorillonite Vermiculite	3-layer sheet silicate, expandable
Sepiolite Attapulgite	Fibrous

Types of Clay

There are three major types of clay:

1. Primary
2. Colluvial
3. Secondary

The formation and location of these types of clay are shown in Fig. 49-1.

Primary Clay

Primary clay is found at its origin. It has not been moved by the natural weathering forces, such as rain. Thus, it is also called residual clay.

The color of the fired clay is white. *Kaolin* or china clay is an example of a primary clay. Kaolin is a white firing clay primarily composed of kaolinite.

Colluvial Clay

Colluvial clay has been transported a short distance from its origin. The color of the fired clay is usually white. A secondary kaolin is an example of a colluvial clay. It is

168 / *Modern Industrial Ceramics*

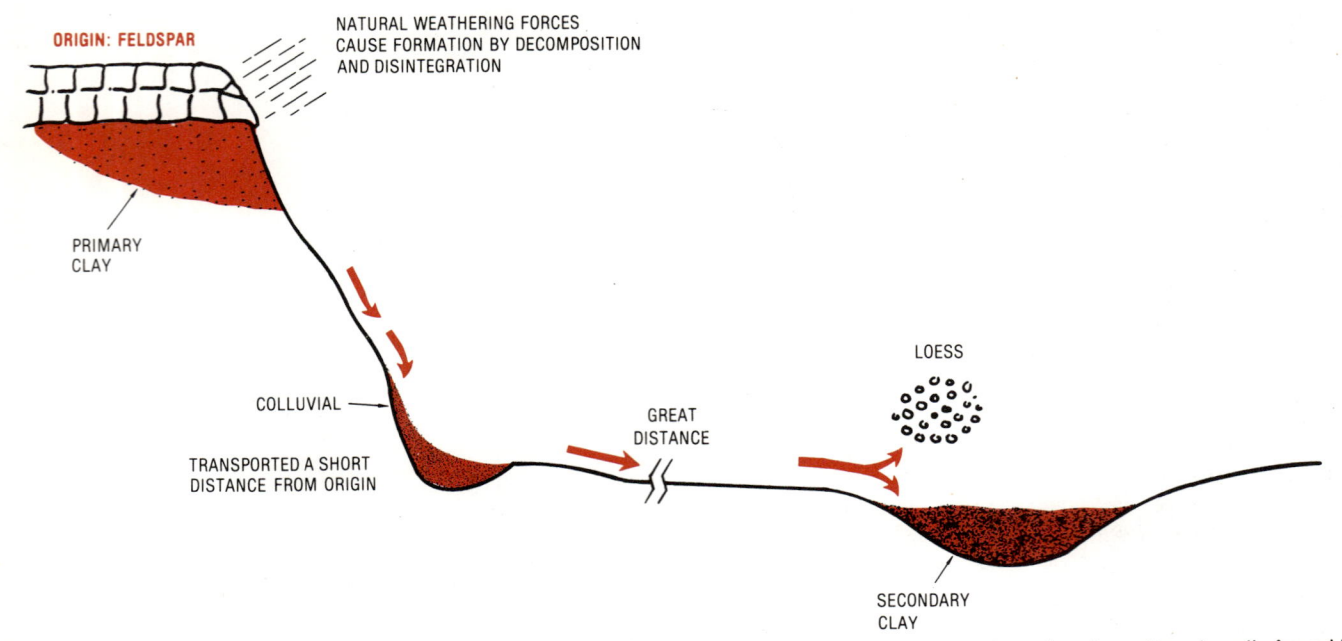

Fig. 49-1. The formation of clay and the three types of clay. Loess is wind-formed from secondary clay deposits primarily found in arid regions.

called secondary kaolin, since it has been moved from its origin.

Secondary Clay

Secondary clay has been transported a great distance from its origin. Water, wind, and glaciers are the major transporters. Thus, the clay usually contains impurities that color the clay (Fig. 49-2).

This clay is also called sedimentary or transported clay. Most of the clay deposits are secondary clays.

The color of the fired clay is white, buff, or red. Limestone ($CaCO_3$) in the clay may cause a buff color. Iron oxide in the clay causes a red color. The color of the raw clay may be caused by organic materials that burn out when fired. Therefore, there is a color change. Only the fired color is permanent.

Ball clay and brick clay are examples of secondary clay. Ball clay is a very plastic sedimentary kaolinite clay that fires white. Brick clay is impure, fires red, and is often used to make bricks and tiles.

Physical Properties

Physical properties describe a material—state of matter, color, etc. The physical properties of clay describe it in terms of the:

1. Properties of the clay particle (Table 49-2)
2. Plastic properties
3. Drying properties
4. Firing properties
5. Properties after firing

Fig. 49-2. Raw sedimentary clay that will fire red (top) and buff (bottom). The color of the raw clay is dull red (top) and gray (bottom).

Material Science of Clay / 169

Table 49-2. Physical Properties of Clay Formulas

A. Water of Plasticity

$$\text{percent water of plasticity} = \frac{\text{weight of water}}{\text{weight of dry clay}} \times 100$$

B. Drying Shrinkage (part of Drying Properties)

$$\text{percent linear drying shrinkage} = \frac{\text{plastic length} - \text{dry length}}{\text{dry length}} \times 100$$

C. Firing Shrinkage (part of Firing Properties)

$$\text{percent linear firing shrinkage} = \frac{\text{dry length} - \text{fired length}}{\text{dry length}} \times 100$$

D. Properties after Firing

1. Total Shrinkage

$$\text{percent total linear shrinkage} = \frac{\text{plastic length} - \text{fired length}}{\text{plastic length}} \times 100$$

2. Water Absorption

$$\text{percent absorption} = \frac{\text{saturated weight} - \text{dry weight}}{\text{dry weight}} \times 100$$

Note: Use metric measurements (grams and centimeters)

Properties of Clay Particles

The clay particle is a *solid*. It is a flat plate, roughly like a hexagonal crystal (Fig. 49-3). The particle size is measured in microns or micrometers. There are approximately 25,000 microns or micrometers in 1 inch. You need an electron microscope to see a crystal.

Plastic Properties

Plasticity is considered to be the most important property of clay. This property allows clay to be formed into many different products.

Plasticity is a property of matter that allows it to be deformed or changed into any shape *and* retain that shape when the deforming force is removed.

There are at least six theories of plasticity. One of these, the plate theory, states that the flat, plate-like clay particles slide over each other when deformed. The clay particles slide easily when enough water is present. Water acts as a lubricant and cohesive force between the flat plate-like particles.

Plasticity is affected by particle size and the amount of surface water. Surface water is not chemically combined with the clay. Smaller particles will hold more surface water. There are more particles to coat with water.

The process of taking on or losing water is called *porosity*. Porosity is the ability of the minute spaces between particles to soak up or give off water. Humidity is the controlling factor.

Plastic clay that lacks water is "short." This clay will develop cracks when deformed. Water is added to correct the condition.

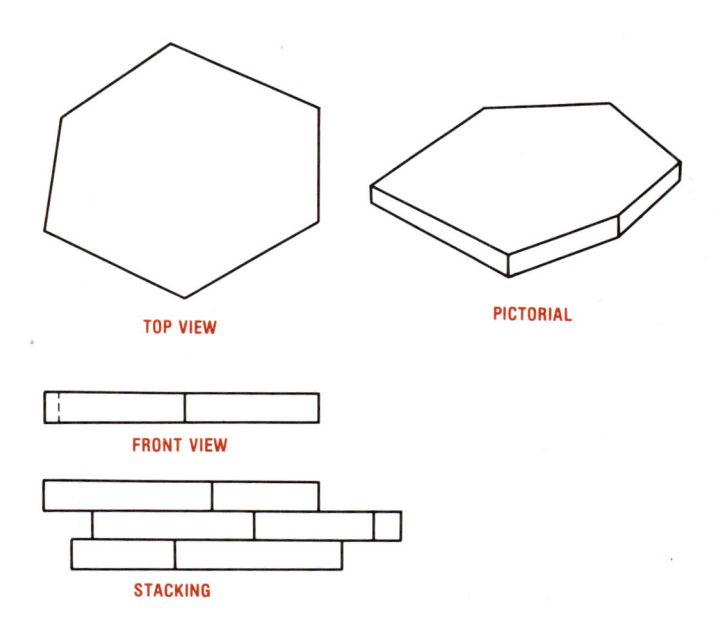

Fig. 49-3. A clay particle is a very small flat plate, roughly like a hexagonal crystal. The above drawings are not to scale. The particle size has been enlarged.

Drying Properties

Plastic clay shrinks as it dries, due to the evaporation of surface water. Shrinkage occurs when the clay particles draw together to take up the space previously occupied by water.

You should be able to carefully handle dry clay. The ability to resist breaking when it is handled is called its dry strength.

Uneven drying and drying too quickly can cause cracking and/or warping. A rack with perforated shelves should be used to permit even drying. The rate of drying can be controlled by reducing the heat or increasing the humidity.

The stages of drying are:

1. Plastic
2. Leatherhard
3. Greenware

Plastic clay has been defined. *Leatherhard* clay is cold and damp, but it is no longer plastic. It can be easily cut. Greenware or *bone dry* is neither cool nor damp when held against your face. It can be scraped, sanded, and sponged. Greenware is ready for firing.

Firing Properties

Clay shrinks when it is fired, due to the loss of surface water and chemically combined water. Shrinkage again occurs when the clay particles draw together to take up the space previously occupied by water.

Vitrifying or fusing also occurs. A glassy bond is formed that fuses the clay particles together.

The two types of firing are:

1. Bisque
2. Glaze

Bisque firing is the first firing of dry clay ware. The usual maximum temperature is *Cone 07* (C/07). This temperature leaves the fired clay porous and easier to glaze. *Glaze firing* is usually the second firing of clay ware. It is often done at a higher temperature (for example, C/04), than the bisque fire. The purpose is to cause the glaze to melt, thereby forming a glassy coating. This firing also causes the clay to become vitreous or glassy.

Properties After Firing

Three properties can be observed after firing. They are water absorption, hardness, and color.

Water absorption is a measure of vitrification. The amount of water absorbed by the fired clay indicates the amount of vitreous or glassy bond. The less water absorbed indicates a nearly vitreous fired clay.

Hardness parallels vitrification. Basically, the higher the firing temperature, the harder the clay.

The fired color is the permanent color of the clay. Organic materials in the raw clay burn out when fired. The primary and colluvial clays fire white. The secondary clays usually contain impurities and fire buff or red.

States of Clay

The three states of clay are:

1. Powdered
2. Plastic
3. Liquid

Powdered clay is beneficiated raw material. Plastic clay is powdered clay and less than 50% water. Liquid clay is either a slurry or casting slip. A slurry is 50–50 powdered clay and water. Casting slip is approximately 70% powdered clay, 30% water, and a very small amount of deflocculant.

Preparation and use of each state of clay will be discussed in Sections XIII and XIV.

SECTION XIII

Structural Clay Products Industry

Unit 50: Material Science

The clay industries were represented by a general word equation:

$$\text{raw materials} + \text{water} \xrightarrow{\Delta} \text{primary or secondary product} + \text{optional processing}$$

The structural clay products (SCP) industry can be expressed by this word equation:

$$\text{raw materials} + \text{water} \xrightarrow{\Delta} \text{primary product}$$

The preceding equation represents most of the structural clay products. However, some products require more processing, as shown by this word equation:

$$\text{raw materials} + \text{water} \xrightarrow{\Delta} \text{bisque} + \text{glaze} \xrightarrow{\Delta} \text{secondary product}$$
(won and processed)

These word equations provide an overview and a logical sequence for the discussion in this section.

Definition

Structural clay products (SCP) are the manufactured clay products primarily used in construction. They are usually used for building on site in their final position of use. There are two major product groups—bricks (Fig. 50-1) and tiles (Fig. 50-2). The British refer to SCP as heavy clay products.

Materials

Both natural raw materials and manufactured materials are used to produce the structural clay products. Most of the materials used are natural raw materials.

Raw Materials

The major raw materials are clay and shale. Secondary clay is used for bricks and most tile. This clay usually contains impurities that cause it to fire a buff or red color

Fig. 50-1. Bricks are a visible and frequently used structural clay product.

171

172 / Modern Industrial Ceramics

(Fig. 50-3). Local names may be used for this clay, such as brick, local, or surface clay. Surface clay is a descriptive name for the clay found near the surface of the earth.

Shale is the sedimentary rock form of clay (Fig. 50-4). The clay has hardened into rock. Shale is fine grained, or has very small particles, like clay. Impurities usually cause shale to fire a buff or red color.

Some of the products, such as floor and wall tile, are white. Thus, primary and/or colluvial clay(s), such as kaolin, and nonplastic materials are needed. The nonplastic materials include flint, feldspars, whiting, talc, and metallic oxides. Due to the number of materials required, several are usually purchased. They are purchased as beneficiated raw materials—powdered.

Fig. 50-2. Ceramic tile may be flat, as shown, or hollow. These glazed wall tile are 3 × 6 in. (8 × 15 cm). Drain tile and sewer pipe are examples of hollow tile. (Courtesy American Olean Tile Company)

Manufactured Materials

Grog is the major manufactured material used in bricks and most tiles. It is made from fired clay. Often, fireclay is used. The fired clay is crushed and ground into particles the size of rice, or finer. Grog is added to clay to reduce shrinkage, prevent warping, and possibly increase strength.

Fig. 50-3. A secondary clay that fires a red color.

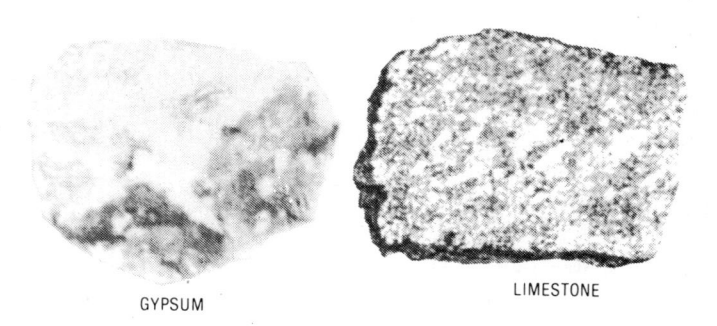

Fig. 50-4. Shale is the sedimentary rock form of clay. It is a clastic sedimentary rock. Clastic means that the rock is composed of particles. Often, the particles can be seen without magnification.

Unit 51: Procurement

Procurement of raw materials for the structural clay products industry is diagrammed in Fig. 51-1. The prospecting and developing processes previously discussed are used by this industry.

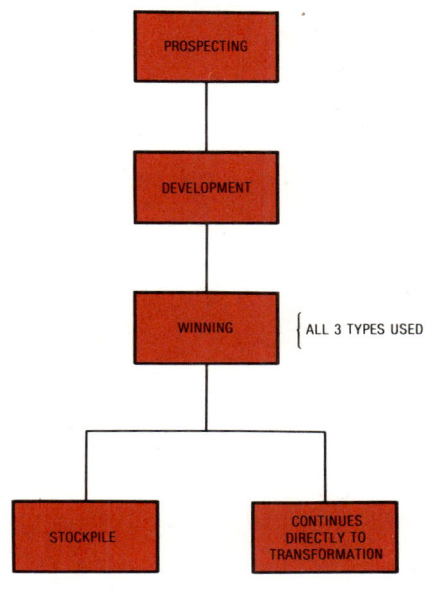

Fig. 51-1. A typical vertical flow sheet for procurement of raw materials used by the structural clay products industry.

Prospecting

Secondary clay and shale are fairly universal materials. Therefore, prospecting is not a difficult task. Locating the large deposits is the difficult part. Large deposits that will provide raw materials for several years are desired. This can reduce the prospecting, developing, transportation, and other costs.

Winning

All three types of winning are used—superficial, subterranean, and subaqueous. Superficial or open-pit winning is the most frequently used (Fig. 51-2). There are two major reasons for this:

Fig. 51-2. Open-pit winning of clay in Ohio. A dipper-stick type of power shovel is used to excavate the deposit. (Courtesy Cedar Heights Clay Co.)

1. Secondary clay and shale often occur near or on the earth's surface.
2. Costs are less, since standard and heavy-duty construction machines can be used.

A *shale planer* may be used in some open pits. This is a specialized machine. Although it is called a shale planer, it can be used to win clay. This machine has a vertical boom with a rotating continuous cutter chain that rakes the material off the bank as it descends. Thus, the bank of the pit is also vertical. The material falls to the bottom of the cutting boom. A conveyor then moves it to a hopper, railroad car, or truck (see Fig. 51-3).

Subterranean winning is used occasionally. Fireclay may be won by this method.

Fig. 51-3. A shale planer is a specialized winning machine used in open pits. The vertical boom has a rotating continuous cutter chain that rakes material off the bank as it descends.

Subaqueous winning is the method least used by this industry (Fig. 51-4). When clay is won by this method, it contains excess water. The clay may be used without removing the excess water, or it may be used after some of the water has been removed.

Stockpiling

Stockpiling the raw materials is common and practical (Fig. 51-5). Enough material for several days of plant operation are usually stored. This insures continuous operations, regardless of weather conditions or breakdown of the winning machinery.

Blending of the raw materials may also be done when they are stockpiled. A material may vary in certain qualities within a deposit. Thus, blending produces a more uniform raw material. A uniform raw material helps insure uniformity in content or quality of the finished products.

174 / *Modern Industrial Ceramics*

Fig. 51-4. A barge-mounted crane equipped with a clamshell. This dredge can lift up to 18 cubic yards of material. (Courtesy Great Lakes Dredge & Dock Company)

Fig. 51-5. A raw materials building protects stockpiles from the weather. (Courtesy American Olean Tile Company)

Unit 52: Transformation

Transformation of materials into structural clay products is diagrammed in Fig. 52-1. There are five basic processes:

1. Beneficiation
2. Material preparation
3. Forming
4. Drying
5. Firing

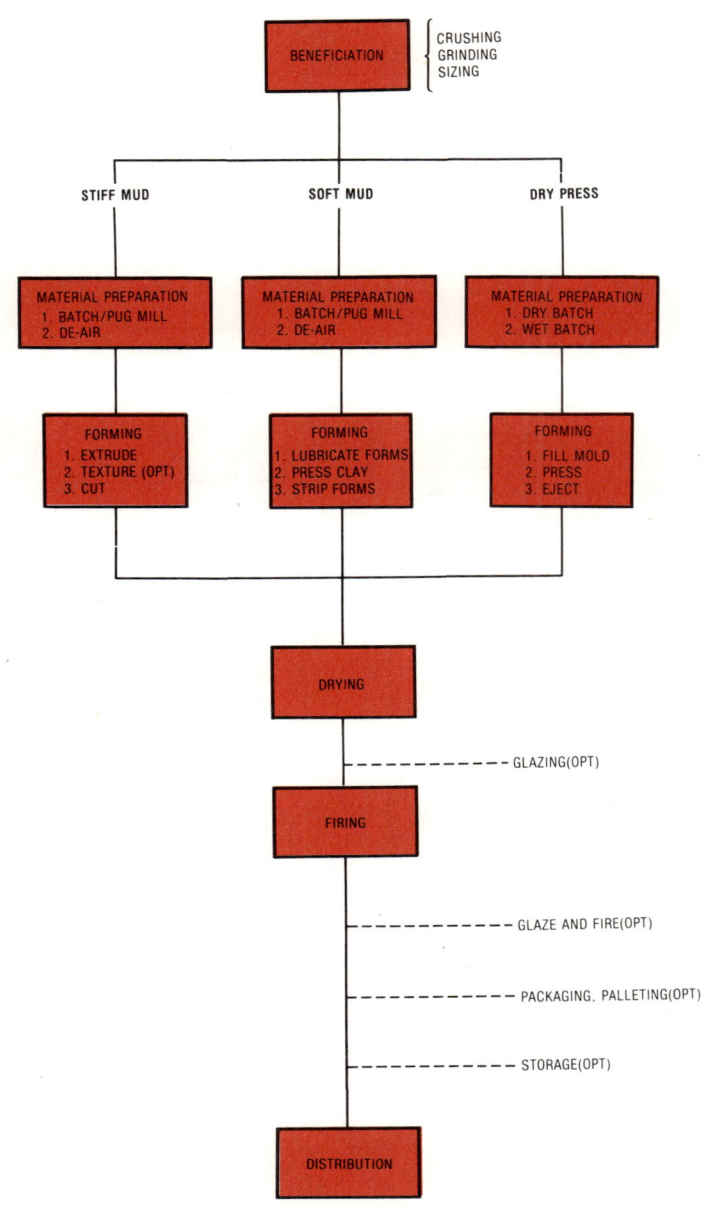

Fig. 52-1. A typical vertical flow sheet for the transformation of materials into structural clay products. Product examples for the three forming methods are: (1) stiff mud—face brick and clay pipe; (2) soft mud—common brick; and (3) dry press—floor and wall tile.

Pictorial flow sheets that parallel the diagram in Fig. 52-1 are provided in Figs. 52-2 and 52-3. They include the five basic processes with machine representations.

Beneficiation

Beneficiation is the process of refining the material into a usable form. This is necessary, since most of the raw materials, especially clay, are in large lumps after winning.

Both *comminution* and *classifying* processes are used. Comminution involves crushing and grinding the large lumps or pieces. Jaw crushers, hammermills, or roll crushers (see Fig. 52-2) are typically used for crushing. Dry pans (see Step 3 in Fig. 52-3) are frequently used for grinding (Fig. 52-4). After grinding, the clay is classified by passing it through an inclined vibrating screen. This controls particle size. The particle size is usually fine or powder-like.

Beneficiation can be omitted, if the shale is soft and has smaller particles. A shale planer often produces shale particles that are ready for the pug mill. Beneficiation is also omitted when the clay is already in a soft-mud or plastic state. This clay would be pugged only before forming.

The condition of the raw material dictates the process necessary. The structural clay products industry does only the minimum required. The materials can usually be coarser, as for bricks and sewer pipe, than those used for the whitewares.

Material Preparation

Material preparation is the mixing of materials. The ceramic term for mixing is *batching*. Clay and/or shale, grog, and water are the typical materials batched. The batching machine, amount of water, and plasticity all vary, depending on the forming methods used.

Batching for Stiff Mud

Batching for a stiff mud is done in a *pug mill*. This is a mixing chamber with a hopper on top for feeding the materials. The chamber contains one or two revolving shafts with blades that cut, mix, and knead the materials into a plastic clay (Fig. 52-5). An auger moves the clay into a *de-airing* chamber that removes the air holes and bubbles. A vacuum pump removes the air from the chamber. De-airing increases the plasticity and workability.

Clay and/or shale, and grog, if used, are mixed with only enough water to produce plasticity. Usually 14 to 20%, by weight, of water is used. Thus, a stiff plastic clay is produced that can only be easily formed by mechanical means.

Batching for Soft Mud

Batching for a soft mud is also done in a pug mill (Fig. 52-5) with a de-airing chamber. Clay and/or shale and grog, if used, are mixed with 20 to 30% water, by

176 / *Modern Industrial Ceramics*

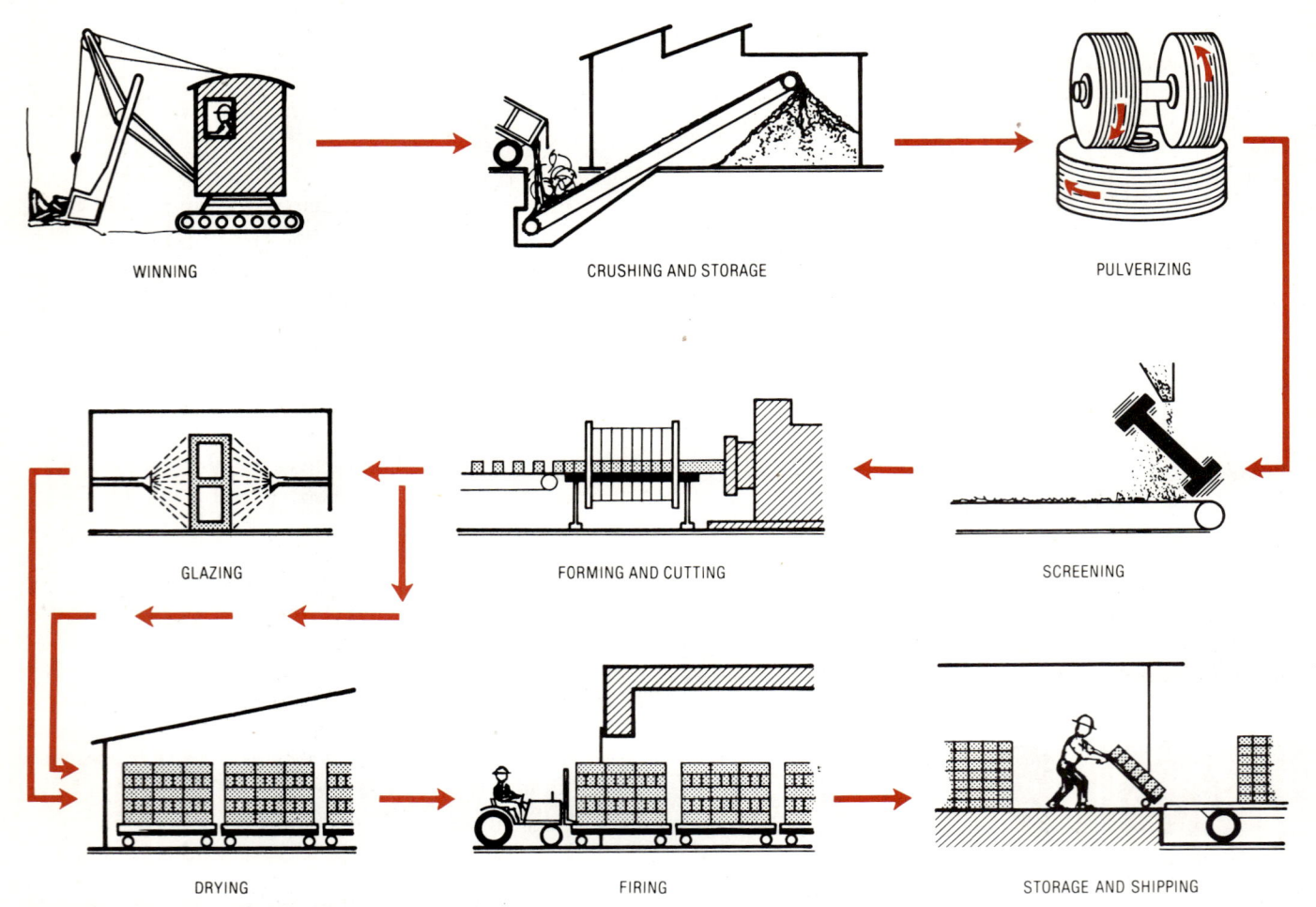

Fig. 52-2. A pictorial flow sheet using stiff-mud forming (see forming and cutting above) for the manufacture of structural clay products. (Courtesy Brick Institute of America)

weight. The available clay may be too wet. Then, only grog would be mixed with the clay. A soft plastic clay is produced that can be easily formed with your hands.

Batching for Dry Press

Batching for a dry press may be done in a dry pan. Clay and other nonplastic materials are mixed with 10% or less water, by weight. This produces a mixture of small pea-size lumps or pellets.

Forming

Forming is the shaping of materials into products. Three major forming processes used to shape structural clay products are:

1. Stiff mud
2. Soft mud
3. Dry press

Today, all these are semiautomatic or fully automatic processes. Each process is capable of producing one thousand or more pieces per hour.

Stiff Mud

An auger usually moves the stiff plastic clay directly from the de-airing chamber (see material preparation) to the extrusion die (Fig. 52-6). The clay is forced through the die in a continuous column by the auger. The die shapes the clay and leaves a hole or holes inside the clay column, if needed (Fig. 52-7).

There are two types of extruders—auger and piston. The auger-type extruder described above is most frequently used today (see Fig. 52-6). It is used to form face brick and many tile shapes, such as structural, drain, roof, and quarry tile.

The piston-type extruder is periodic. It forces the clay through the die only when the piston moves forward. This type of extruder has been used to form the sewer pipe that has a bell at one end. See steps 5-10 in Fig. 52-3.

The continuous-clay column is supported as it leaves the auger-type extruder die. An optional process *texturing* may occur as the clay leaves the die. The die leaves a smooth surface on the clay. Many different textures may

1. MINING THE CLAY Vitrified Clay Pipe is made of selected clays and shales. Laboratory tests determine the correct properties of all raw materials for maximum strength and other physical characteristics.

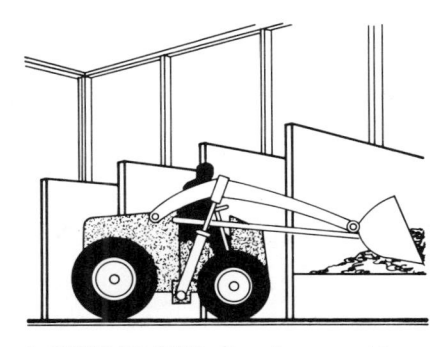

2. MIXING THE CLAYS Many clays are aged to various degrees in storage bins and then blended in the proper combinations.

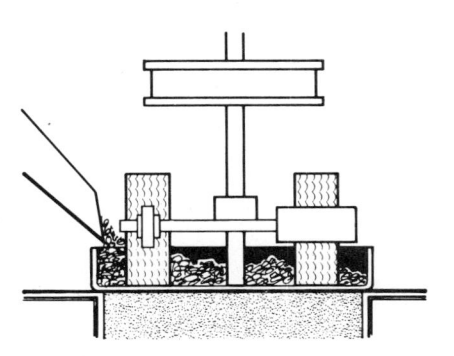

3. GRINDING THE CLAYS Clays are ground in heavy, perforated metal pans by large crushing wheels. The clay is ground fine enough to fall through the perforations in the metal pan.

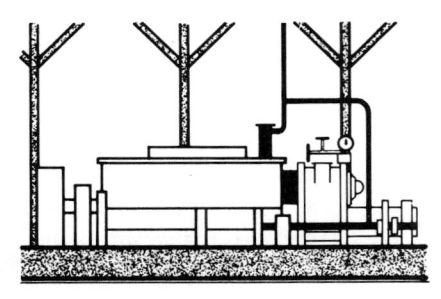

4. PUGGING THE CLAYS Ground raw materials are mixed with water in a pug mill. This material is forced through a vacuum, de-airing chamber to produce a smooth, dense mixture.

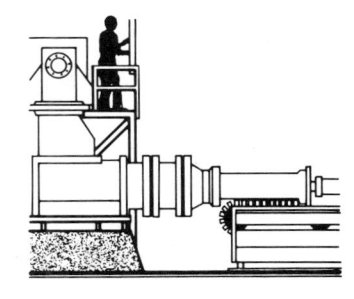

5. FORMING THE PIPE The moistened clay is extruded under tremendous pressure to form the barrel and socket as an integral unit.

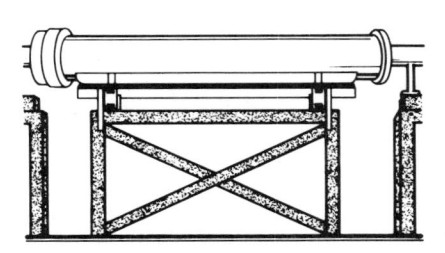

6. FINISHING THE GREEN PIPE Automatic machines trim and finish the moist pipe.

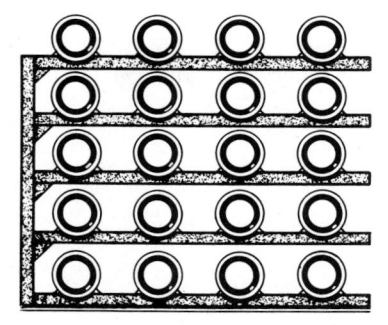

7. DRYING THE PIPE The pipe is stored in large, heated drying rooms until all moisture is driven off.

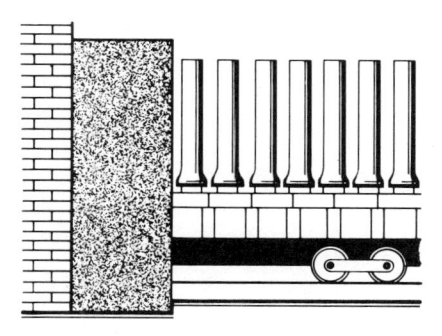

8. SETTING THE KILN The pipe is then set on tunnel kiln cars, as illustrated, or in the familiar beehive kiln.

9. FIRING THE KILN The temperature in the kiln is gradually increased to the intense heat required for vitrification of the pipe, which is approximately 2000°F (1100°C).

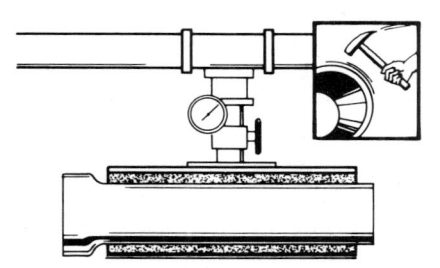

10. TESTING THE PIPE Soundness is determined by the hammer test. Samples from all batches are tested for physical property performance.

Fig. 52-3. A pictorial flow sheet using stiff-mud forming (see Step 5—*Forming the Pipe* above) for the manufacture of vitrified clay pipe. This flow sheet diagrams the manufacture of sewer pipe in Steps 5-10 above. (Courtesy National Clay Pipe Institute)

Fig. 52-4. An industrial-size dry pan that is 19 feet high. Compare the size of the machine with the size of the man standing at the left-hand side. (Courtesy J. C. Steele & Sons, Incorporated)

Fig. 52-5. Top view of a double-shaft pug mill. Pug mills are used to batch plastic clay. (Courtesy J. C. Steele & Sons, Incorporated)

be applied by various attachments that cut, scratch, roll, brush, or otherwise roughen the surface. Only three surfaces can be textured if the column is a rectangular shape, as for bricks. Face bricks are frequently textured.

Texturing occasionally takes the form of imprinting the manufacturers name and/or trademark. Imprinting may be done by a roller. Drain tile may be imprinted.

The continuous-clay column is then moved through an automatic *cutter* (Fig. 52-8). There are two types of cutters—side-cut and end-cut. Bricks are usually side-cut. This machine resembles a series of revolving egg cutters (Fig. 52-2). Most extruded tile shapes, such as structural, drain, and quarry tile, are end-cut. This machine resembles a series of revolving cheese cutters. Taut piano wire does the cutting in both machines.

After leaving the cutting machine, the individual units are deposited on a continuous conveyor belt. The conveyor moves the units through inspection to the drying area or dryers.

Two factors should be noted—extruder die size and cutter wire spacings. Both of these are larger than the final product. This allows for drying and firing shrinkage.

Soft Mud

Soft-mud forming is done in molds or forms. The common brick are soft-mud formed. Originally, this was done by hand, using wood molds that made one or more bricks. The process was similar to that used by the soft-mud machines today. The basic process includes:

1. Lubricate molds/forms
2. Press clay into molds
3. Cut off excess clay
4. Strip or dump mold
5. Bricks conveyed to drying

The molds are lubricated with sand or water to prevent the clay from sticking to them. When sand is used, the bricks are called "sand struck." They are called "water struck," if water is used.

The molds or forms may be made of hardwood, faced with steel. There are multiple molds. Each mold may have six to ten compartments. The compartments are made oversize to allow for shrinkage.

The soft plastic clay is forced or pressed into a lubricated mold. The excess clay is cut off by a taut piano wire. Then, the mold is immediately dumped and the new bricks are moved to drying.

Dry Press

Dry pressing is done in steel molds under pressure (Fig. 52-9). A hydraulic press that exerts 500 to 1500 psi (35 to 105 kg/sq cm) is often used. The basic process includes:

1. Fill mold
2. Press
3. Eject

Usually, the steel dry press mold has three parts—male, female, and ejector. The male part presses the clay into a compact and dense shape in the female or cavity part. The bottom of the cavity is a separate part that ejects the formed piece after pressing. Molds may have multiple cavities. Thus, one or more pieces may be made at each pressing. Note the place where dry pressing occurs in the manufacturing sequence of glazed wall tile in Fig. 52-10.

Drying

Drying removes the excess water. The bricks or tile contain up to 30 percent excess surface water after forming. The amount of water depends on the forming process. Most of this excess surface water is evaporated before firing.

Air drying and dryers are used. Temperature and humidity affect both methods. Air drying is a slower process.

Dryer temperatures range from 100–400°F (38–204°C) and time varies from 24 to 48 hours. Frequently, waste heat from the kiln(s) is used for the dryers. Time,

Fig. 52-6. An industrial-size pugging, de-airing, and extrusion machine. Pugging is done in the large open hopper. De-airing is done in the smaller chamber with the gauge at the right-hand end of the pugging hopper. Extrusion occurs at the right-hand end of the machine. An auger pushes the clay through the die. This machine is capable of forming 15–22,000 bricks per hour. (J. C. Steele & Sons, Incorporated)

Fig. 52-7. An unassembled extrusion die for the machine in Fig. 52-6. Note how the die housing (left) tapers down to the brick shape. (Courtesy J. C. Steele & Sons, Incorporated)

heat, and humidity must be controlled to avoid cracking and/or warping the unfired clay products.

Glazing

Glaze forms a glassy coating when fired onto the clay. It protects and beautifies the product. Glaze materials include fluxes, alumina, and flint. Feldspars, whiting, talc, kaolin, flint, metallic oxides, and water are the glaze materials most frequently used.

Glazing is an *optional* process. Some face brick, structural tile, and most wall tile are glazed. The glaze is usually applied to only one surface—the face. Some face brick may have two glazed surfaces.

Liquid glaze is usually applied to structural clay products. The four methods of glaze application are:

1. Brushing
2. Dipping
3. Pouring
4. Spraying

All of the methods may be used. Spraying is the most frequently used. Glazing, when used, is usually an automatic process that includes conveyor, spray guns, exhaust unit, and dryer.

Fig. 52-8. Two views of an automatic brick and tile wire cutter, showing front view (top) and rear view (bottom). Bricks would be side-cut, and the tile would be end-cut by this type of machine. The bricks or tile are cut when the reel revolves. This machine is pneumatically and electrically operated. (Courtesy J. C. Steele & Sons, Incorporated)

Fig. 52-9. Raw materials feed from storage bins (upper right) into automatic dry presses which form the shape of the tile. Note the flexible tubing which draws off all loose dust, thus keeping the air in the plant clean and pure. (Courtesy American Olean Tile Company)

Firing

Firing causes a glassy bond to form that fuses the clay particles together. This is called vitrifying. *Vitreous* means glassy. When a glaze is used, firing also causes the glassy coating to fuse to the body.

Kilns

Both periodic and continuous kilns are used by the structural clay products industry. Large kilns are usually used, due to the quantity and size of some of the products.

The periodic kilns used are electric, updraft, downdraft, and shuttle (Fig. 52-11). Updraft and downdraft kilns use combustible fuels, usually oil or natural gas. Large, round beehive-shaped updraft and downdraft kilns have been used by this industry. The shuttle kiln may use electricity or combustible fuels. Combustible fuels are necessary for salt glazing and flashing. A discussion of these processes follows.

The continuous kilns used are straightline tunnel kilns (Fig. 52-12). Kiln cars are used to move most of the structural clay products through the three firing zones—preheat, firing, and cooling. Roller-hearth and pusher-slab methods may be used for small items, such as floor, wall, and mosaic tile. Tunnel kilns are frequently used by this industry.

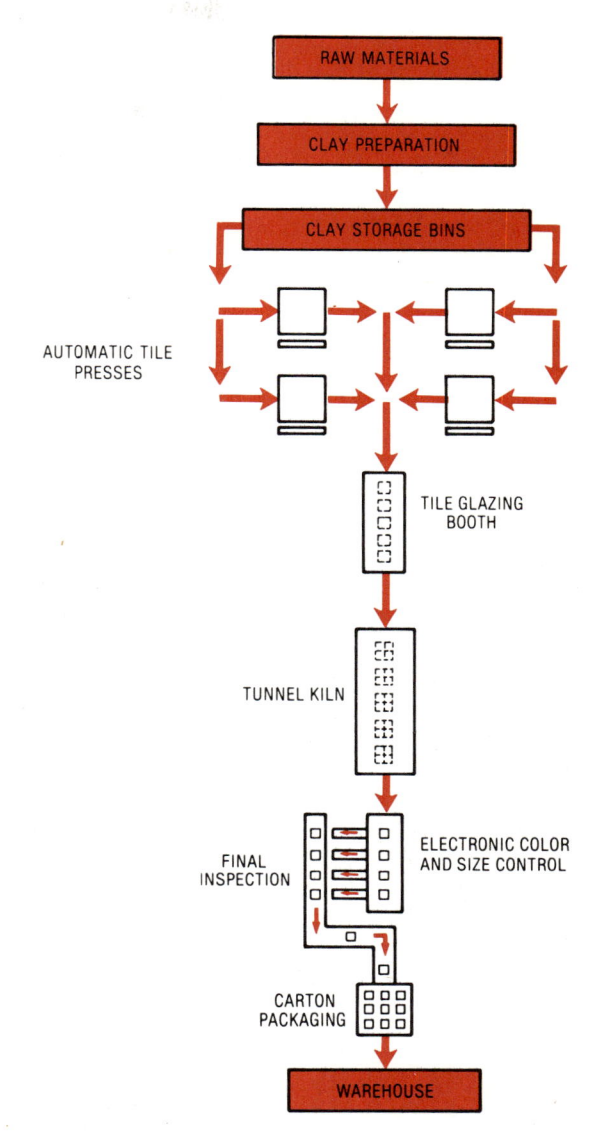

Fig. 52-10. A pictorial vertical flow sheet diagramming the manufacturing sequence for glazed wall tile. (Courtesy American Olean Tile Company)

Temperatures

Firing temperatures depend on the product. Temperatures range from 1700–2700°F (927–1482°C). Firing time also depends on the product and the kiln size. The large periodic kilns may require 150 hours to fire, plus cooling time. Tunnel kilns usually require less time.

Firing time, temperature, and cooling time have to be carefully controlled to avoid defects. A typical firing of a clay and/or shale product is outlined in Table 52-1.

See Figs. 52-1, 52-2, 52-3, 52-10, and Section VI—*Kiln Industry* for firing and other kiln diagrams.

Types of Firing

The two types of firing (discussed in Section XII—*Clay Industries*) are bisque and glaze. Bisque firing is the first firing of the dried clay product. A low bisque temperature leaves the fired clay porous and easier to glaze. This is done only when a glaze is to be applied and the product is to be refired. Structural clay products without a glaze, such as quarry tile, are fired to a higher bisque temperature that causes vitrification.

The glaze fire is usually the second firing. It is often done at a higher temperature than bisque firing. Glazed structural clay products may be fired once or twice. The traditional firing is done twice—bisque and glaze. Wall and mosaic tile may have been fired twice.

However, a single firing is also used. Five reasons for its use are:

1. Slow firing is used. This means the temperature is slowly increased over a period of several hours.
2. Glaze is usually applied to only one surface. Thus, moisture can still escape from the clay via unglazed surfaces.
3. One firing reduces time and cost.
4. Efficiency is increased.
5. Conserves fuel.

Face brick, structural tile, wall tile, and mosaic tile may be single fired with a glaze.

Fig. 52-11. A group of four large periodic shuttle kilns with a transfer-car system. Large structural clay products can be fired in these kilns. The kilns are 18 ft (5.49 meters) in width by 50 ft (15.25 meters) in length. They can be fired in 36 hours, cold-to-cold. (Courtesy Bickley Furnaces, Inc.)

Fig. 52-12. A straight-tunnel kiln, using cars to convey ware through its length. These cars are stacked with saggars. Saggars are used to hold small ware, such as floor and wall tile, for firing. (Courtesy Bickley Furnaces, Inc)

Table 52-1. Typical Firing of a Clay and/or Shale Structural Clay Product

Firing Stage	Temperature*
1. Watersmoking	up to about 400°F
2. Dehydration	300° to 1800°F
3. Oxidation	1000° to 1800°F
4. Vitrification	1600° to 2400°F
5. Flashing (optional)	1600° to 2400°F
6. Cooling	2400°F down to normal

* (Courtesy Brick Institute of America)

Salt Glaze

Sewer pipe (see Fig. 52-3) are also single fired. They may be vitrified without a glaze, or they may be salt glazed. *Salt glazing* is a special single glaze firing that eliminates bisque firing and glaze application. It is usually done in a periodic kiln fired with a combustible fuel and used *only* for salt. The salt will also coat the inside of the kiln. The kiln is stacked with greenware sewer pipe. Then, it is fired to the maturing temperature. Salt thrown into the kiln at this temperature vaporizes. The salt, sodium chloride (NaCl), vapors react with the clay to form a glaze. Chemically, the sodium (Na) in the salt combines with silica (SiO_2) in the clay to form a simple glaze. The chlorine (Cl) in the salt combines with water (H_2O) to form hydrochloric acid (HCl) that leaves the kiln in the fumes. Thus, the fumes are dangerous.

Flashing

Flashing may be done near the end of the firing process to produce color variations. This is an optional process usually done to only unglazed structural clay products. Face brick and quarry tile are the two products usually flashed. This process is best accomplished in a kiln fired with a combustible fuel. The normal kiln atmosphere has sufficient oxygen. This is called *oxidizing*. When the flow of air into the kiln is reduced, the atmosphere lacks oxygen. This is called *reduction* or flashing when used on structural clay products. The term *flashing* is unique to the SCP industry.

Red clay provides a typical example of the effect flashing can achieve. Clay containing iron in practically any form will fire red in an oxidizing atmosphere, due to the formation of ferrous oxide (FeO). The same clay fired in a reducing atmosphere will take on a purple cast.

Flashing may be done only on every fourth firing. Then, the flashed bricks mixed with those from a normal or oxidation firing will produce a variegated color effect.

Defects

Some defects that can occur in structural clay products (especially bricks) are bloating, black coring, and scumming. Bloating and black coring are firing defects. These are primarily caused by firing too fast or increasing the heat too rapidly. Moisture trapped inside the bricks causes both defects. Cored brick has practically eliminated the trapping of moisture, since the cores provide exposure of the interior.

Scumming is a body defect. It is usually caused by soluble salts, such as calcium sulfate in gyspum, in the raw material(s). The soluble salts are carried out of the brick by water during drying and firing and deposited on the exterior surfaces, due to evaporation. A small amount of barium carbonate ($BaCO_3$) mixed with the raw material(s) before forming renders the soluble salts insoluble.

Storage and/or Distribution

Final inspection and, if needed, packaging and/or palleting are done before storage and/or distribution. Today, packaging and palleting may be semi- or fully automatic. Storage or immediate distribution completes transformation (Fig. 52-13).

Fig. 52-13. Pallets of new bricks at the job site. Pallets allow easy handling during storage and distribution.

Unit 53: Utilization

Structural clay products are used in both manufacturing and construction. Their utilization is discussed in three divisions—manufacturing, construction, and product information. The third division contains information that is useful to all SCP consumers.

Manufacturing

The traditional structural clay products were not used in manufacturing, due to their weight. However, at least three products are used today in manufactured products—face brick, floor tile, and wall tile.

Prefabricated brick panels are now manufactured. The manufacturing process involves assembling face bricks, concrete reinforcing rods, and mortar into large panels. The panels are usually assembled on the horizontal. A crane is used to lift the panels. They are transported on heavy trucks. Two or more panels are usually used for an exterior wall.

Flexible pregrouted ceramic tile is another manufactured product. Sheets of floor and wall tile are now assembled with a flexible silicone rubber grout and are ready to be glued in place. The sheets are about 24 × 24 in. (61 × 61 cm). Traditionally, the floor and wall tile were individually placed and grouted. Thus, the pregrouted sheets save installation time.

Prefabrication variations using structural clay products will become available. You can be aware of these new products as they become available by reading the magazines and newspapers.

Construction

Traditionally, the structural clay products have been used in construction. They have been and still are used for building on site in their final position of use. They may be individually placed or prefabricated. Usually, the construction is an assembling process.

Terms

Bricks have two surfaces that may be exposed in a wall. They are the header and stretcher. The *header* is the end of a brick with only the height and width exposed. The *stretcher* is the face of a brick with only the height and length exposed (Fig. 53-1).

Course and bond are two terms that apply to the structural units in a wall or floor. A *course* is one horizontal row of structural units. A *bond* is the overall pattern that is made by several courses.

Mortar is the bonding agent that adheres or fastens the structural units together. Its composition was discussed earlier in Section IX—*Lime*. A variety of mortar joints are used, and most have decorative qualities.

Uses

Bricks are the well-known and visible clay products. Some construction uses of bricks are in:

1. Veneer walls—both interior and exterior (Fig. 53-2)
2. Fireplaces—both interior and exterior
3. Planters
4. Patios (Fig. 53-3)

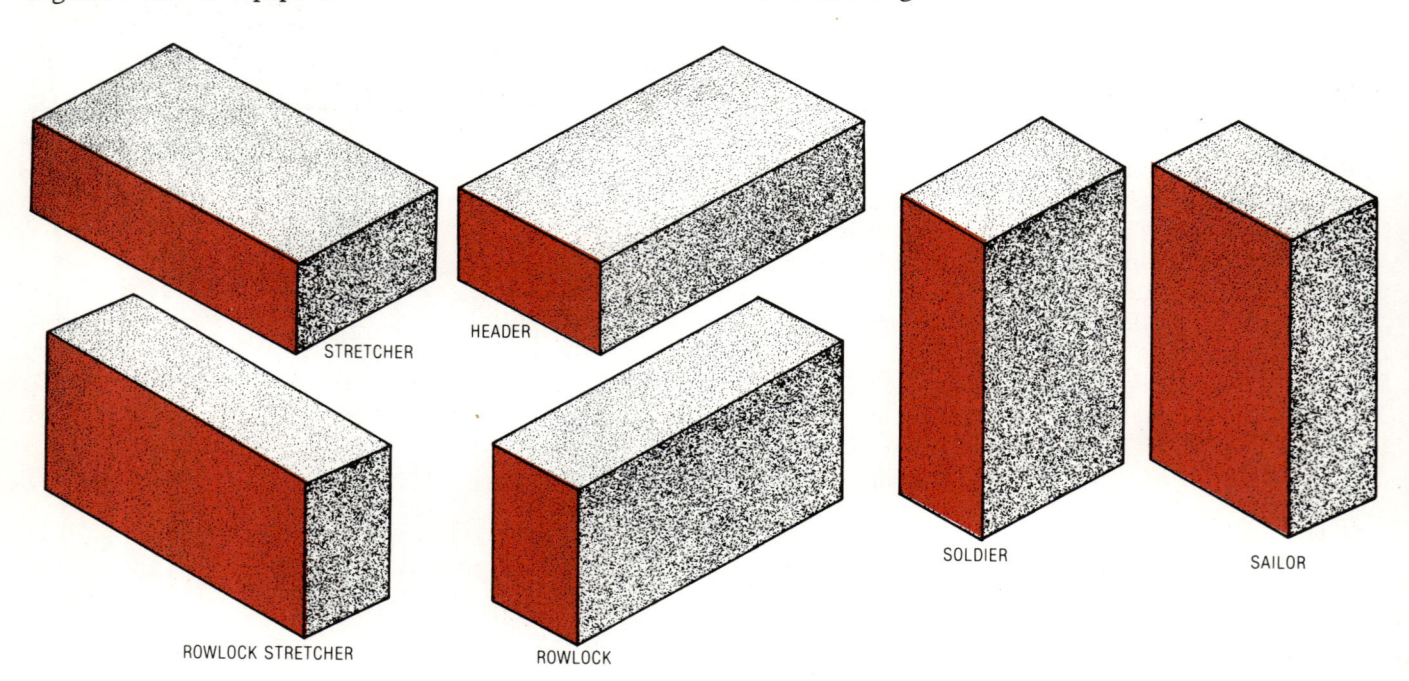

Fig. 53-1. Six positions of a brick in a wall. The shaded area is the one surface exposed, except at corners where two surfaces are visible. (Courtesy Brick Institute of America)

184

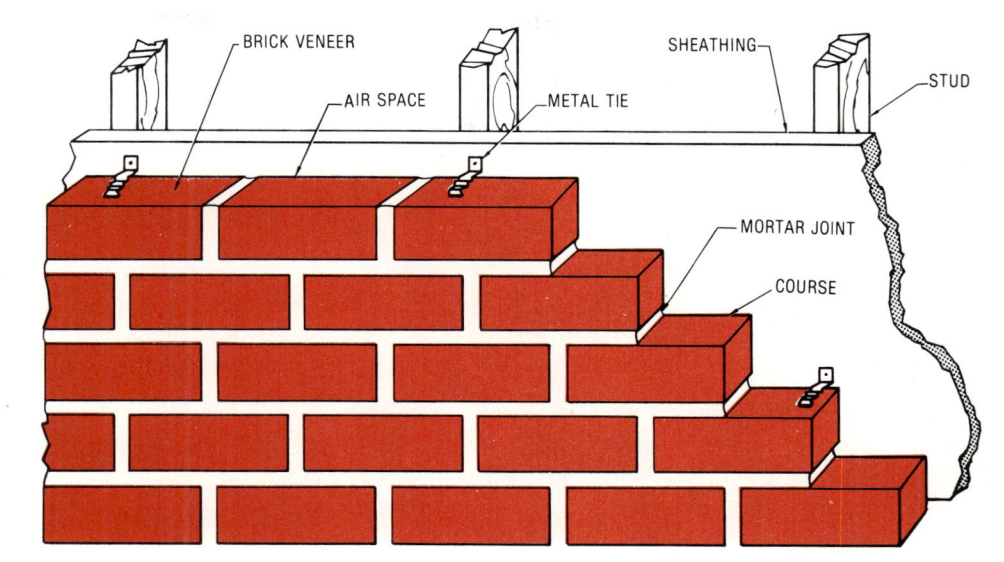

Fig. 53-2. Construction details of a brick veneer exterior wall. Brick veneer means that the wall is only one brick in thickness.

Fig. 53-3. Paving brick used for a patio. These bricks are also used for walks, driveways, and streets.

Fig. 53-4. Drain tile made in Canada. These are 3 inches in diameter by 12 inches in length. They are used to carry water away from gutter downspouts and to drain wet land.

5. Walks
6. Pavements—driveways and streets

Structural tile may be either exposed or used as backup units in walls. The glazed structural tile are usually used when they are exposed, especially as interior walls. The unglazed structural tile have also been used for farm silos.

Drain tile and sewer pipe are usually used underground (Fig. 53-4). Thus, they are hidden from normal view. The drain tile are used to carry water away from gutter downspouts, from a septic system leach field, and from wet land. Sewer pipe carries liquid waste to a septic tank or sewage plant. They are also used for sanitary and storm sewer systems, drainage, irrigation, culverts, conduits, and heating ducts.

Floor, wall (Fig. 53-5), and quarry tile are used on both interior and exterior surfaces. Probably, more of them are used for interiors than for exteriors. All these tile protect and beautify the surface to which they are applied. Quarry tile are used on floors (Fig. 53-6), patios, and walks. Traditionally, quarry tile were used only for commercial and institution applications. Now they are used in homes.

Roof tile usage is obvious. They often last longer than the building.

Mosaic tile are used for countertops, tabletops, a back splash, and novelty uses.

Product Information

The two major types of structural clay products are bricks and tiles.

Bricks

There are three types of bricks—face, common, and

Fig. 53-5. This plaid pattern was made by using scored and plain 4¼-in. × 4¼-in. glazed wall tile. Unglazed floor tile are used more frequently than these glazed tile. Glazed tile tend to be slippery when wet. (Courtesy American Olean Tile Company)

Fig. 53-6. This laundry room floor was made of 6-in. × 6-in. quarry tile. (Courtesy American Olean Tile Company)

specialty. Face bricks usually have one surface exposed when in use. A large variety of face brick are available. Variations occur in their color, texture, unglazed or glazed, and size. Color (flashing), texture, and glazing were discussed in Unit 52—*Transformation*. Usually, face brick are semi-vitreous. This allows the mortar to achieve a stronger bond.

Common brick are designed and used as backup units in walls. They are softer and are not vitreous. These bricks are soft-mud formed. They can be identified by the recessed panel that often contains the name of the manufacturer. Also, they are a standard-size solid brick without cores.

Paving brick are the major specialty brick. The current sizes available are listed in Table 53-1. Paving brick are used for patios, walks, driveways, and streets. These bricks are vitreous.

Table 53-1. Paving Brick Sizes

Face Dimensions (actual size in inches)		Thickness
w x	l	
4	8	The unit thickness of brick varies. The most popular thicknesses are 2¼ in. and 1⅝ in. The range of thickness is generally from ¾ in. to 2½ in.
3¾	8	
3¾	7½	
3⅝	7⅝	
3½	7½	
6	6	
8	8	
6	6 Hexagon	
8	8 Hexagon	

(Courtesy Brick Institute of America)

Face-brick sizes are either modular or nonmodular. Modular means that a standard size is used. Modular size is based on 4 in. (10.2 cm), or a division or multiple of 4 in. (10.2 cm). Modular brick height is 2¼ in. (5.7 cm), so that the height of three courses with mortar is equal to 8 in. (20.3 cm). Nonmodular bricks are not based on the 4-in. (10.2 cm) standard. The modular face-brick sizes are diagrammed in Fig. 53-7 and in Table 53-2. The nonmodular brick sizes are diagrammed in Fig. 53-8 and in Table 53-3.

The brick grades are listed in Table 53-4. These should be considered for correct brick usage. For example, SW bricks can be used on the ground or under the ground; MW bricks can be used above the ground; and NW bricks can be used as backup bricks inside a wall.

Tiles

Tiles include the following:

1. Structural
2. Drain
3. Sewer pipe
4. Wall
5. Floor
6. Quarry
7. Roof
8. Mosaic

Structural tile are used for walls. They may be exposed or backup units for brick or wall tile. They are made glazed and unglazed.

Drain tile are usually cylindrical (Fig. 53-4). They are 3 or 4 in. (7.6 or 10.2 cm) in diameter and 12 in. (30.5 cm) in length.

Sewer pipe is manufactured with a bell or socket on one end. The bell or socket is used to join the successive pieces together. This is also known as vitrified clay pipe. It is made unglazed or salt glazed. The pipe diameters range from 4 to 42 in. (10.2 to 107 cm). The lengths begin at 24 in. (61 cm) and increase in 1-ft (30.5 cm) intervals as the diameter increases.

Wall tile are available in a variety of shapes, sizes, colors, and either glazed or unglazed. Traditional wall tile are glazed 4¼ in. (10.8 cm) square. Most of the wall tile have a white body. Color is usually achieved by using glazes.

Floor tile are also available in a variety of shapes, sizes, and colors. Traditional floor tile are unglazed and 1-in. (2.5 cm) square. Most of the floor tile are unglazed. Color is achieved by adding metallic oxides to the raw materials used for the clay body.

Quarry tile are usually larger and thicker than floor tile. They are also available in a variety of shapes, sizes, and colors. The traditional quarry tile is a red unglazed 6-in. (15.2 cm) square. Color is due to the clay(s) used and flashing if used.

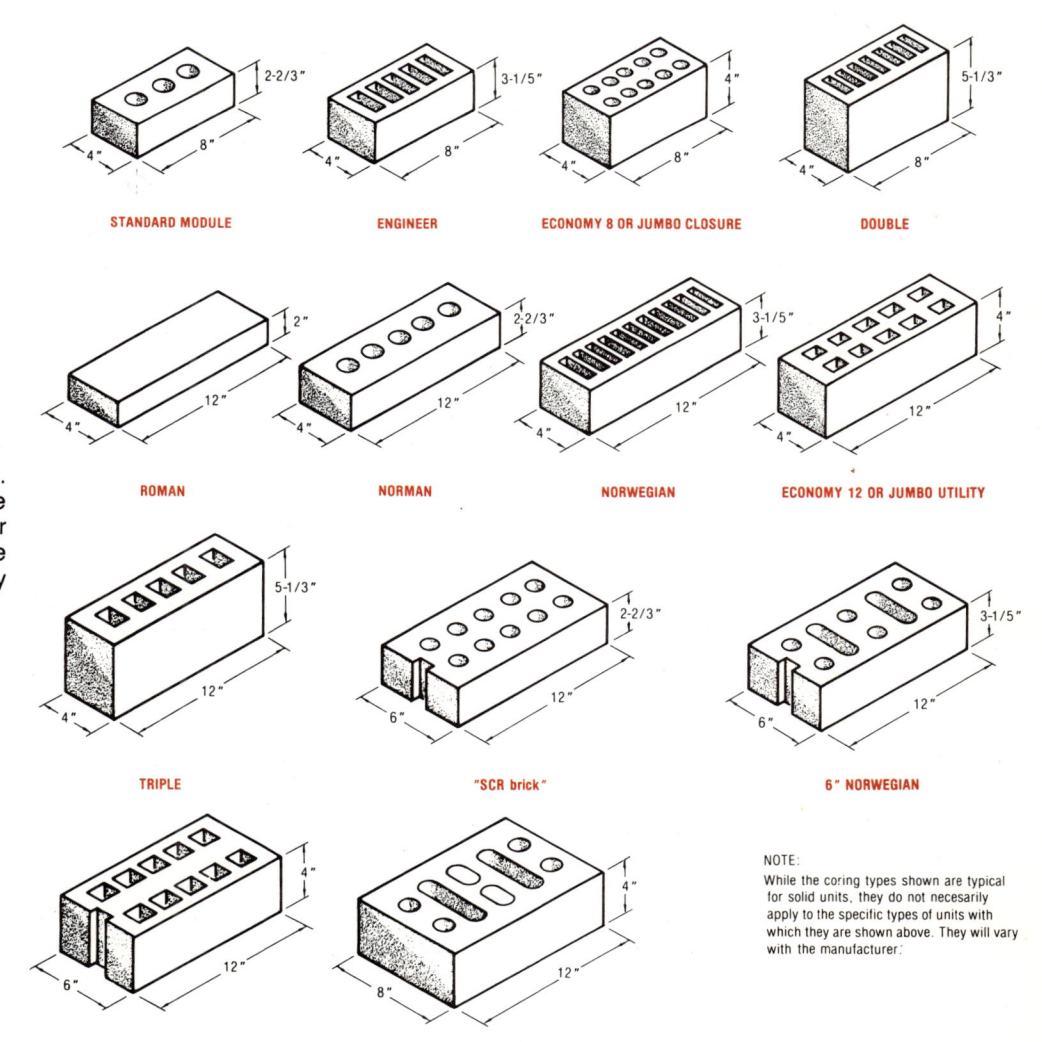

Fig. 53-7. Modular brick sizes. Nominal sizes are used for the brick dimensions. Actual or manufactured dimensions are given in Table 53-2. (Courtesy Brick Institute of America)

NOTE:
While the coring types shown are typical for solid units, they do not necessarily apply to the specific types of units with which they are shown above. They will vary with the manufacturer.

Table 53-2. Modular Face Brick Sizes.

Unit Designation	Nominal Dimensions, in. t	Nominal Dimensions, in. h	Nominal Dimensions, in. l	Joint Thickness in.	Manufactured Dimensions in. t	Manufactured Dimensions in. h	Manufactured Dimensions in. l	Modular Coursing in.
Standard Modular	4	2 2/3	8	3/8	3 5/8	2 1/4	7 5/8	3C = 8
				1/2	3 1/2	2 1/4	7 1/2	
Engineer	4	3 1/5	8	3/8	3 5/8	2 13/16	7 5/8	5C = 16
				1/2	3 1/2	2 11/16	7 1/2	
Economy 8 or Jumbo Closure	4	4	8	3/8	3 5/8	3 5/8	7 5/8	1C = 4
				1/2	3 1/2	3 1/2	7 1/2	
Double	4	5 1/3	8	3/8	3 5/8	4 15/16	7 5/8	3C = 16
				1/2	3 1/2	4 13/16	7 1/2	
Roman	4	2	12	3/8	3 5/8	1 5/8	11 5/8	2C = 4
				1/2	3 1/2	1 1/2	11 1/2	
Norman	4	2 2/3	12	3/8	3 5/8	2 1/4	11 5/8	3C = 8
				1/2	3 1/2	2 1/4	11 1/2	
Norwegian	4	3 1/5	12	3/8	3 5/8	2 13/16	11 5/8	5C = 16
				1/2	3 1/2	2 11/16	11 1/2	
Economy 12 or Jumbo Utility	4	4	12	3/8	3 5/8	3 5/8	11 5/8	1C = 4
				1/2	3 1/2	3 1/2	11 1/2	
Triple	4	5 1/3	12	3/8	3 5/8	4 15/16	11 5/8	3C = 16
				1/2	3 1/2	4 13/16	11 1/2	
SCR brick[2]	6	2 2/3	12	3/8	5 5/8	2 1/4	11 5/8	3C = 8
				1/2	5 1/2	2 1/4	11 1/2	
6-in. Norwegian	6	3 1/5	12	3/8	5 5/8	2 13/16	11 5/8	5C = 16
				1/2	5 1/2	2 11/16	11 1/2	
6-in. Jumbo	6	4	12	3/8	5 5/8	3 5/8	11 5/8	1C = 4
				1/2	5 1/2	3 1/2	11 1/2	
8-in. Jumbo	8	4	12	3/8	7 5/8	3 5/8	11 5/8	1C = 4
				1/2	7 1/2	3 1/2	11 1/2	

[1] Available as solid units conforming to ASTM C 216-or ASTM C 62-, or, in a number of cases, as hollow brick conforming to ASTM C 652-.
[2] Reg. U.S. Pat. Off., SCPI.

(Courtesy Brick Institute of America)

Fig. 53-8. Nonmodular brick sizes. Actual or manufactured sizes are used for the brick dimensions. (Courtesy Brick Institute of America)

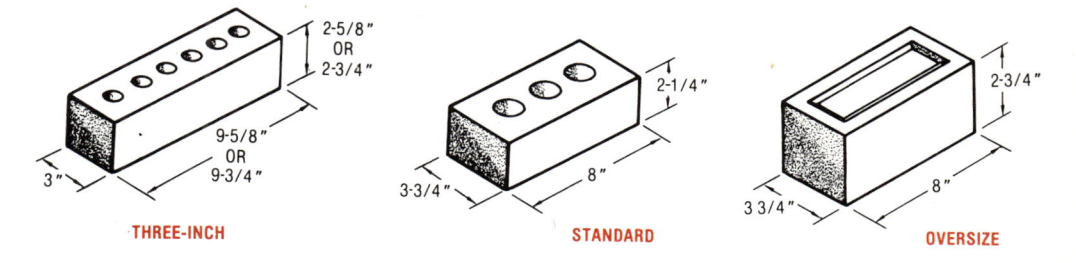

Table 53-3. Nonmodular Face Brick Sizes

Unit Designation	Manufactured Size (in.)		
	thickness	height	length
Three-inch*	3	2⅝	9⅝
	3	2¾	9¾
Standard**	3¾	2¼	8
Oversize**	3¾	2¾	8

* In recent years, the so-called "three-inch" brick has gained popularity in certain areas. The term "three-inch" designates its thickness or bed depth. The sizes shown in the table are the ones most commonly produced under the designation "Kingsize." Other sizes of 3-in. brick are also produced under such designations as "Big John," "Jumbo," "Scotsman," and "Spartan." Originally developed primarily for use as a veneer unit, it is also used to construct 8-in. cavity walls and 8-in. grouted walls.

** The manufactured thickness of standard or oversize nonmodular brick will vary from 3½ to 3¾ in. Therefore, if other than a running bond is desired, the designer should check with the manufacturer of the brick selected.

(Courtesy Brick Institute of America)

Table 53-4. Brick Grades*

Grade	Explanation	Type of Brick
SW	Severe weathering	Paving brick
MW	Moderate weathering	Face brick
NW	No weathering	Common brick

* ASTM C62-58

Roof tile have been available in several shapes and in one color. The tile shape is usually a half-cylinder with horizontal surfaces that are overlapped by the adjoining tiles. They are usually red and unglazed.

Mosaic tile are usually small, 1 in. (2.5 cm), or less, in length. They are available in a variety of shapes, colors, and glazed. These tile are usually sold in 1-ft (30.5 cm) square sheets. The tile are arranged in a pattern and mounted with glue on paper or plastic mesh.

Unit 54: Disposition

The disposition of used structural clay products presents a problem. All are solid waste. However, many of these products can be reclaimed.

Reclaiming

The materials reclaimed are either recycled or reused. Both are positive uses for existing resources.

Recycle

Many of the structural clay products can be recycled by crushing. The coarser crushed fired clay could be used in place of crushed stone and gravel that was to be covered by another material, such as the base for concrete. Another recycling use of crushed fired clay could be concrete aggregate. A 1-in. (2.5 cm) crushed clay might be used as a mulch for shrubbery and other formal plantings.

Reuse

All the structural clay products can be reused. However, mortar or other bonding agents often make their reuse difficult. Removal of the bonding material is a problem that may require manual labor. Manual labor takes longer and thus increases cost. Therefore, the reuse of many of the structural clay products is never attempted.

The reuse of old brick is popular. Mortar and weathering give old brick the unique coloring that is not available in new brick (Fig. 54-1). Typical uses are in the brick veneer walls for homes, planters, and fireplaces. The demand has been so large for old brick that some manufacturers have started making a new brick that looks used.

Some products, such as drain tile and roofing tile, are not mortared or bonded together. Therefore, reuse of these products is easy and encouraged. They can be utilized again for draining and roofing.

Quarry tile are reused after mortar removal. They are often laid without mortar for patios, walks, and garden borders.

Broken brick and tile can be used as a riprap (Fig. 54-2). Hard fill is another use for broken pieces.

Discarding

All too frequently, used structural clay products are discarded. Discarding is the least time consuming. Time is money, so there is also less cost involved. These reasons are given when brick buildings are being dismantled for urban renewal and the brick are being discarded.

Often, the huge quantity of material from an old brick building makes discarding the only feasible method of disposition. Also, many of the bricks may be common bricks that are too soft and porous for reuse.

Discarding can become beneficial, if the material is used as hard fill for low areas or old dumps. Fill is actually a way of reusing used materials. Normally, the natural materials, such as stone or gravel, would be used as hard fill. Thus, natural resources can be conserved by intelligent use of discarded material.

Fig. 54-1. Reuse of old brick for an interesting river bank wall.

Fig. 54-2. Broken brick and tile can be reused as riprap. The riprap prevents erosion of the river bank.

SECTION XIV

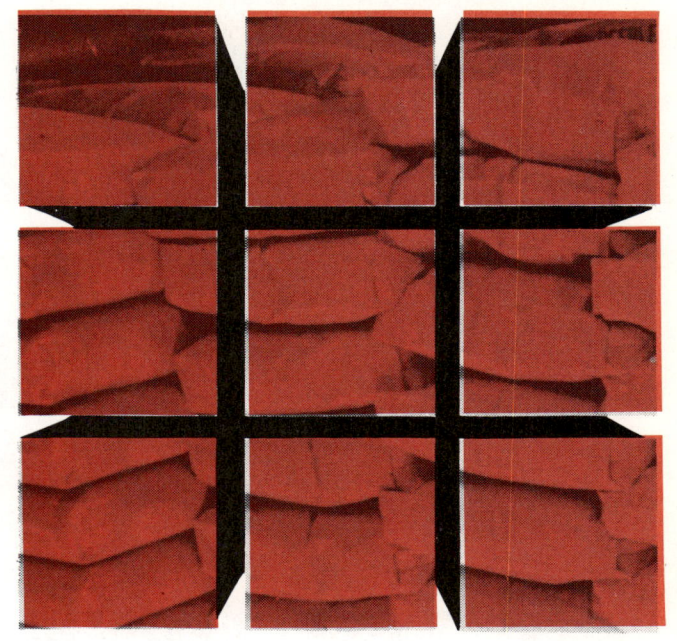

Whiteware Industry

Unit 55: Material Science

Whitewares are fired ware, glazed or unglazed, that usually have a fine-texture white-clay body. These products include handmade ware (Fig. 55-1) and manufactured ware (Fig. 55-2). This is consistent with the *American Society of Testing and Materials,* ASTM (C242) definition.

A general word equation has been used in the preceding sections to express the clay industries:

$$\text{raw materials} + \text{water} \xrightarrow{\Delta} \text{primary or secondary product} + \text{optional processing}$$

Most of the structural clay products were expressed by a simplified word equation:

$$\text{raw materials} + \text{water} \xrightarrow{\Delta} \text{primary product}$$

However, some structural clay products (SCP) require more processing as follows:

$$\text{raw materials} + \text{water} \xrightarrow{\Delta} \text{bisque} + \text{glaze} \xrightarrow{\Delta} \text{secondary product}$$
(won and purchased)

A few whiteware products, such as flower pots, are manufactured similar to the second or simplified structural clay products (SCP) equation. Most of the whiteware products require additional processing, similar to the third word equation. This word equation provides an overview of the content of this section.

Similarities

There are several similarities and differences between the structural clay products and the whitewares.

Some of the *similarities* between the structural clay products and the whitewares are:

1. Overall processing is basically the same.
2. Both products require one or more raw materials and usually some water.
3. Both products require at least one high-temperature heat treatment to make a hard durable end product.

Differences

Some of the *differences* between the structural clay products and the whitewares are:

1. The clay and/or shale used for SCP may have coarser particles than the clay used for whitewares.
2. SCP manufacturers of brick and sewer pipe, for example, use limited material preparation, only enough to make plastic clay.
3. Most of the SCP have a buff or red body color, while the whitewares are usually white.
4. Forming processes are different. Soft-mud forming only is used by SCP. Whitewares use only limited stiff-mud and dry-press forming.
5. Most SCP are only formed and dried, without any cleanup or finishing work. Most whitewares are cleaned up or fettled before firing.

Fig. 55-1. Typical handmade whiteware products made by art potters and hobby ceramists. The forming methods are pinch, coil, slab, and thrown (left to right).

Fig. 55-2. Typical manufactured whiteware products that are formed by machines.

Fig. 55-3. Raw sedimentary clay that will fire red (top) and buff (bottom). The color of the raw clay is dull red (top) and gray (bottom).

6. Most SCP are fired once, while most whitewares are fired twice. Some whitewares are fired three or more times.
7. Most SCP are unglazed, while most whitewares are glazed.
8. Most SCP are heavy, while most whitewares are lightweight and/or delicate.

Materials

Both natural raw materials and manufactured materials are used to manufacture the whitewares. Basically, most of the materials used are natural raw materials.

Raw Materials

The raw materials are either plastic or nonplastic. The plastic materials or clay have plasticity. Clay is used for the body and in glazes. The primary, colluvial, and sedimentary clays are used (Fig. 55-3). However, when a company uses any of these clays from its own area, the clay is called a native or local clay. When a company purchases one type of clay and makes products from only this one clay, it is called *singular-purchased* clay.

For example, in Oak Hill, Ohio there is a well-known red sedimentary clay deposit. When a local Oak Hill, Ohio company uses the clay, it is a native or local clay. When a Buffalo, New York company buys the same clay and has it shipped to Buffalo, it is then a singular-purchased clay. Many educational institutions, potters, and hobbyists use a singular-purchased clay.

The *nonplastic* materials do not have plasticity. They are primarily glass former, fluxes, and colorants. These materials are used for the body and in glazes. Flint (SiO_2) is the major *glass former*. *Fluxes* lower the melting point of the glass former. Feldspars, whiting, and talc are typical fluxes. Metallic oxides and carbonates are used for *colorants*.

Manufactured Materials

Most of the whitewares are made from clay bodies. A *clay body* is made from one or more clays and/or nonplastic materials. For example, a typical clay body might contain kaolin, ball clay (for plasticity), flint, (glass former and refractory), feldspar (flux), and whiting (flux). Clay bodies are used, since their physical properties can be controlled by the addition of one or more materials.

Four major whiteware clay bodies used today (Fig. 55-4) are:

1. Earthenware or semivitreous ware
2. Stoneware
3. Porcelain
4. China

Each of these wares has distinct physical properties that distinguishes it. The properties are:

1. Absorption
2. Firing range
3. Color
4. Firing technique

These properties are listed in Table 55-1.

Grog finds limited use in the whiteware clay bodies. Its major use is in the stoneware clay bodies used by the art potters. Grog was discussed in Section XIII—*Structural Clay Products Industry*.

Fig. 55-4. The four major whiteware clay bodies are earthenware, stoneware, procelain, and china (left to right).

Table 55-1. Whiteware Clay Bodies*

Clay Body	Absorption	Firing Range	Color	Firing Technique and Notes
Earthenware or semi-vitreous ware	<10%	C/06-C/6	White Buff Red	Low bisque fire, such as C/07 and higher glaze fire, such as C/04. Usually requires a glaze to be watertight.
Stoneware	<5%	C/4-C/10	Off-white Buff Brown Metallic oxides occasionally added for blue, green, and black body	Lower bisque fire, such as C/07, and higher glaze fire, such as C/10. Can fire watertight without a glaze. Can salt glaze.
Porcelain	<2%	C/6-C/20	White only	Usually, one fire technique. Body and glaze matured together. Frequently dry footed. Occasionally, body is translucent.
China	<2%	C/6-C/20	White Metallic oxides occasionally added to body	High bisque fire, such as C/10, and lower glaze fire, such as C/05. Stilts usually used for glaze fire. Occasionally, fired three or more times, such as for overglaze. Fine and bone china usually have translucent bodies.

* Note: See Section XII—*Clay Industries* for formulas

Unit 56: Procurement

Procurement of raw materials for the whiteware industry is diagrammed in Fig. 56-1. Clay bodies are used for most of the whiteware products. Two or more materials are used in the clay bodies. Therefore, most if not all the clay body materials are purchased. This is the reason why few if any of the whiteware manufacturers do their own procurement.

Purchasing

Procurement is a purchasing job, rather than an extraction process, for most of the whiteware companies. Even though a company owns a deposit, such as clay, one or more materials still have to be purchased. Purchasing assumes a more important role than has been evident in the ceramic industries previously discussed.

Ceramic Raw Material Industry

The flow sheet in Fig. 56-1 also represents the ceramic raw material industry. This industry specializes in the extraction type of procurement, beneficiation processes, and selling materials to other manufacturers.

The extraction processes of prospecting, development, and winning were discussed earlier in Section III—*Ceram-*

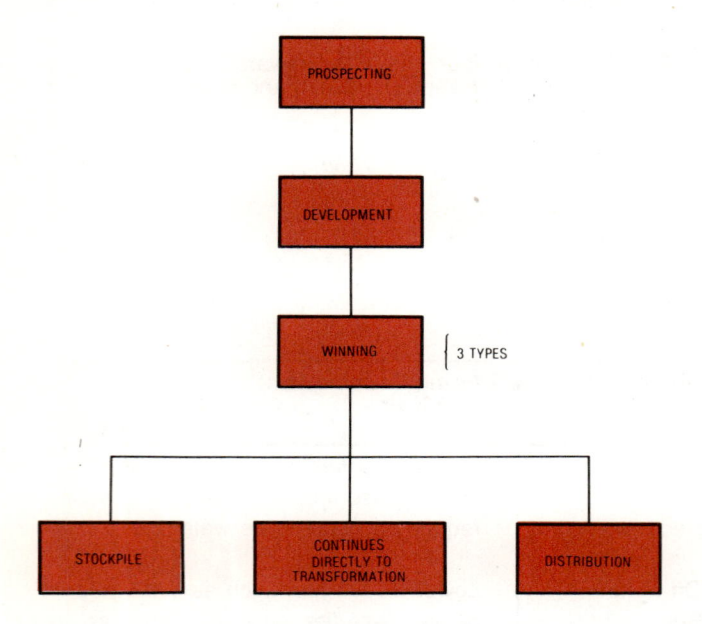

Fig. 56-1. A typical vertical flow sheet for procurement of raw materials used by the whiteware industry.

Fig. 56-2. A crawler-type tractor equipped with a blade and ripper. This machine is used to loosen the clay in a deposit for easy removal by other machines. (Courtesy Cedar Heights Clay Co.)

Fig. 56-3. Superficial or open-pit winning with a dipper-stick power shovel. This machine is used to load trucks that transport the clay to storage for beneficiation. (Courtesy Cedar Heights Clay Co.)

Fig. 56-4. A closeup of the bucket of the dipper stick power shovel in Fig. 56-3. (Courtesy Cedar Heights Clay Co.)

ic Raw Materials Industry. All three types of winning are used, due to the variety of raw materials used in the clay bodies and glazes. However, superficial winning is used more frequently than subterranean and subaqueous winning. Subaqueous is the least used of the three methods. Typical winning processes are shown in Figs. 56-2 through 56-5.

Stockpile

Whiteware manufacturers seldom stockpile the won raw materials. This type of stockpiling is done by the ceramic raw materials industry (Fig. 56-6). The beneficiated raw materials are stockpiled by the whiteware companies. However, beneficiation is a transformation process.

Fig. 56-5. Mining clay in South Carolina with a scoop on a crawler-type tractor. This is an open-pit operation. (Courtesy R. T. Vanderbilt Company, Inc.)

Fig. 56-6. A raw materials building protects the stockpile from the weather. (Courtesy Cedar Heights Clay Co.)

Unit 57: Transformation

Transformation of materials into whiteware products is diagrammed in Fig. 57-1. This flow sheet applies to both handmade ware and manufactured ware. Manufactured ware are machine-made products. The machines are frequently semi- or fully automated. A typical pictorial flow sheet for manufactured ware is shown in Fig. 57-2.

The five basic processes involved in all whiteware transformation are:

1. Beneficiation
2. Batching
3. Forming
4. Drying
5. Firing

Numerous optional processes may also be used to manufacture the whiteware products. The processes used depend on the product. These processes also are shown in Fig. 57-1 and include:

1. Fettling
2. Decoration
3. Glazing
4. Two or more firings
5. Packaging

Beneficiation

Beneficiation is the refining of a raw material into a usable form. Also, beneficiation includes the transformation processes of comminution, classifying, and separating. *Comminution* involves crushing and grinding. *Classifying* involves particle sizing. The *separation* processes frequently used are drying and magnetic separation. The typical beneficiation processes are shown in Figs. 57-3 and 57-4.

These processes are usually performed by the ceramic raw material industry that was discussed earlier in Section III—*Ceramic Raw Material Industry*. Most of the whiteware materials are purchased already beneficiated, as shown in Figs. 57-5 and 57-6.

Batching

The ceramic term for mixing is *batching*. Batching is the mixing of materials for the whiteware body. The two types of batching are dry and wet. Batching can be done either manually or mechanically.

Manual Batching

The educational institutions, art potters, and hobby ceramists may use manual batching. All the three states of clay (powdered, plastic, and liquid) can be mixed by hand. Two special manual processes are used—wedging and slaking.

Wedging is a kneading process that gives the plastic clay a uniform consistency and removes the air and/or lumps to produce a homogeneous mass. The three types of wedging are:

1. Cutting and pounding
2. Jelly roll
3. Oriental or spiral

A wedging board (Fig. 57-7) is frequently used for this process.

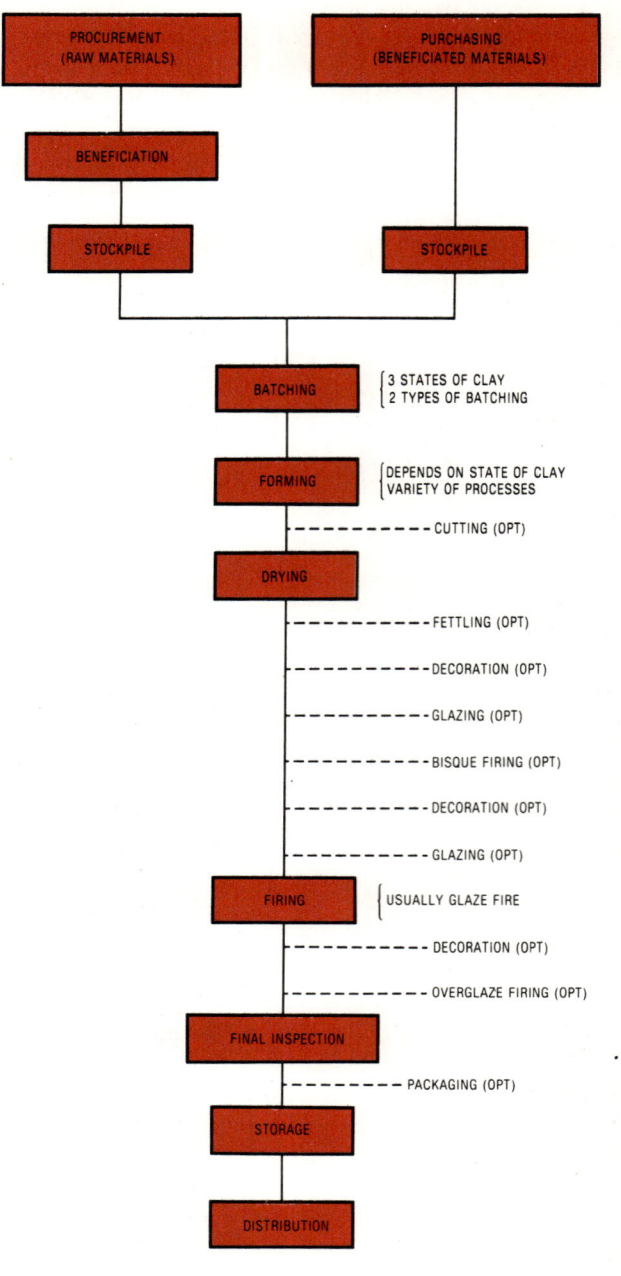

Fig. 57-1. A typical vertical flow sheet for the transformation of materials into whiteware products. These are handmade and manufactured ware. Examples are: handmade—art pottery and hobby ceramics; and manufactured—flowerpots and dinnerware.

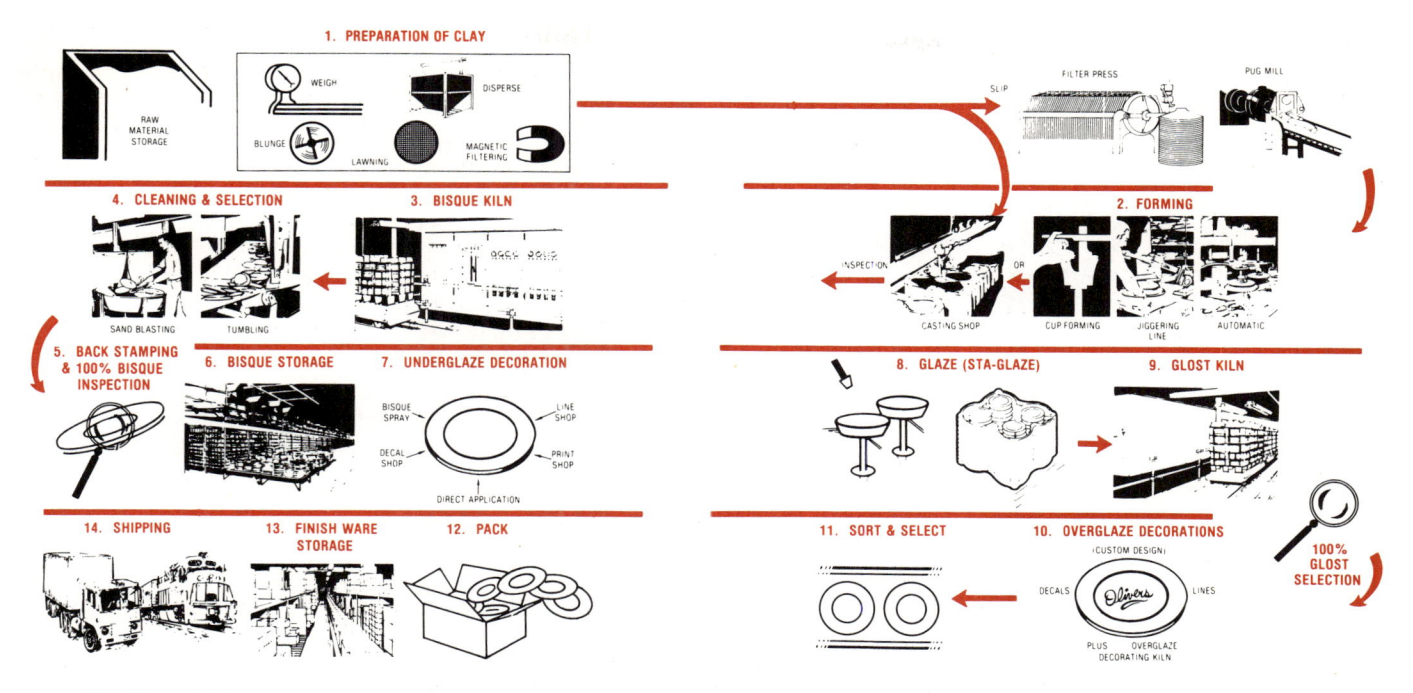

Fig. 57-2. Pictorial flow sheet for the manufacture of dinnerware. Glost is another term for glaze fire. (Courtesy Shenango China Products)

Slaking allows a clay body to take on water at its own rate. When making the casting slip by hand, the clay body is sifted into the water and deflocculant. Then, it is allowed to slake for 24 hours or more before mixing. This should remove all the lumps, but if any lumps remain, the casting slip can be poured through a sieve.

Mechanical Batching

The batching machine and the amount of water, if used, depend on the state of the clay for the forming method used. The three states of clay are powdered, plastic, and liquid. Typical batching machines for the three states of clay are listed in Table 57-1. Some of the machines are shown in Figs. 57-8 through 57-11.

Forming

Forming is the shaping of the materials into products. The whiteware products are either hand formed or machine formed.

Hand Forming

The hand forming processes are listed according to the state and condition of the clay in Table 57-2.

Pinch forming is the simplest and probably the oldest method of hand forming. However, all the plastic clay hand forming methods are old when compared to machine methods. Only your hands are required when pinch forming. Small dishes can be pinch formed.

Some hand tools are used when coil and slab forming. *Coil forming* is done with coils that are rolled out, cut, or extruded. The usual shape of the coils is round. The coils

Fig. 57-3. This is clay after winning. It is being fed into a crusher by the truckload. (Courtesy R. T. Vanderbilt Company, Inc.)

198 / *Modern Industrial Ceramics*

are welded together with water or a slurry. Bowls and vases can be coil formed.

Slab forming is done with flat slabs of clay. The slabs can be made by pounding, hand rolling, or manual machine rolling. Templates are often used when cutting the slabs prior to shaping. The slabs may be welded together or shaped over a stone or drape mold. The drape molds were discussed in Section VIII—*Plaster Mold Industry*. A large variety of slab forming methods are used. Dishes, vases, and plaques can be slab formed.

Throwing is done on a *potter's wheel*. This plastic clay forming process requires your hands, a few hand tools

Fig. 57-4. An industrial-size ball mill. This machine is 10 feet (3 meters) in diameter. The revolving and tumbling of media inside the mill apply continuous pressure and impact for size reduction. (Courtesy R. T. Vanderbilt Company, Inc.)

Fig. 57-5. Loading a hopper car for bulk transport of beneficiated material. Whiteware manufacturers often purchase their materials in bulk, especially when used in large quantities. (Courtesy R. T. Vanderbilt Company, Inc.)

Fig. 57-6: Loading a boxcar with bagged beneficiated material for shipment. Whiteware manufacturers usually purchase materials bagged when only a small quantity is needed. (Courtesy R. T. Vanderbilt Company, Inc.)

Table 57-1. Typical Batching Machines Used to Prepare Whiteware Clay Bodies

Powdered*	Plastic*	Liquid**
A. 1-step process if no water used. If 10% water is used, it is a 2-step process—dry and wet batch. B. Typical machines used: 　1. Dry batch—ball mill, dry pan, V mixer, cone or similar mixer 　2. 10% water—ball mill, V mixer, cone or similar mixer	A. Usually a 5-step process 　1. Dry batch 　2. Wet batch—with excess water into slurry 　3. Filter press—remove excess water 　4. Pug mill } conditioning 　5. De-air B. Typical machines used 　1. Dry batch—any in powdered list 　2. Wet batch—blunger 　3. Filter press—same 　4. Pug mill—same 　5. De-air—vacuum pump	A. Usually a 2-step process 　1. Dry batch 　2. Wet batch B. Typical machines used 　1. Dry batch—any in powdered list 　2. Wet batch—blunger 　3. Optional—screens and/or magnetic operation

* State of clay
** Specifically casting slip

(optional), and a potter's wheel. Your hands do the actual forming while the machine turns the clay. The various types of potter's wheels are listed in Table 57-3. One type of Potter's wheel is shown in Fig. 57-12. The basic parts common to all the potter's wheels are:

1. Head (horizontal)
2. Splash pan (optional)
3. Shaft (vertical)
4. Mechanism to rotate head counterclockwise

The basic throwing process after the material preparation or wedging involves:

1. Centering
2. Opening
3. Raising a cylinder
4. Shaping
5. Finishing
6. Allow to dry to leatherhard
7. Finishing—turn foot

200 / *Modern Industrial Ceramics*

Plastic clay is also hand formed by *pressing*. Usually, the plaster press molds are used. These molds were discussed earlier in Section VIII—*Plaster Mold Industry*. The general procedure for using a one-piece press mold includes:

1. Dust the mold cavity with powdered clay.
2. Press-wedge the plastic clay into the mold cavity.
3. Cut off the excess clay—with taut wire or fettle knife.
4. Allow to dry if necessary—till leatherhard.
5. Invert the mold to remove the piece.

Technically, *slip casting* is not true hand forming, since the plaster does the actual shaping. However, it is hand or manual work. The drain-type and solid casting molds that are used were discussed in Section VIII—*Plaster Mold Industry*. This type of forming is extensively used by hobby ceramists and industrial firms. Manual slip casting is still used in the industry, although casting machines are now available. Small- and medium-size drain-type plaster molds are primarily used with slip-casting machines. Thus, the large drain castings, such as sanitary ware, and solid castings are usually poured by hand. Industrial slip casting is shown in Fig. 57-13. The slip-casting procedure is diagrammed in Figs. 57-14 and 57-15.

Fig. 57-7. A small portable wedging board for laboratory or studio use. It is made of plaster and has a taut wire for cutting. Plastic clay is wedged on the right-hand half of the board when excessively wet or sticky. (Courtesy American Art Clay Co., Inc.)

Fig. 57-8. This is a laboratory-size ball mill with two 1-gallon jars. Ball mills are frequently used for dry batching or when less than 10 percent water is added to the powdered materials. (Courtesy American Art Clay Co., Inc.)

Fig. 57-9. A filter press removes the excess water from a slurry to make plastic clay. This is a laboratory-size filter press. Note the size of the filter cake of plastic clay at the top of the press.

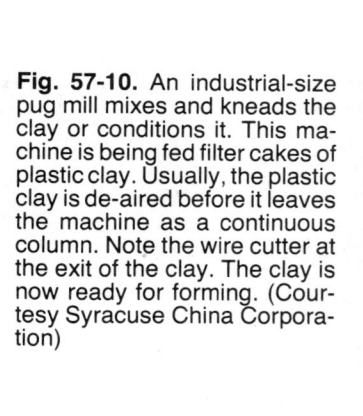

Fig. 57-10. An industrial-size pug mill mixes and kneads the clay or conditions it. This machine is being fed filter cakes of plastic clay. Usually, the plastic clay is de-aired before it leaves the machine as a continuous column. Note the wire cutter at the exit of the clay. The clay is now ready for forming. (Courtesy Syracuse China Corporation)

Slip-casting molds are used when the:

1. Product has draft in two or more directions.
2. Product has handle(s), knob, and/or spout.
3. Irregular product shape is to be duplicated or mass produced.

See Fig. 57-16 for typical slip-cast shapes.

Cutting is not a frequently used hand forming process. Turning a foot on a thrown piece is an example of cutting. Carving, incising, and piercing may be used to form and/or decorate clay.

Machine Forming

The machine forming processes for clay are listed in Table 57-4 according to the state and condition of the clay.

Isostatic pressing is done inside a high-pressure chamber (Fig. 57-17). No water or other liquid is used with the clay body. The basic process involves the following steps:

1. Pour the powdered body into a rubber mold.
2. Place inside a chamber that usually contains water-soluble oil.

Fig. 57-11. Slaking the clay body in water and deflocculant prior to blunging. This is an initial step in the manufacture of casting slip. Casting slip is one of the two liquid states of clay. (Courtesy Syracuse China Corporation)

202 / *Modern Industrial Ceramics*

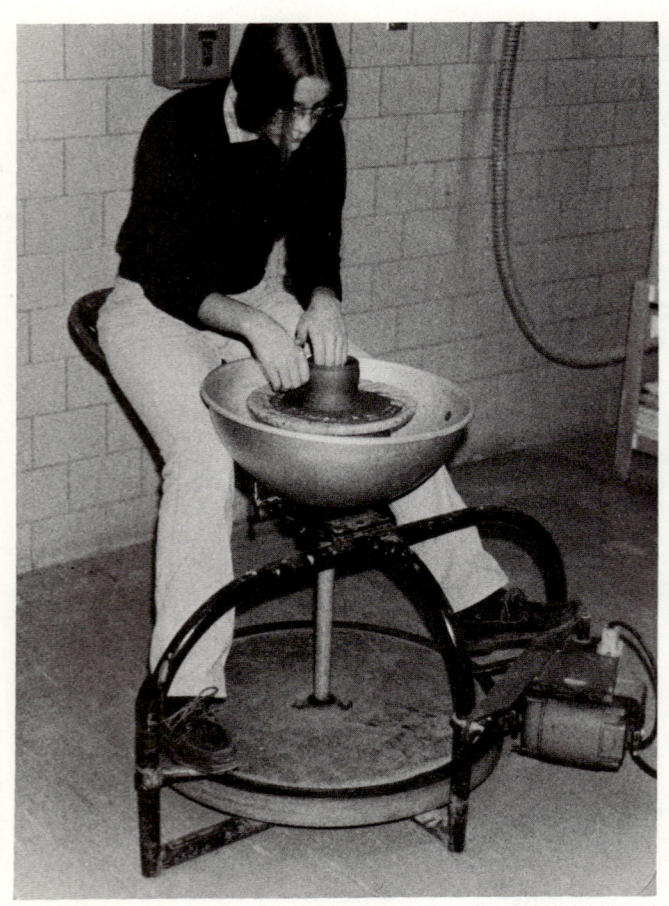

Fig. 57-12. A combination type of potter's wheel that is powered by kicking and/or an electric motor. This machine is used to form plastic clay by throwing. The head rotates counterclockwise.

3. Seal the chamber.
4. Pump fluid into the chamber until 5,000 to 20,000 psi (350 to 1,050 kg/sq cm) is reached.
5. Release the pressure.
6. Remove the rubber mold from chamber.
7. Strip the rubber mold off the pressed product.

This process allows solid pieces with a draft in two directions, such as a spark plug insulator, to be formed.

Fig. 57-13. An industrial casting shop. Dumping slip from a double-drain mold (top); removing the moist clay piece, a cream pitcher from the opened mold (center); and cup handles being removed from a two-piece gang mold (bottom). (Courtesy Syracuse China Corporation)

Table 57-2. Hand Forming Processes for Clay

Powdered*	<10% Water Added**	Plastic*	Liquid* Casting Slip	Leatherhard or Greenware**
none	none	Pinch forming Coil forming Slab forming Throwing Pressing (molds)	Drain casting Solid casting	Cutting—carving Turning

* State of clay
** Condition of clay

Transformation / 203

Table 57-3. Types of Potter's Wheels

Type Wheel	Sit	Stand
Kick	x	
Treadle	x	x
Electric	x	x
Combination (kick and electric)	x	

Dry pressing was discussed earlier in Section XIII—*Structural Clay Products Industry*. Less than 10% water is used with the body. Hydraulic or pneumatic pressure ranges from 5,000 to 15,000 psi (350 to 1,050 kg/sq cm) are used. Multiple-cavity steel molds are frequently used that allow high-speed mass production. Semi- and fully automatic machines are often used today. Pieces formed by this process usually have vertical sides.

Extruding (Fig. 57-18) was also discussed in Section XIII—*Structural Clay Products Industry* where it was called "stiff-mud" forming. Wire cutters are used to cut the continuously extruded column of clay. Extrusion permits a variety of external and internal shapes. Solid and cored shapes are possible. Only a few whiteware products are extruded. The process is frequently used to make blanks. Then, the plastic clay blanks are jiggered or pressed. Leatherhard blanks are turned (for example, large electrical insulators) to final shapes.

Jiggering produces cylindrical ware with draft in one direction from a plastic clay. All sizes of round plates, cups, and most bowls are jiggered. The basic process uses

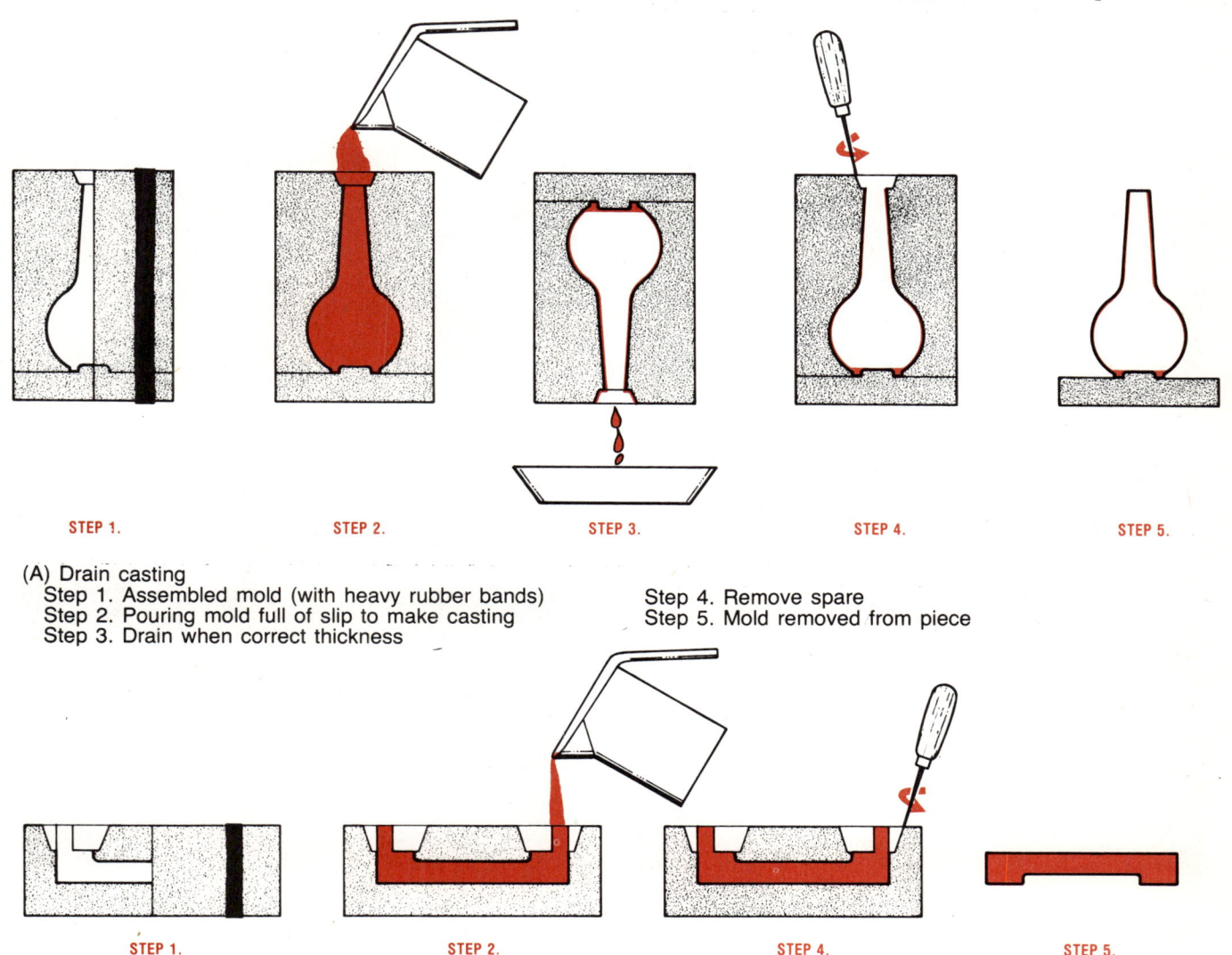

STEP 1.   STEP 2.   STEP 3.   STEP 4.   STEP 5.

(A) Drain casting
   Step 1. Assembled mold (with heavy rubber bands)
   Step 2. Pouring mold full of slip to make casting
   Step 3. Drain when correct thickness
   Step 4. Remove spare
   Step 5. Mold removed from piece

STEP 1.   STEP 2.   STEP 4.   STEP 5.

(B) Solid casting
   Same procedure, except no draining (Step 3 is omitted).

Fig. 57-14. Basic procedure for drain- and solid-type slip castings.

204 / *Modern Industrial Ceramics*

a template that forces the plastic clay against a revolving plaster mold (Fig. 57-19). The mold revolves counter-clockwise on a machine that was originally a potter's wheel.

The template forces the plastic clay against the inside of the mold to make hollow ware such as cups (Fig. 57-19) and bowls. The template forces the clay against the outside of the mold to make flatware such as plates. Typical jiggering molds and templates are shown in Fig. 57-20.

Today, jiggering is highly automated. Many companies have automated production lines that revolve (merry-go-round style) or are straight. The automated production lines may use a template, pressing and template, or roller forming (Fig. 57-21).

Ram pressing is done between two dense plaster molds that contain imbedded air lines and air passages. Oval shapes and tray shapes with draft in one direction, such as platters, are usually made by a *Ram* press. The process was patented and named by Ram, Inc. Hydraulic pressure is used for the pressing sequence illustrated in Fig. 57-22.

Machines are in use today that make *drain castings*. Small- and medium-size, drain-type plaster molds, such as for mugs with handles, are primarily used with these machines. Manually controlled and semiautomatic machines are being used. Large molds, such as for sanitary

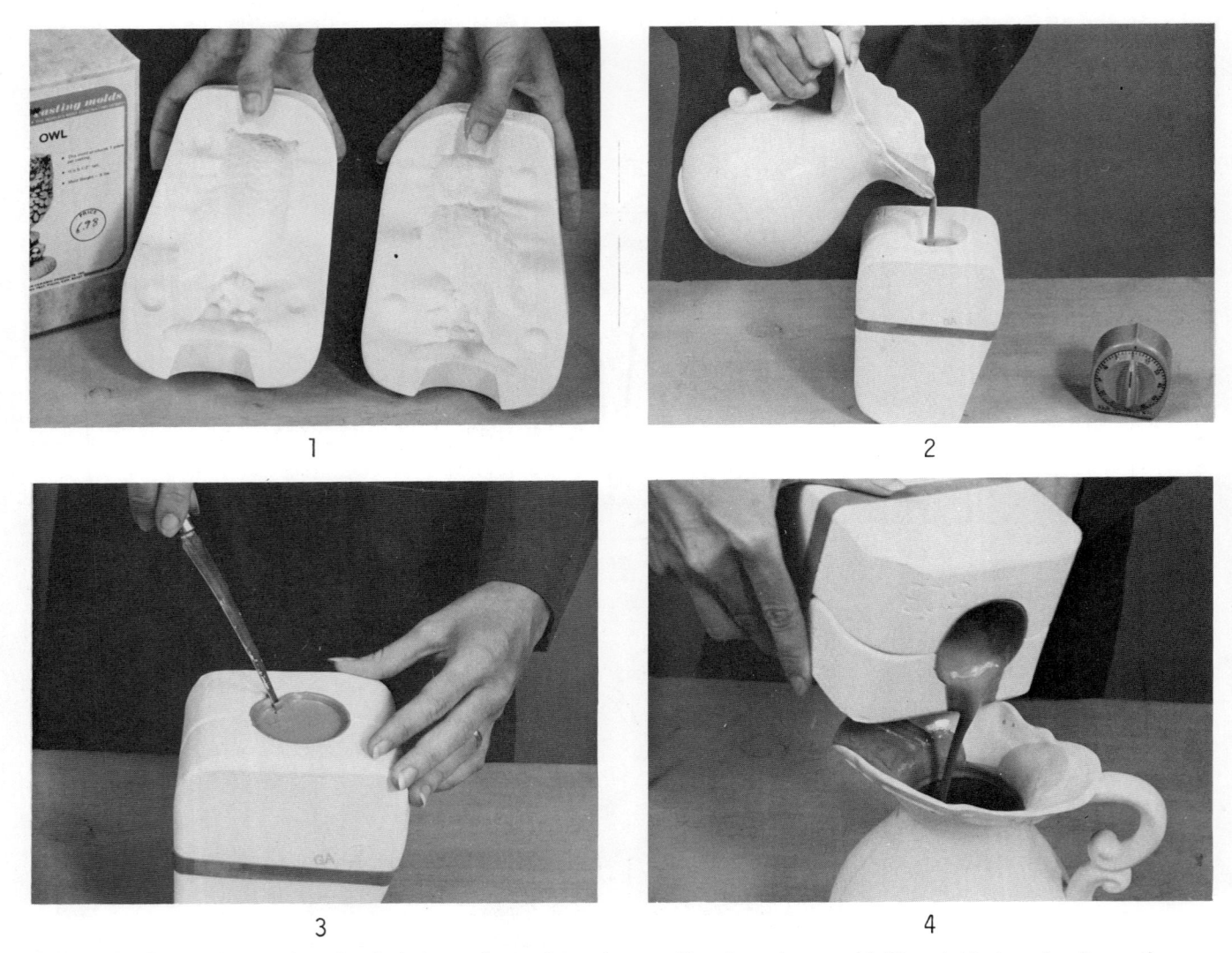

1

2

3

4

Fig. 57-15. Detailed procedure for drain-type slip casting, shown with a two-piece mold. The eight steps in slip casting are:

1. Open the mold to check for dust or clay, and reassemble with heavy rubber bands.
2. Pour the slip slowly into the mold until full. Then, vibrate mold to release any trapped air. Refill spare as needed.
3. Check wall thickness of casting in spare by cutting notch. Usually, thickness should be 1/8 to 3/16 in.
4. Drain slip after wall of casting has achieved the correct thickness.

ware, are too heavy for the machines. Solid cast molds are too time consuming.

Cutting machines are specialized. For example, they may cut a cup handle from a blank. *Turning* the extruded blanks into electrical insulators is done on a horizontal or vertical lathe, depending on the size. Leatherhard clay is usually used for these operations.

Usually, cutting is a part of the forming process—for example, the excess clay is trimmed away when jiggering or spares and fettle lines are removed from the slip castings.

Drying

Drying removes the excess water, if used, from the product. The amount of excess water depends on the forming process. Most of the excess water is removed before firing.

Air drying and dryers are used. The dryers use forced drying by controlling the time, heat, and humidity. Time, heat, and humidity must be controlled to avoid cracking and warping. The dryer temperatures range from 100°–400°F (38°–204°C). Waste heat from the kiln(s) is frequently used for the dryers.

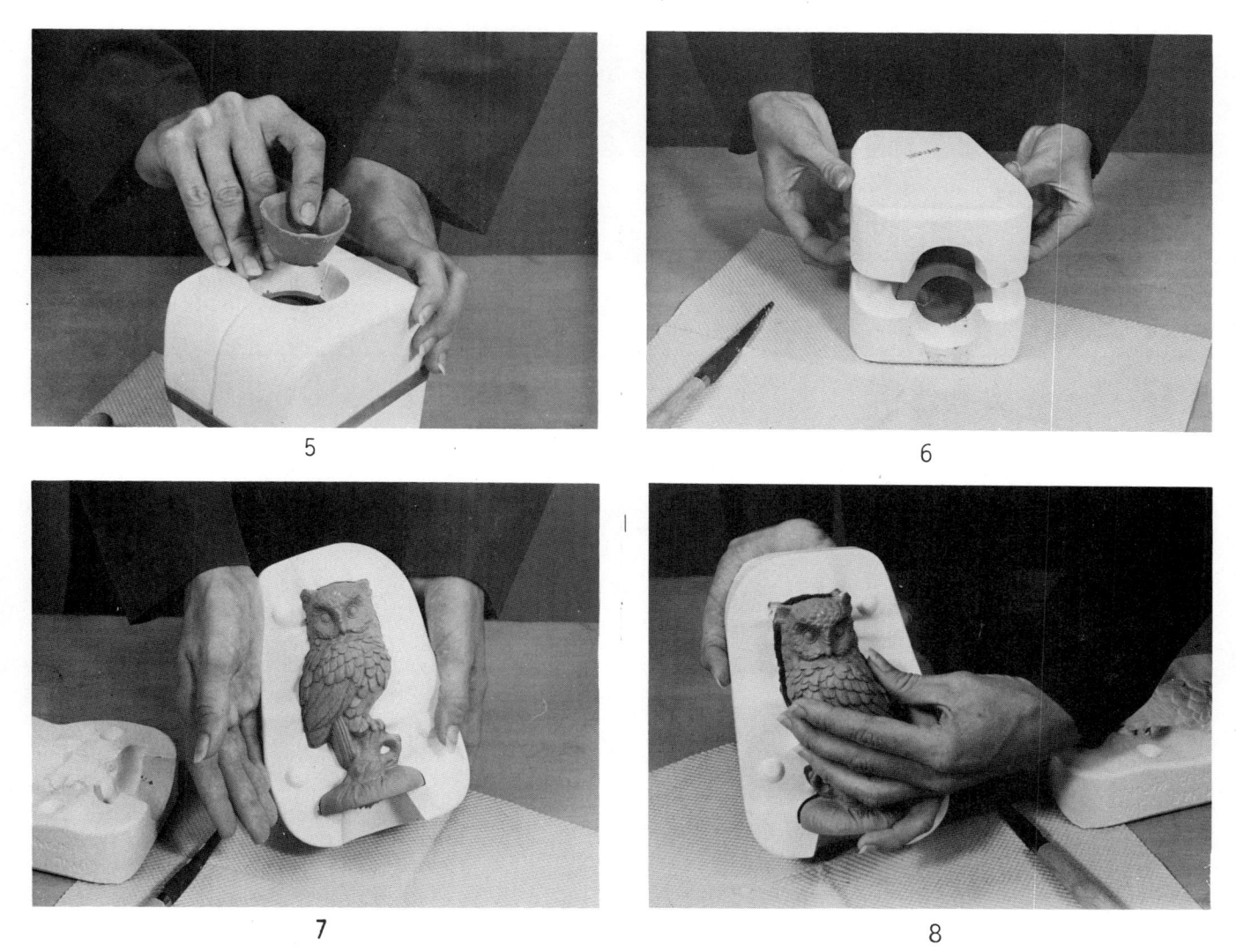

5.
6.
7.
8.

5. When the water sheen has left the surface, cut out and remove spare.
6. Allow casting to continue drying in mold until leather-hard. Then slowly and carefully open mold.
7. One half of the mold has been removed, revealing the casting. The casting may require further drying before removal.
8. Remove the casting by tilting the mold forward, and allow it to fall into your hand. Then allow the casting to continue drying until greenware.

(Copyright © Duncan Ceramic Products, A Division of Duncan Enterprises, Fresno, California)

206 / *Modern Industrial Ceramics*

Fig. 57-16. Typical mass-produced product shapes that can only be formed by using drain- and solid-type slip casting molds. Note draft in two or more directions and handles. Knobs and spouts also limit a shape to being duplicated only by slip casting. (Courtesy R. T. Vanderbilt Company, Inc.)

Table 57-4. Machine Forming Processes for Clay

Powdered*	<10% water Added**	Plastic*	Liquid* (casting slip)	Leatherhard or Greenware**
Isostatic pressing	Dry pressing	Extruding Jiggering Ram press	Drain casting	Cutting—cup handle Turning

* State of clay
** Condition of the clay

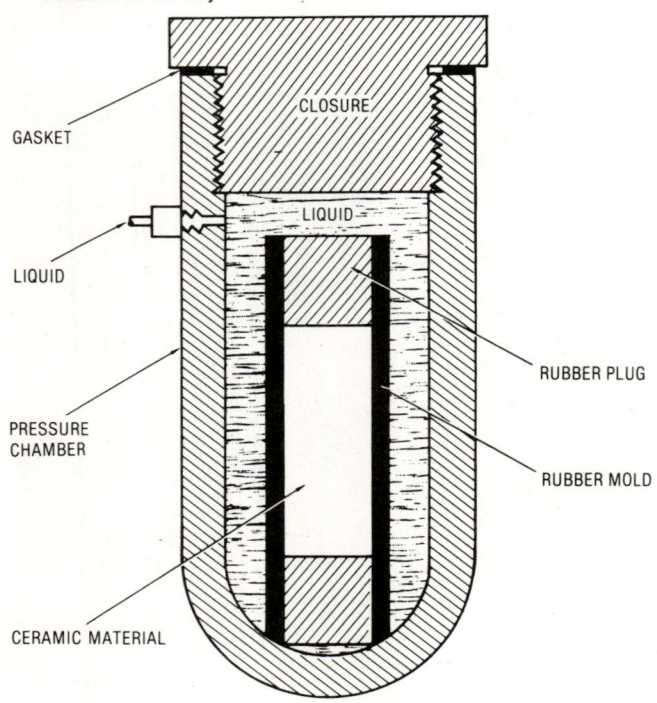

Fig. 57-17. Isostatic pressing is done inside a high-pressure chamber.

Air drying is usually used for the hand forming processes (see Table 57-2). Drying processes for the machine-formed clay products are shown in Table 57-5.

Optional Processes

Several optional processes may be used after drying. The process(es) used depends on the product (Table 57-6). For example, none of them are used for flowerpots, while all of them might be used for dinnerware. These processes and the optional processes that may occur after firing are listed in Fig. 57-1. Some of the processes are shown in Fig. 57-23. Decoration, types of glazes, and glaze application are discussed in Section XV—*Glaze Industry*.

Firing

Firing causes a glassy bond to form that fuses the clay particles together. When a glaze is used, firing also causes the glassy coating to fuse to the clay.

Kilns

Both periodic and continuous kilns are used by the whiteware industry for bisque and glaze firings. They were discussed in Section VI—*Kiln Industry*. The kiln type, size, and the number used depend on the product.

The periodic kilns used today are listed in Table 57-7. This table also lists the people who frequently use each of

Transformation / 207

these types of kilns. Typical periodic kilns are shown in Figs. 57-24 and 57-25.

The tunnel kilns are the only continuous type of kilns used. Both straightline and circular tunnel kilns are in use. The straightline tunnel kiln is more common (Fig. 57-26). The ware is moved through the tunnel kiln by one of three methods—roller hearth, pusher slab, or kiln cars. The size and quantity of products helps determine the type of conveyance. Kiln cars are the method most frequently used by industrial firms for production (Fig. 57-27). Industrial research and development departments and educational institutions use all three methods of conveyance. Usually, their tunnel kilns are the smaller, or laboratory-size kilns.

Temperatures

Firing temperatures depend on the type of firing and the product. The temperatures usually range from C/06 to C/20, as shown in Table 57-8.

The firing time depends on the type of firing, the

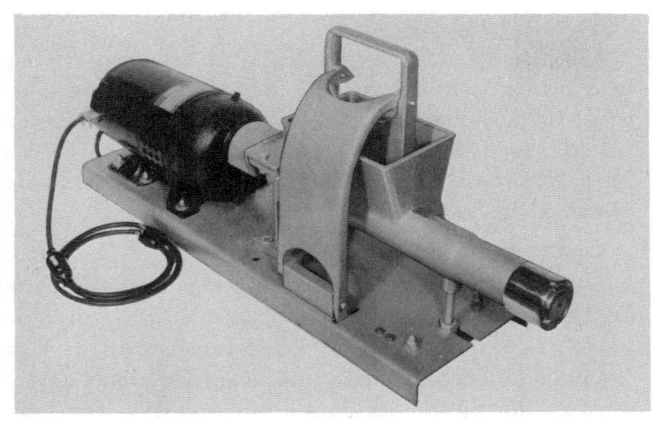

Fig. 57-18. A laboratory-size portable pug mill that can also be used as an extruder. Note the tamper and hopper for feed. Various die shapes and sizes can be used at the exit for extruding. (Courtesy American Art Clay Co., Inc.)

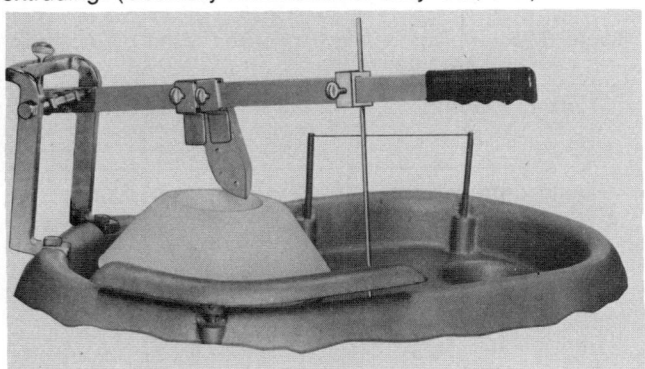

Fig. 57-19. A laboratory-size manually controlled jiggering machine setup to make cups. Cups are hollow ware. The template is lowered into the revolving plaster mold to form the plastic clay. (Courtesy American Art Clay Co., Inc.)

PLATE MOLD = PLATE TEMPLATE

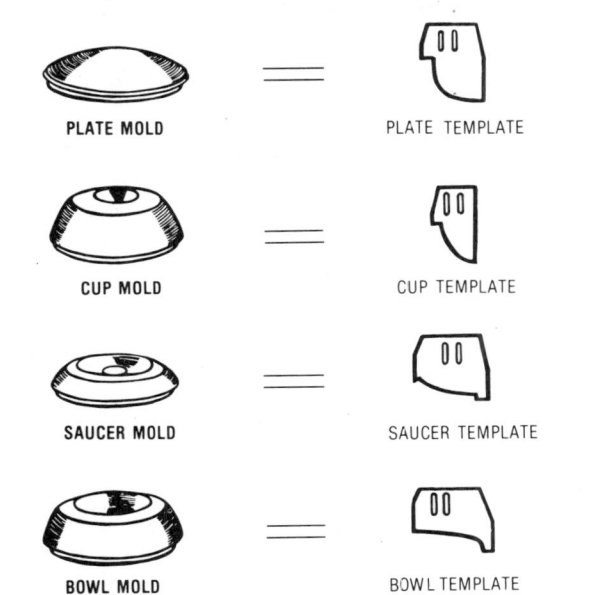

CUP MOLD = CUP TEMPLATE

SAUCER MOLD = SAUCER TEMPLATE

BOWL MOLD = BOWL TEMPLATE

Fig. 57-20. Typical jiggering molds and templates. Cups and bowls are hollow ware. Plates are flatware. (Courtesy American Art Clay Co., Inc.)

Fig. 57-21. A roller-head jiggering machine. A slug of clay has been placed on the plate mold. Then, the roller will drop down and form the clay to the contour of the mold. Both the roller head and the mold revolve during the forming step. (Courtesy Syracuse China Corporation)

208 / *Modern Industrial Ceramics*

product, and the kiln size. The large periodic kilns require the most time, especially if they are wood or coal fired. Tunnel kilns usually require the least amount of time.

Firing time, temperature, and cooling time have to be carefully controlled to avoid defects. A typical firing curve for whiteware products is shown in Fig. 57-28.

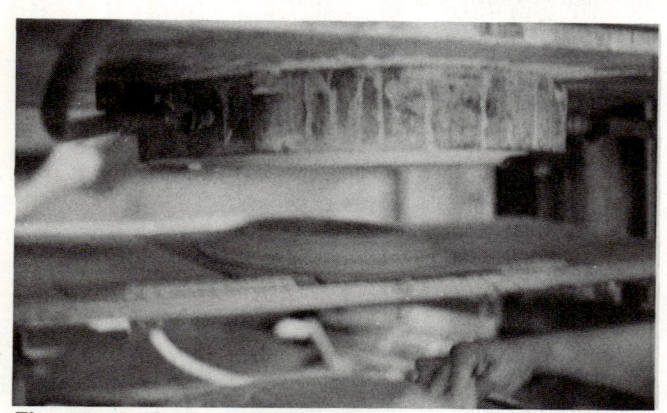

Fig. 57-22. Ram pressing of plastic clay uses two dense plaster molds that contain imbedded air lines. The pressing sequence is: plastic clay blank placed in mold (top); pressing (center); and air pressure blows piece out of bottom mold as top mold is raised (bottom). (Courtesy Buffalo China, Inc.)

Special Firings

Salt glazing and reduction firing are used, but they are used primarily by art potters. Some industrial firms use these firings (Fig. 57-29). Salt glazing and reduction firings, such as flashing, were discussed in Section XIII—*Structural Clay Products*. Reduction firing was also discussed in Section VI—*Kiln Industry*.

Completion

Final inspection, packaging (optional), storage, and distribution complete the manufacture of the whiteware products. Several of the processes are shown in Figs. 57-30 and 57-31. Actually, inspection occurs at several intervals during the transformation processes. Typically, inspection occurs after each of the basic processes listed at the beginning of this unit and in Fig. 57-1.

Table 57-5. Typical Drying Processes Used by Whiteware Manufacturers

Type of Machine	Air	Forced*	Other
Isostatic pressing			Not necessary
Dry pressing	x	x	
Extruding		x	
Jiggering		x	
Ram press		x	
Drain casting	x	x	
Cutting	x	x	
Turning		x	

* Dryers are used for forced drying by controlling time, heat, and humidity.

Table 57-6. Optional Processes Used for Whiteware Manufacturing

Process*	Description	Typical Product(s)
Fettling or trimming**	Clean up greenware	Dinnerware Sanitary ware
Decoration	Texturing, incising or piercing greenware	Art pottery and/or hobby ceramics
Glazing**	Applying underglaze or glaze to greenware	Hobby ceramics Dinnerware Sparkplug insulator
Bisque firing**	First firing of greenware. Typically a low fire (C/07) for earthenware and stoneware, a high fire (C/10) for china, and none for porcelain	Dinnerware Flower pots
Decoration**	Applying engobe, underglaze, or decals to bisque ware	Dinnerware
Glazing**	Applying glaze to bisque ware	Dinnerware Art pottery
Decoration	Applying overglaze or ceramic decals	Dinnerware
Overglaze firing	Done at a lower temperature than glaze fire, typically C/010–C/017	Dinnerware
Packaging	Protect individual pieces. Modular (4 pieces), packaging frequently used.	Dinnerware

* Listed in sequence used in Fig. 57-1.
** May be done by semi- or fully automatic machine

Table 57-7. Typical Use of Periodic Kilns for Whiteware Products

Type Periodic Kiln	Frequently Used By	Bisque	Glaze	Overglaze
Electric	Educational institutions	x	x	x
	Art potters	x	x	
	Hobby ceramics	x	x	x
	Industry*	x	x	x
Updraft	Educational institutions		x	
	Art potters		x	
Downdraft	Educational institutions		x	
	Art potters		x	
Muffle	Industry	x	x	
Shuttle	Industry	x	x	x

(Type of Firing: Bisque, Glaze, Overglaze)

* Electric periodic kilns primarily used for industrial research and development and overglaze firings. Smaller firms may also use them for production.

210 / *Modern Industrial Ceramics*

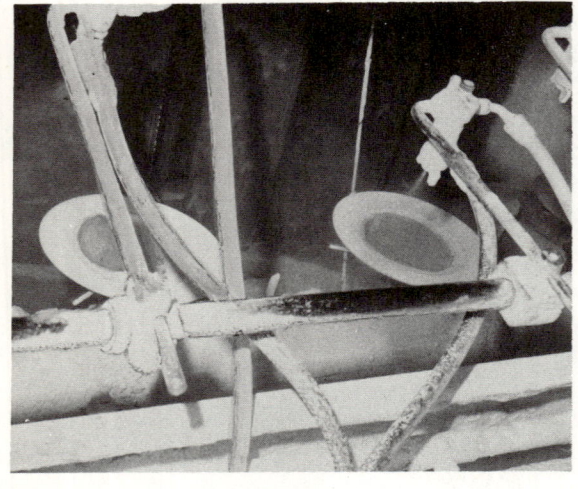

Fig. 57-23. Industrial decoration of bisque fired dinnerware with underglaze. Hand decorator applying a line decoration (A); automatic decorating machine applying a line decoration (B); and a closeup of the automatic glaze application by spray gun on bisque fired dinnerware. Pieces revolve through the machine similar to a merry-go-round (C). (Courtesy Shenango China Products)

Fig. 57-24. A typical glaze firing in an electric front-loading periodic kiln. This type of kiln is frequently used by art potters, hobby ceramists, and educational institutions. Note the use of kiln shelves, posts, and stilts. (Courtesy American Art Clay Co., Inc.)

Fig. 57-25. Shuttle kilns are frequently used by whiteware manufacturers. This is a small-size bottom-loading elevator-type shuttle kiln. It is fired with natural gas to 3002°F (1650°C) using a 24-hour, cold-to-cold, firing cycle. The saggers hold small electronic components for firing. Note the instrument control panel on the left-hand side. (Courtesy Bickley Furnaces, Inc.)

Fig. 57-26. Entrance to straightline tunnel kilns used for industrial manufacture of whiteware. Two kiln-car type tunnel kilns are used for bisque firing (left). Note the line for stacking cars (left foreground) and the transfer car. A roller-hearth type of tunnel kiln is used for fast firing (right). Note the "one-high" stacking of cups and the roller system. The stacking line is at the left. (Courtesy Shenango China Products)

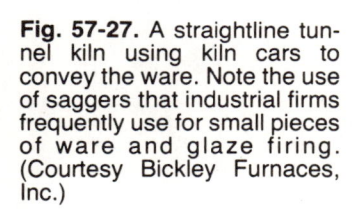

Fig. 57-27. A straightline tunnel kiln using kiln cars to convey the ware. Note the use of saggers that industrial firms frequently use for small pieces of ware and glaze firing. (Courtesy Bickley Furnaces, Inc.)

Table 57-8. Typical Whiteware Firing Temperature

Type of Firing	Typical Temperature			
	Earthenware	Stoneware	Porcelain	China
Bisque	C/07	C/07	Usually none	C/10
Glaze	C/04	C/10	C/6-C/20	C/5
Overglaze	C/016	C/016	C/016	C/016

212 / *Modern Industrial Ceramics*

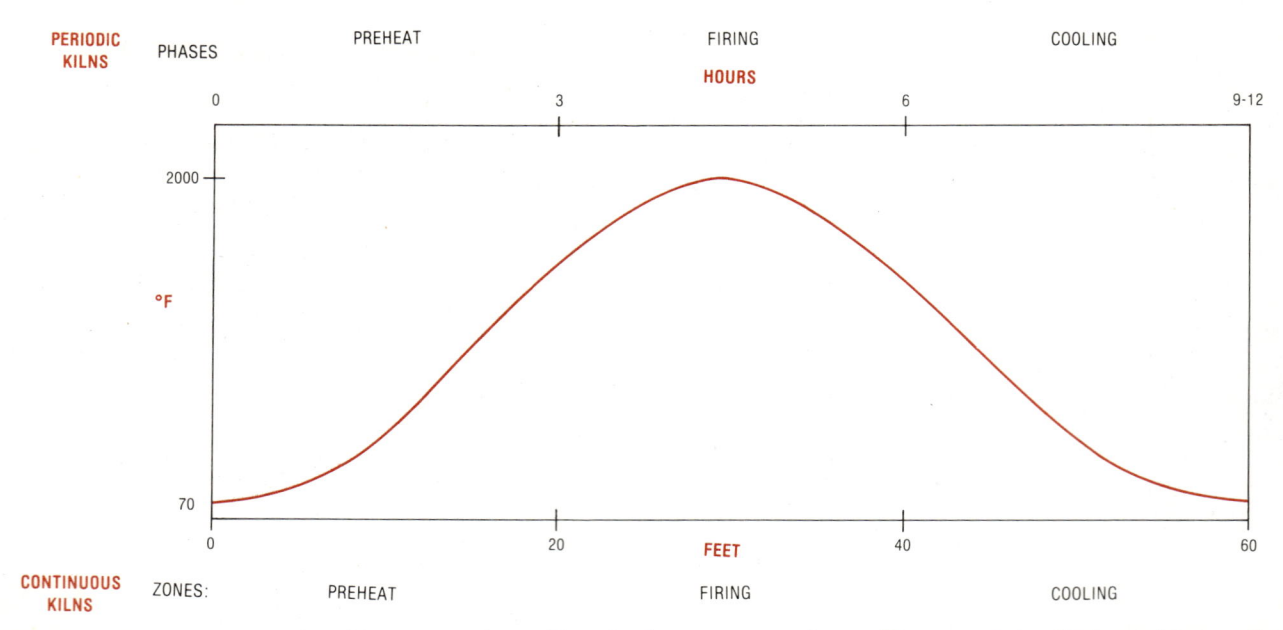

Fig. 57-28. A typical firing curve for whiteware products. Note the three zones or phases. These apply to all firings—bisque, glaze, and overglaze.

Fig. 57-29. An instrument control panel for automatic programming of the kiln's firing cycle. This panel includes recording temperature regulations for six temperature control zones, plus automatic control of the reduction firing period and diffusion air (excess air) firing periods. Typically, they are used for shuttle kilns. Note the indicating pyrometer (top center), recording pyrometer (top left), and cam programmer (top right). (Courtesy Bickley Furnaces, Inc.

Fig. 57-30. A machine that removes stilt and plate pin marks after glaze firing. This can be semi- or fully automatic machine process. (Courtesy Shenango China Products)

Fig. 57-31. A carton-sealing machine for dinnerware. This is usually an automatic machine process. (Courtesy Shenango China Products)

Unit 58: Utilization

Whiteware products are manufactured products. Some of them are primary products. The majority of whiteware products are secondary products.

Some whiteware products are used in the manufacture of other products, such as spark-plug insulators and high-voltage electrical insulators, and in construction—sanitary ware, for example. Sanitary ware includes bathroom sinks, urinals, and water closets.

Products

Typical manufactured whiteware products are listed in Table 58-1. Some of the products, such as art pottery, are only handmade; while others such as sanitary ware, are only machine made. A few products, such as dinnerware, are made using both methods.

Handmade ware is usually produced by the art potters, hobby ceramists, and educational institutions (Fig. 58-1). Very little handmade ware is produced by industrial firms. Manufactured or machine-made ware is produced by industrial firms (Fig. 58-2). Educational institutions also use machine forming processes.

Product Information

The stoneware clay bodies used by the art potters frequently contain grog. They use the grog for texture and to reduce shrinkage. This is the only real use for grog in the whiteware products.

Sometimes, art potters and educational institutions reduction-fire the stoneware. Reduction firing is used by a few industrial firms. This type of firing was discussed in Unit 57—*Transformation*.

Table 58-1. Typical Manufactured Whiteware Products

Type of Clay	Art Potter	Hobby Ceramics	Ed. Inst.	Industrial	Typical Product(s)
A. Natural clay					
Native clay	x		x		Art pottery, such as plates, pots, mugs, Flower pots
Singular purchased clay	x	x	x		Art pottery, Hobby ceramics, such as decorative pieces
B. Clay bodies					
Earthenware	xx	x	x	x	Art pottery, Hobby ceramics, Flower pots, Dinnerware
Stoneware	x		x	x	Art pottery, Dinnerware
Porcelain	xx		xx	x	Art pottery, Spark-plug insulator, High-voltage insulator, Crucibles, Ball-mill jars and media, Dinnerware, Sanitary ware
China		xx		x	Dinnerware, Sanitary ware
C. Type of ware					
Handmade	x	x	x	xx	
Manufactured/machine made			x	x	

Key: x—Frequently used
xx—Not frequently used

214 / *Modern Industrial Ceramics*

Fig. 58-1. Handmade whiteware is usually produced by art potters, hobby ceramists, and educational institutions.

Fig. 58-2. Manufactured or machine-made whitewares. All these products were produced by industrial firms.

Fig. 58-3. *Basaltware* is an unglazed black stoneware made by Wedgewood of England.

Fig. 58-4. *Jasperware* is an unglazed, colored body with white decoration called *sprigging*. It is made by Wedgewood of England.

An unglazed, colored stoneware body is not common. This is produced by an industrial firm, Wedgewood, in England with the trade names *Basaltware* and *Jasperware*. *Basaltware* is an unglazed black stoneware that is practically watertight (Fig. 58-3). *Jasperware* is usually an unglazed, colored body with white decoration that is practically watertight (Fig. 58-4). The colored body may be blue, black, green, or lavender. These colors are made by adding metallic oxides and/or carbonates to the body raw materials. Thus, the body color is homogeneous throughout the entire piece. The white decoration is the same clay body without colorants. The white low-relief decoration called *sprigging* is made in shallow press molds and welded onto each piece by hand. The white sprigging contrasts nicely with the colored body. A few pieces of glazed *Jasperware* may be seen. The glaze would be on food contact surfaces.

China

There are at least five types of china. All except the sanitary china are used to make dinnerware. The five types of china are:

1. Bone china (Fig. 58-5)
2. Fine china (Fig. 58-5)
3. Everyday china
4. Hotel or institutional china (Fig. 58-5)
5. Sanitary china

Bone china was originally made only in England. Today, some United States manufacturers also make it. As its name implies, it contains bone ash. The bone ash is calcined bone containing calcium phosphate that acts as a flux. Bone china may contain up to 50% bone ash. A typical chinaware body, such as British bone china, contains 40-50% bone ash, 25-30% china clay, and 25-30% china stone or Cornwall stone. The chinaware body may contain only 25% bone ash, if it is made in the United States.

Usually, bone china is whiter and slightly heavier than fine china, due to the bone ash. Both bone china and fine china are translucent.

Fine china is primarily made in the United States. Kaolin, feldspars, alumina, and flint are used in the typical body. Alumina makes a whiter and stronger body. Fine china was not quite as white, but thinner, than bone china. However, some fine china is now made as white as bone china, due to the use of alumina.

Basically, the only difference between fine, everyday, and hotel china is the body thickness. *Everyday china* has a medium body thickness. It is not translucent. This china is intended for everyday or daily use—thus, its name. It has not become as well known as the fine china. Of course, it has been manufactured only for 10-15 years. Fine china has been manufactured for nearly one hundred years in the United States.

Hotel or *institutional china* has a thicker body than the fine and everyday china. Also, it is not translucent. It is made thicker to withstand the abuse of rough handling and dishwashers.

Sanitary china has the thickest body of the chinas. The name is derived from its usage. The sanitary ware includes sinks (usually bathroom sinks), urinals, and water closets.

Fig. 58-5. Three types of china—(left to right) fine, bone, and hotel or institutional china.

Unit 59: Disposition

Disposition of the whiteware products usually does not occur until a piece is actually broken. When the whiteware products are disposed of in large quantities, they can be a problem. All are solid waste. However, most of them can be and are reclaimed.

Reclaiming

In fact, probably more whiteware products are reclaimed than are discarded. This is encouraging, since reclaiming involves recycling and reusing.

Recycling

Some of the products can be and are recycled. Spark plugs that are cleaned and regapped are probably one of the best examples of recycling. High-voltage insulators may be stripped of usable parts before they are discarded. The broken pieces, especially antiques, are repaired and reused (Fig. 59-1). Dinnerware and vases are typical examples.

Reuse

Most of the whiteware products are reused. Many of these products become collectibles or antiques (Fig. 59-2). The age of a product increases its value. Art pottery and dinnerware are excellent examples of collectible and/or antique whiteware.

Flea markets, antique shows and shops, secondhand stores, junk dealers, garage sales, and collectors all help to resell the used whiteware products. Most of the products are again used for their intended purposes after resale. Others become part of a collection and are used only for display purposes.

Frequently, when we discontinue the use of a whiteware product, such as dinnerware and sinks, in our homes, it is reused in our vacation home. Many camps, cottages, summer homes, and hunting lodges have been and still are equipped with the discards from homes (Fig. 59-3).

When the industries discontinue the use of a piece of equipment due to its age, they often donate the equipment to educational institutions. Crucibles, ball mills, and media are often reused in this manner.

Discarding

Occasionally, discarding is necessary. For example, if a vase or piece of dinnerware is dropped and breaks into many pieces, it can only be discarded, which does not cause any great problems.

Discarding of whiteware products is minimal. No real problems occur if the item(s) is properly discarded. Improper discarding causes problems—for example, litter, which disfigures our country.

Fig. 59-1. Broken stoneware mugs and a kit for repairing them.

Fig. 59-2. Antique stoneware jugs and bottles. These are typical whiteware products that are collected.

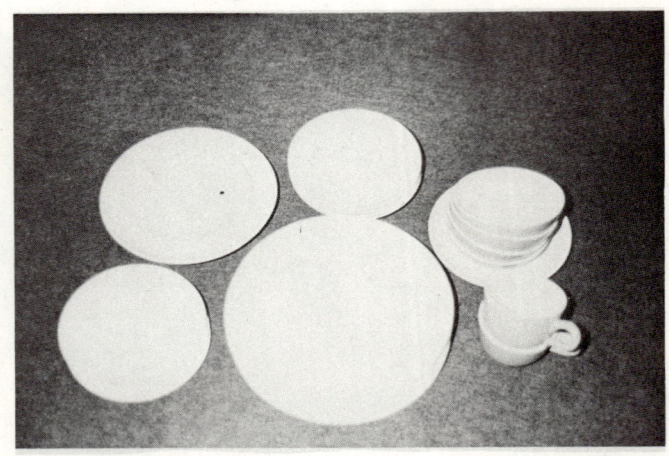

Fig. 59-3. These pieces are all that remain from two different sets of dinnerware. They can still be used at camp or a summer cottage.

SECTION XV

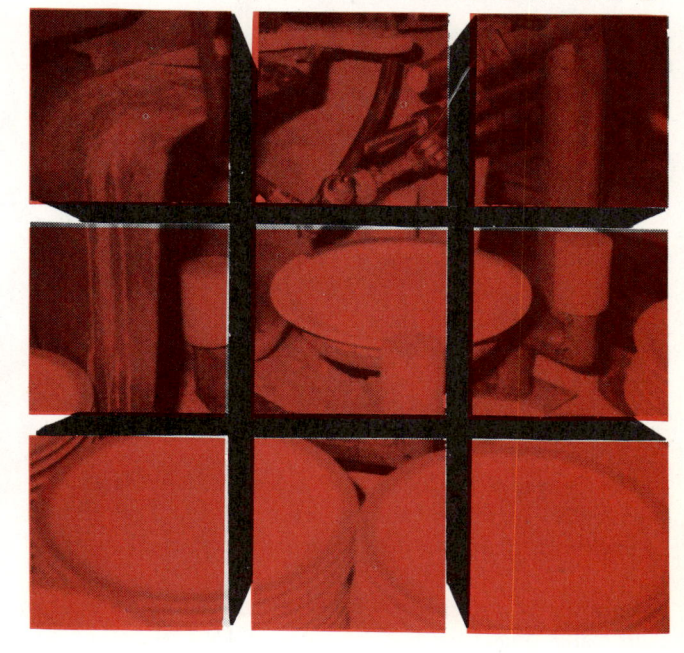

Glaze Industry

Unit 60: Material Science

Four glassy materials are used in ceramic products (Fig. 60-1). They are:

1. Glassy bond
2. Glaze
3. Porcelain enamel
4. Glass

A glassy bond is formed when the clay products are fired to a high temperature. This makes them durable solids. Glassy bond was discussed in Section XII—*Clay Industries*; Section XIII—*Structural Clay Products Industry*; and Section XIV—*Whiteware Industry*.

Fig. 60-1. Four glassy materials in ceramic products are glassy bond, glaze, porcelain enamel, and glass (left to right).

Glassy Material Industries

Glaze, porcelain enamels, and glass are the major glassy material industries. Glazes will be defined, discussed, and illustrated here. Porcelain enamel and glass will be discussed in later sections.

There are distinct differences between the glassy materials. These can be seen in the definition of each material.

A *glassy bond* is formed when clay is fired. It is found in all the structural clay products and in nearly all the whiteware products. A glassy bond also occurs when some refractories and abrasive products are fired.

A *glaze* is a glassy coating fired onto clay products. It is an exterior coating for clay (Fig. 60-2).

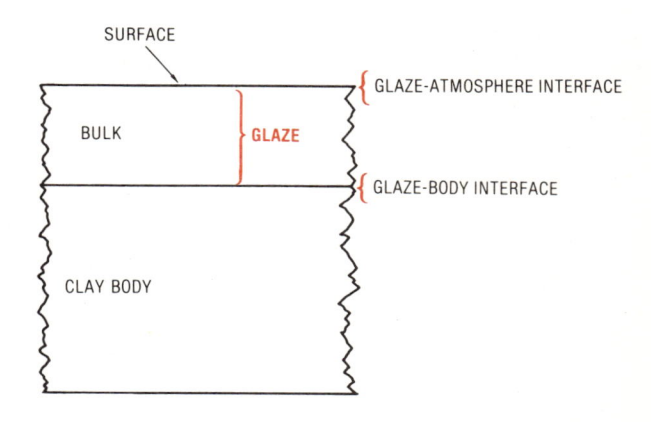

Fig. 60-2. Glaze is a glassy coating fired onto clay products. Relationship of the glaze and clay body are diagrammed above.

Porcelain enamel is also a glassy coating. It is fired onto some glass and metal products as an exterior coating.

Glass is thicker than the preceding glassy coatings. Thus, the glass products are made of only a single material, and are homogeneous.

Glazes

Glazes are glassy coatings fired onto clay products to protect and/or beautify them. High temperatures are used for the firing process. The high temperatures melt the glaze materials. The glaze flows when it is melted. It forms a glassy coating on the surface of the product. The glaze becomes an exterior coating when it is cooled (Fig. 60-3).

Properties

The properly designed and compounded glazes are chemically inert to almost everything. The major exception is hydrofluoric acid, which attacks all types of glass.

Improperly designed and/or compounded glazes can be chemically unstable. For example, a lead glaze can be dissolved by acetic acid. Lead is soluble in acetic acid. Acetic acid is found in vinegar, cider, and similar common foods. Thus, only lead-free glazes should be used on dinnerware.

The major physical properties of glazes are:

1. Hard surface
2. Nonabsorbent
3. Easily cleaned
4. Decorative

Technically, all the glazes should have these properties, except to be decorative. However, decoration is a relative thing that depends upon your point of view. You could say that all glazes are also decorative.

Basically, physical properties can be reduced to the two words used in the earlier definition—*protect* and/or *beautify*. Protection by waterproofing results from a hard surface that is nonabsorbent, and easily cleaned. To state that a surface is easily cleaned implies a smooth or reasonably smooth surface. A glaze surface texture, such as gloss or mat, varies, but generally, all the glazes are reasonably smooth and therefore easily cleaned.

Materials

Both the natural raw materials and the manufactured materials are used to manufacture glazes. The majority of the materials used are the natural raw materials.

Raw Materials

Two types of materials are found in all glazes. They are:

1. Glass former
2. Flux(es)

Most of the glazes also contain one or more of the following materials:

1. Stiffener
2. Opacifier
3. Matting agent
4. Colorant
5. Binder

The number of these materials used can make a glaze simple or complex, as follows:

1. The *glass former* is silica or silicon dioxide (SiO_2). The major source is flint (SiO_2). Clay ($Al_2O_3 \cdot 2SiO_2 \cdot 2H_2O$) and the feldspars ($K_2O \cdot Al_2O_3 \cdot 6SiO_2$) also provide silica when they are used in a glaze. Silica (SiO_2) has one major disadvantage. It melts at about 3110°F (1710°C). This excessively high temperature is impractical for a glaze.
2. Thus, one or more *fluxes* are used to lower the melting temperature of the silica. This is the purpose for including the lead compounds such as lead oxide (PbO), whiting ($CaCO_3$), feldspars ($K_2O \cdot Al_2O_3 \cdot 6SiO_2$), and similar materials in a glaze.

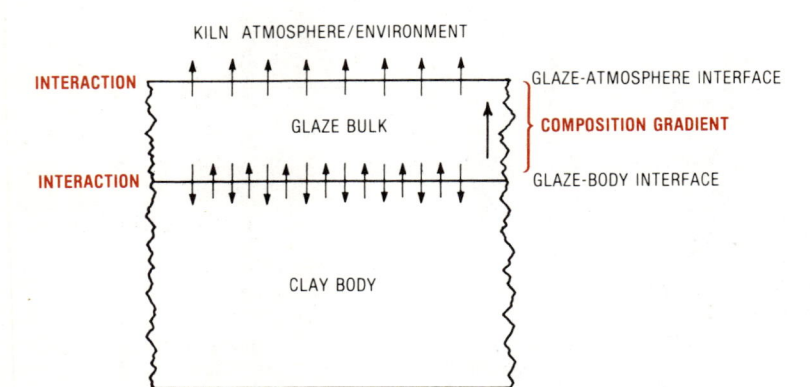

Fig. 60-3. The interactions of the two glaze interfaces during firing, as shown in the diagram above.

The fluxes either contain lead or are leadless. Thus, glazes are classified as either lead or leadless. The *leadless glazes are preferred and recommended*, since the lead can be dissolved out of a glaze by any material containing acetic acid.

The oxides that function as fluxes are listed in Table 60-1. This table also lists the typical raw materials used to provide the necessary fluxes.

One or more fluxes are used in all glazes. The number of fluxes used is limited by specific rules that govern the designing of a glaze. The designing of a glaze is usually called *glaze calculation*. This technical subject is not discussed here.

1. The glass former and the fluxes usually make a clear transparent-base glaze when correctly fired. Any additional materials simply act as controllers or modifiers of the base glaze or two basic components. A base glaze is just that—it is the base upon which a glaze is built or developed.
2. A *stiffener* prevents the glaze from flowing off the piece during firing. Alumina or aluminum oxide (Al_2O_3) is usually used to stiffen a glaze. Clay and/or feldspars are sources of the alumina (see Table 60-1).
3. An *opacifier* reduces the transparency of the glaze, until it is either translucent or opaque. The more opacifier used, the less the translucence or the greater the opacity. Tin oxide is the best opacifier, but it is very expensive. Four other oxides listed in Table 60-1 can also be used. Transparency and opacity are two types of glaze texture.
4. The *matting agent* reduces the glossiness of the glaze. The more matting agent used, the less glossiness or the greater matness. A semigloss glaze is the same as a mat glaze. Gloss or matness are two types of

Table 60-1. Glaze Materials

Type of Glaze Material	Fired Oxide	Chemical Name	Typical Source[1]	Example of Raw Material[1]
Glass former	SiO_2	Silicon dioxide	Flint	Same (SiO_2)
Flux	PbO	Lead oxide[3]	Lead compounds[2,3]	Litharge (PbO)[3]
	Na_2O)	Sodium oxide[2]	Feldspar[2]	Soda feldspar ($Na_2O \cdot Al_2O_3 \cdot 6SiO_2$)
	K_2O	Potassium oxide[2]	Feldspar[2]	Potash feldspar ($K_2O \cdot Al_2O_3 \cdot 6SiO_2$)
	ZnO	Zinc oxide	Zinc oxide	Same
	CaO	Calcium oxide[2]	Whiting	Same ($CaCO_3$)
	MgO	Magnesium oxide[2]	Magnesium carbonate	Same ($MgCO_3$)
	BaO	Barium oxide[3]	Barium carbonate[3]	Same ($BaCO_3$)
	SrO	Strontium oxide	Strontium carbonate	Same ($SrCO_3$)
	Li_2O	Lithium oxide	Lithium carbonate	Same (Li_2CO_3)
Stiffener	Al_2O_3	Alumina or aluminium oxide	Clay or feldspar	Kaolin ($Al_2O_3 \cdot 2SiO_2 \cdot 2H_2O$) Potash feldspar ($K_2O \cdot Al_2O_3 \cdot 6SiO_2$)
Opacifier	SmO_2	Tin oxide	Tin oxide	Same
	ZrO_2	Zirconium oxide[2]	Zirconium oxide	Same
	ZnO	Zinc oxide	Zinc oxide	Same
	TiO_2	Titanium dioxide	Titanium oxide	Same
	Sb_2O_3	Antimony oxide	Antimony oxide	Same
Matting agent	BaO	Barium oxide[3]	Barium carbonate[3]	Same ($BaCO_3$)
Colorants	(See Table 60-2)			

[1] The most frequently used raw materials are listed
[2] Two or more raw material sources are available
[3] Should not be used for glazes on tableware, since they might release toxic compounds

220 / Modern Industrial Ceramics

glaze texture. Barium carbonate (see Table 60-1) is added to cause matness. *A mat glaze containing barium carbonate (BaCO$_3$) should not be used on tableware, since it might release toxic compounds.*

5. *Colorants* give the glaze color. Usually, 0.1% to 10% of a metallic oxide and/or carbonate are used, as indicated in Table 60-2. Glaze stains are also used for colorants (see Table 60-2). They are synthetic materials. A stain contains two or more materials, one of which is always a metallic oxide or a carbonate. The materials are calcined to produce the stain.

Two or more metallic oxides, metallic carbonates, and/or glaze stains can be combined to produce colors not listed in Table 60-2. For example, 0.5% cobalt carbonate and 2% iron oxide will produce a blue-gray color; or, 0.5% cobalt carbonate, and 2% copper carbonate will produce a blue-green color.

6. *Binders* help to keep a liquid glaze in suspension. This means the materials do not settle out. They also improve a liquid glaze that is to be applied by brush. The binders are optional. Organic materials that burn out during firing are used for this purpose. The typical binders are:

a. Gum arabic
b. Gum tragacanth
c. CMC or cellulose gum (sodium cellulose glycolate or sodium carboxymethylcellulose—synthetic)

Often, a binder is not necessary. When used, use as little binder as possible. Up to 1% of binder by weight is recommended.

Manufactured Materials

Two major manufactured materials may be used in glazes—glaze stains and frits. *Glaze stains* are used to color glazes. They were included and discussed with raw material colorants (see 5 above). Stains were discussed with the other colorants to make that subject complete.

Frits are ground-up glass. There are three types of frits:

1. Glaze frits
2. Body frits
3. Porcelain enamel frits

Table 60-2. Common Glaze Colorants

Raw Material	Formula	Fired Oxide	Color[1]	Recommended Amount
Cobalt carbonate	CoCO$_3$	CoO	Blue	0.1–1%
Copper carbonate	CuCO$_3$	CuO	Blue to green	1–8%
Chrome oxide	Cr$_2$O$_3$	Cr$_2$O$_3$	Green	2%
Ilmenite[2]	Fe·TiO$_2$	Fe·TiO$_2$	Brown[3]	1–3%
Iron chromate	FeCrO$_4$	FeCrO$_4$	Grey	2%
Iron oxide (black)	FeO	FeO	Tan to dark brown	2–6%
Iron oxide (red)	Fe$_2$O$_3$	Fe$_2$O$_3$	Brown	2–6%
Manganese carbonate	MnCO$_3$	MnO	Purple[3]	4–6%
Nickel oxide (green)	NiO	NiO	Grey to brown	2%
Rutile[2] (impure titanium dioxide)	TiO$_2$	TiO$_2$	Tan[3]	5%
Uranium oxide	UO$_2$	UO$_2$	Yellow	6–10%
SELECTED GLAZE STAINS[4]				
Cadmium and selenium stain	Approx 20% selenium and 80% cadmium sulphide[5]		Red[5]	2–10%
Naples yellow[4]	Antimony and lead[5]		Yellow[5]	2–6%
Vanadium stain[4]	Vanadium pentoxide and (V$_2$O$_5$) and tin oxide (SnO$_2$)		Yellow	6%

[1] Oxidation firing.
[2] An ore containing iron and titanium oxides.
[3] Will produce specks or spots in glaze when using coarse particles, such as 80 mesh.
[4] Manufactured.
[5] Should not be used for glazes on tableware. They might release toxic compounds.

We are interested only in the glaze frits at this time. Glaze frits are used to make:

1. Soluble compounds insoluble
2. Toxic compounds nontoxic

Soluble and toxic glaze materials are listed in Tables 60-3 and 60-4.

Frits contain glass former, one or more fluxes, and occasionally a stiffener and/or opacifier. The frits are made by melting the dry, powdered raw materials in a crucible. The molten glass is poured into cold water. This causes the molten glass to shatter into small pieces. The pieces are then ground into a fine powder.

A glaze can be made entirely of frit, but this is not common. Usually frit, clay, colorant (optional), and a binder (optional) are combined with water to make a fritted glaze.

Table 60-3. Typical Soluble Glaze Materials*

Type of Glaze Material	Fired Oxide	Chemical Name	Other Name	Raw Material
Flux	$Na_2O \cdot 2B_2O_3$	Borax	None	Same ($Na_2O \cdot 2B_2O_3 \cdot 10H_2O$)
Flux	B_2O_3	Boric acid	None	Same ($B_2O_3 \cdot 3H_2O$)
Flux	K_2O	Potassium nitrate	Nitre	Same (KNO_3)
Flux	K_2O	Potassium carbonate	Pearl ash	Same (K_2CO_3)
Flux	Na_2O	Sodium carbonate	Soda ash	Same (Na_2CO_3)

* These materials dissolve in water.

Table 60-4. Typical Toxic Glaze Materials*

Type of Glaze Material	Fired Oxide	Chemical Name	Other Name	Raw Material
Flux	PbO	Lead carbonate	White lead	Same ($2PbCO_3 \cdot Pb(OH))_2$
Flux	$3PbO \cdot 2SiO_2$	Lead monosilicate	None	Same ($3PbO \cdot 2SiO_2$)
Flux	PbO	Lead oxide	Litharge	Same (PbO)
Flux	PbO	Lead oxide	Red lead (oxide)	Same (Pb_3O_4)
Matting agent	BaO	Barium oxide	None	Barium carbonate ($BaCO_3$)
Colorant	CdO	Cadmium oxide	None	Same (CdO)

* Should not be used for glazes on tableware, since they might release toxic compounds.

Unit 61: Procurement

Procurement of raw materials for the glaze industry is diagrammed in Fig. 61-1. Glazes are made of two or more materials. In fact, six or more materials may be used in a glaze.

The number and variety of materials used in glazes, as discussed in Unit 60, usually makes purchasing, rather than procurement, more feasible. Thus, few if any glaze manufacturers and whiteware firms that make their own glaze(s) are involved in procurement. Procurement, for these companies, becomes a purchasing job rather than an extraction process. Purchasing is an important role (Fig. 61-2). This is true for any of the whiteware manufacturers, as well as for the glaze manufacturers.

Ceramic Raw Material Industry

The flow sheet (see Fig. 61-1) also represents the ceramic raw material industry. This industry specializes in the extraction type of procurement, beneficiation processes, and selling materials to other manufacturers. Ceramic industries that frequently purchase these materials are: whiteware, glaze, porcelain enamel, glass, and abrasives.

The extraction processes of prospecting, development, and winning were discussed earlier in Section III—*Ceramic Raw Material Industry*. All three types of winning are used, due to the variety of the raw materials needed. Superficial winning is used most often, followed by subterranean, and finally subaqueous winning. These have been discussed in preceding sections.

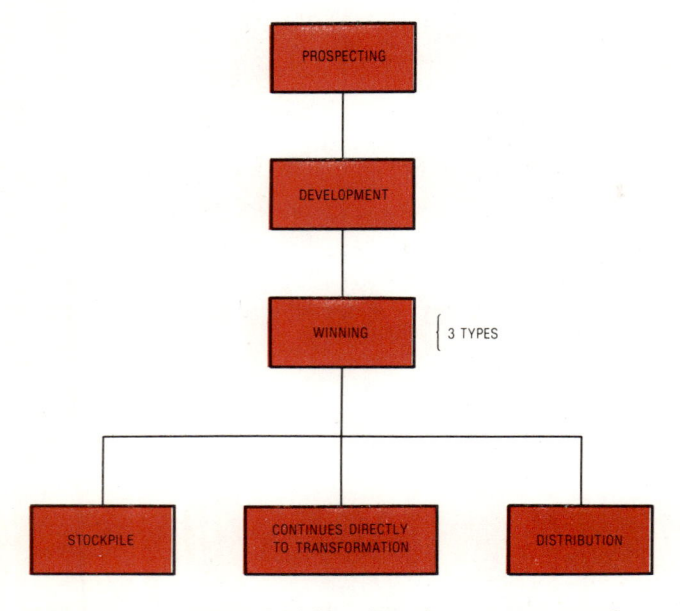

Fig. 61-1. A typical vertical flow sheet for procurement of raw materials used by the glaze industry. These processes may be done by the glaze industry, but few companies do their own procurement. Usually, the ceramic raw material industry does the procurement, some transformation (such as beneficiation), and sells powdered raw materials to the glaze industry.

Stockpile

Glaze manufacturers, including the whiteware firms that make their own glazes, seldom stockpile the won raw materials. This type of stockpiling is done by the ceramic raw material industry (Fig. 61-3). The beneficiated raw materials are stockpiled by the glaze manufacturers (Figs. 61-4 and 61-5). Beneficiation is the first transformation process.

Fig. 61-2. Large quantities of raw materials are frequently purchased in bulk. (Courtesy R. T. Vanderbilt Company, Inc.)

Procurement / 223

Fig. 61-3. A typical storage building for large quantities of bulk raw materials. (Courtesy American Olean Tile Company)

Fig. 61-4. Stockpiling bagged beneficiated raw materials. (Courtesy R. T. Vanderbilt Company, Inc.)

Fig. 61-5. Transporting bagged talc, beneficiated raw material, to storage at Gouveneur, New York. Note how the wood pallet allows the forklift truck to easily transport palletized material. (Courtesy R. T. Vanderbilt Company, Inc.)

Unit 62: Transformation

There are two major subdivisions in this unit—glaze manufacturing and glaze use. Technically, only glaze manufacturing should be discussed here. However, after you have made the glaze, its use becomes important. Thus, glaze use has been included.

Manufacture of Glazes

Basically, there are two types of glaze manufacturers: (1) firms that make glazes for wholesale and/or retail sale, and (2) firms, institutions, and individuals who make glazes only for their own use. The second group includes the clay industries (SCP and whitewares), educational institutions, and art potters.

Flow Sheets

Transformation of materials into glaze products is diagrammed in Fig. 62-1. The products manufactured by this industry are either powdered or liquid. Usually, only water is added to the prepared powdered glaze before its application to a clay product. Prepared liquid glazes are ready for application.

Many whiteware manufacturers, educational institutions, and art potters make their own glazes (Fig. 62-2). Some of the structural clay product manufacturers also make their own glazes. Usually, their glazes are made from purchased materials.

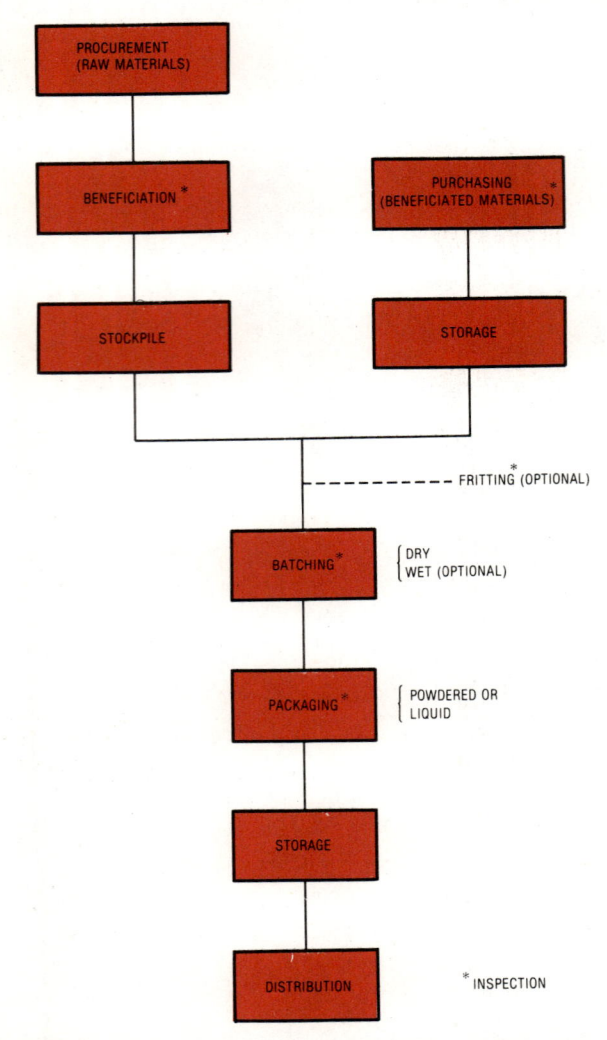

Fig. 62-1. A typical vertical flow sheet for the transformation of materials into glaze products. Prepared glaze products are in either a powdered or liquid form. Both forms are available in a variety of quantities.

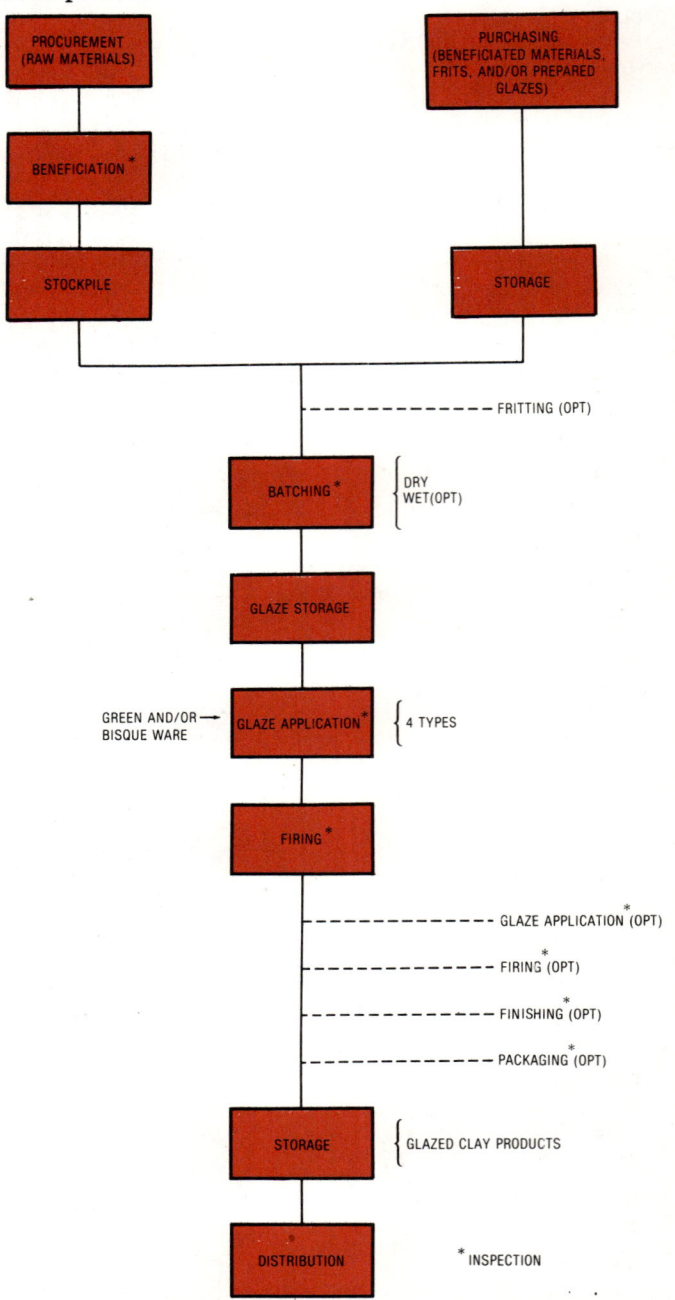

Fig. 62-2. A typical vertical flow sheet for the transformation of materials into glazes by whiteware manufacturers that make their own glaze. Glaze transformation ends at glaze storage. The remainder of the flow sheet shows how whiteware manufacturers use the glazes they have prepared.

Only a few glaze manufacturers or others that make their own glazes do their own procurement or beneficiation of the raw materials. The number and variety of glaze materials required does not make these processes feasible for a single firm. Procurement and beneficiation are usually done by the ceramics raw material industry.

Usually, four basic processes involved in making glazes are:

1. Purchasing
2. Batching
3. Packaging
4. Storage

Purchasing

Nearly everyone that makes glazes purchases one or more of the necessary raw and/or manufactured materials. The reason for this practice is the number and variety of glaze materials required.

Beneficiated raw materials and frits are purchased. Beneficiation is done by the ceramic raw material industry. They use the transformation processes of comminution, classifying, and separating.

Frits are ground-up glass. The glaze frits are the type used to make glazes. The types and manufacture of frits were discussed in Unit 60. At least three firms specialize in the manufacture of frits. Therefore, a large variety of frits are available. They are easily purchased. Occasionally, a company, institution, or art potter will make its own frits.

Batching

Each individual glaze material is carefully weighed before use. You should use a metric balance—for example, a triple-beam balance. Measurement is easier with the metric system.

All the glaze materials are then *dry batched* to insure uniform distribution. This is especially important when colorants are used. The process is complete when there is a uniform color to the dry materials.

Dry batching may be done manually or mechanically. A mortar and pestle is usually used for manual dry batching. This is only feasible for quantities up to 1 quart or 1 liter of liquid glaze.

A ball mill or V-type mixer can be used for mechanical dry batching. Both machines are available in a variety of sizes. They range upward in size from a 1-quart ball mill or a 2-gallon V-type mixer. The largest quantity either type of machine can handle is several tons.

Wet batching involves adding water to the dry materials. The usual ratio is 50–50 by weight—100 gm of dry materials, plus 100 ml of water. A consistency similar to thick cream is desired.

Wet batching is also done manually and mechanically. The mortar and pestle are used for small batches. The ball mills are usually used for mechanically mixing both small and large batches.

Packaging and/or Storage

Commercially manufactured glazes are packaged in paper, plastic, and glass containers. The dry glazes are packaged in paper and plastic bags. The liquid glazes are packed in glass and plastic jars.

Firms, institutions, and individuals who make glazes for their sole use store them in stainless steel or glass-lined tanks, in porcelain-enamel canisters, and in glass or plastic jars, depending on the quantity.

Small-mouth bottles and jugs should not be used. It is difficult to completely remove the glaze from these containers. Zinc-coated (galvanized) tanks or containers should not be used. The glaze materials may react with this coating.

The liquid glazes that do not contain an organic binder have an infinite shelf line. The glazes that contain binders, such as commercially prepared liquid glazes, also have a long shelf life if they are not allowed to freeze. Freezing and thawing causes these glazes to congeal.

Liquid glazes stored for a long period of time tend to lose water and thicken. These glazes may be restored to a heavy cream consistency by slowly adding water to the contents while hand mixing. A glaze that has completely dried out will require time to allow the additional water to soak into and soften the materials. Then it can be easily remixed.

General Procedure

The general procedure or steps for glaze preparation includes:

1. Select a leadless glaze.
    a. Base glaze—gloss or mat and transparent or opaque
    b. Color, if desired
2. Determine amount to prepare (Table 62-1).

Table 62-1. Determining the Amount of Base Glaze

Number of Pieces to be Glazed	Amount of Dry Base Glaze Needed
1 small piece or glaze test	50 gm
1 medium-size piece (4-in. diameter)	100 gm
2 medium or larger pieces	300 gm
6 or more medium or larger pieces	750 gm*
24 or more medium or larger pieces	3000 gm**

* 750 gms of dry base glaze will make approximately 1 quart (0.946 liters) of liquid glaze.
** 3000 gms of dry base glaze will make approximately 1 gallon (3.785 liters) of liquid glaze.

226 / *Modern Industrial Ceramics*

3. Carefully weigh each glaze material, using a metric balance—triple-beam balance.
4. Dry batch until a uniform color is achieved.
5. Wet batch until the lumps disappear and the mixture is a heavy cream consistency. Color at this point seldom resembles the fired color.
6. Store the glaze in a fully labeled container. Do not use galvanized or tin containers.

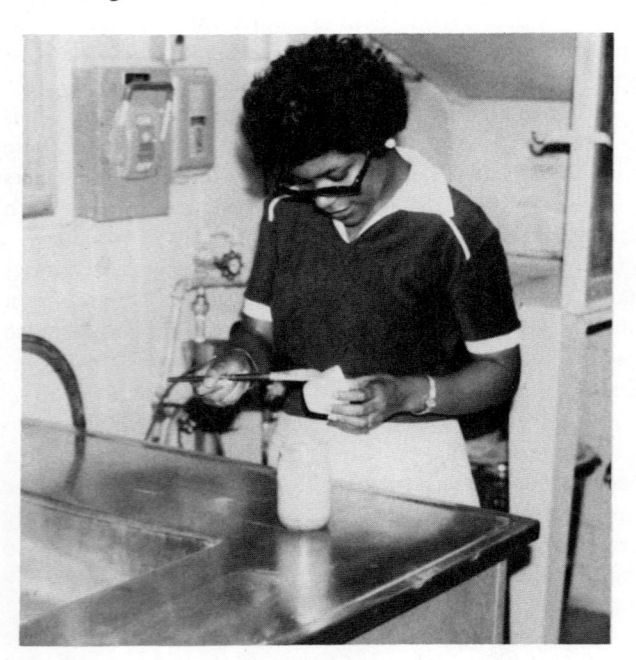

Using Glazes

Using the glazes involves: (1) glaze application; (2) glaze firing; (3) finishing fired glaze; and (4) glaze defects.

Glaze Application

Four major methods of glaze application are:

1. Dipping (Fig. 62-3)
2. Pouring (Fig. 62-4)
3. Brushing (Fig. 62-3)
4. Spraying (Fig. 62-5)

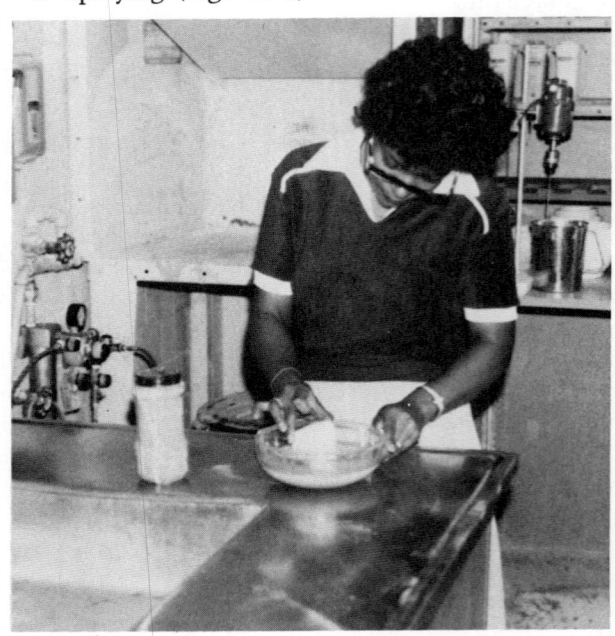

Fig. 62-3. Two manual glaze application methods are brushing (left) and dipping (right). These methods are frequently used when only a few pieces are to be glazed.

Fig. 62-4. Pouring is another method of manually applying glaze. This method is used to glaze the interior of tall pieces, such as bottles and vases.

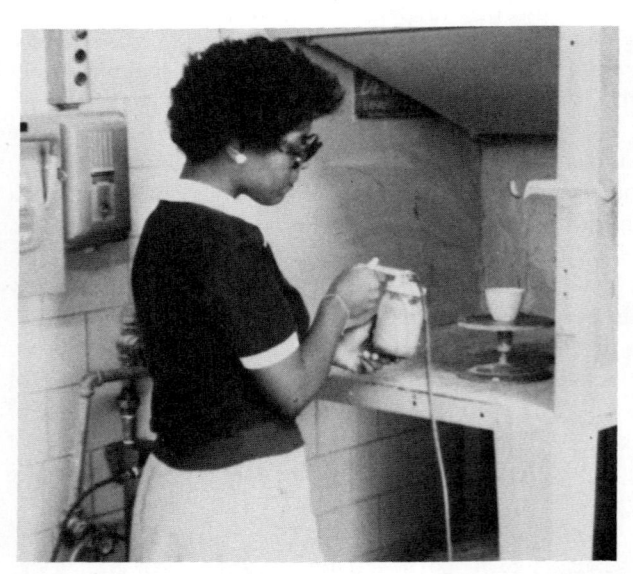

Fig. 62-5. Spraying is an easy way to apply glazes. Note the use of the spray gun, whirler, and spray booth. Also note that this worker is not wearing safety equipment. The booth should be equipped with an exhaust fan that is vented outdoors.

All these methods can be done by hand. Industry has developed machines that will dip, pour, or spray glazes. Automatic spraying of the ware on a conveyor line is used extensively by the whiteware manufacturers in the United States.

Certain guidelines are followed for each method of glaze application. A liquid glaze is use for all these methods.

Dipping—One coat is usually applied for bisque ware (see Fig. 62-3). A piece may be partially dipped in one glaze color and the remainder dipped in another glaze color for the decorative effect. Glazes with the same base glaze should be used. Partial dipping is a method used by the art potters. Dipping tongs are usually used. Touching up with a brush is required after the glaze dries. Dry the dipped pieces on a wire rack.

Pouring—One coat is usually applied to bisque ware, (see Fig. 62-4). This method should be used to glaze the interior of tall pieces, such as a vase. Glaze the interior first and allow the piece to dry before pouring the exterior. When pouring the interior, fill the piece about two-thirds full of glaze and *immediately* pour out the glaze while revolving the piece. Revolving the piece causes all the remaining interior to be glazed. Use a large clean pan when pouring the glazes. Place a wire rack over the pan to support a piece for pouring the exterior. Dry the poured pieces on a wire rack. Touch up with a brush any unglazed areas.

Brushing—Three coats applied in opposite directions are usually used on either green or bisque ware (see Fig. 62-3). Fewer glaze defects occur when glazing bisque ware. Brushing works best when the liquid glaze contains organic binder.

Do not brush a glaze, as you would brush paint. *Dab* the glaze onto the piece. The brush is only a device for conveying the glaze. Allow each coat to dry before applying the next coat. Always apply the second and third coats in opposite directions to the preceding coat (apply first coat the length of the piece, second coat the width, and third coat diagonally). Brushing uses the least amount of glaze.

Spraying—Three to six coats of glaze are usually applied to either green or bisque ware (see Fig. 62-5). A coarse granular surface indicates sufficient glaze has been applied. The surface will resemble flint paper or "sandpaper." Use a whirler or banding wheel and a spray booth with an exhaust fan (Fig. 62-6). The banding wheel allows you to rotate the piece while spraying. *Always* use the exhaust fan and a clean spray booth. Also, always pour the glaze through a strainer when filling the spray gun (Fig. 62-7). This removes any lumps that could clog the gun.

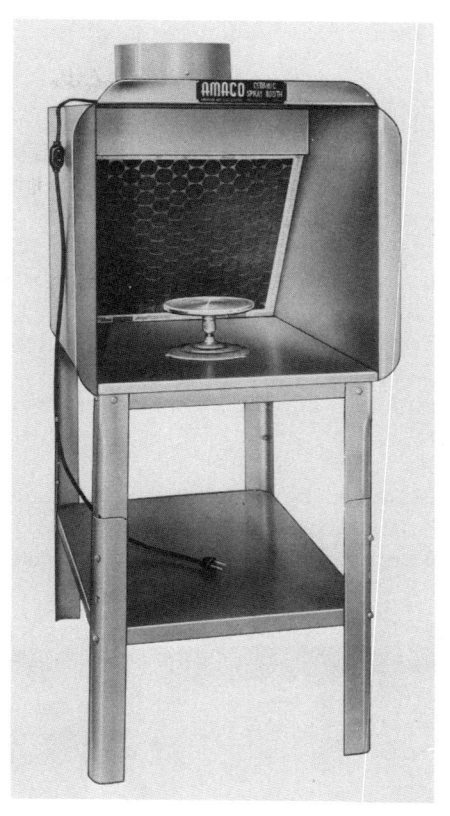

Fig. 62-6. A spray booth equipped with a whirler, air filter, and exhaust fan. The fan should be ducted outdoors. (Courtesy American Art Clay Co., Inc.)

Fig. 62-7. An inexpensive and small spray gun for glaze application. This gun uses 40–60 psi (2.8–4.2 kg/sq cm). Caution must be taken not to break the glass jar. The glass allows you to easily see the amount of glaze remaining in the jar. (Courtesy American Art Clay Co., Inc.)

The following principles apply to all the methods of glaze application:

1. Use less glaze on the bottom third of the pieces. Remember that glaze flows during firing.
2. You can lightly rub a dry glaze to remove the lumps, pinholes, or other blemishes. Allow the glaze dust to remain on the piece.
3. Melted paraffin wax or wax resist can be applied to the foot of a piece. This prevents glaze from adhering. The wax will burn off in firing, leaving a piece with a dry foot.
4. The color of a liquid glaze seldom resembles its fired color!

Glaze is applied by spray guns in semiautomatic or fully automatic machines. These are used by many manufacturers of whiteware and structural clay products (Fig. 62-8).

The general procedure for glaze application includes:

1. Prepare the glaze, or remix if the liquid glaze has been stored. Be sure its consistency is similar to heavy cream.
2. Dampen the piece to remove the dust. Caution if greenware.
3. Apply the glaze.
4. Allow the piece to dry on a wire rack.
5. Stack in a kiln with stilts, or place in a dust-free storage for later firing.
6. Glaze fire.
7. Finishing—grind off the stilt marks.

Glaze Firing

The general procedure for stacking a periodic kiln for glaze firing was discussed earlier. Please refer to Unit 24 in Section VI—*Kiln Industry*. A correctly stacked periodic kiln is shown in Fig. 62-9.

Caution should be observed when unstacking the fired glaze ware. Sharp points may remain from the stilts on the bottom of the piece. Do not rub your fingers over these stilt marks. Use a screwdriver to gently pry off the stilts

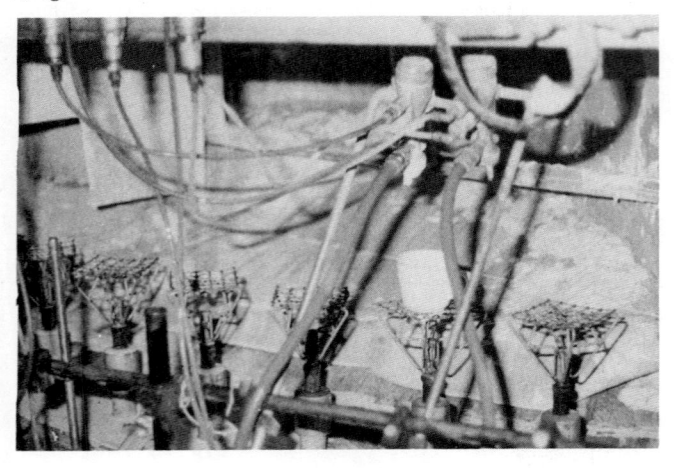

Fig. 62-8. Automatic glazing machines using spray guns on coffee cups (top) and on plates (bottom). Both machines move the pieces of ware past stationary spray guns on holders that also revolve. These machines resemble a merry-go-round. (Courtesy Buffalo China, Inc.)

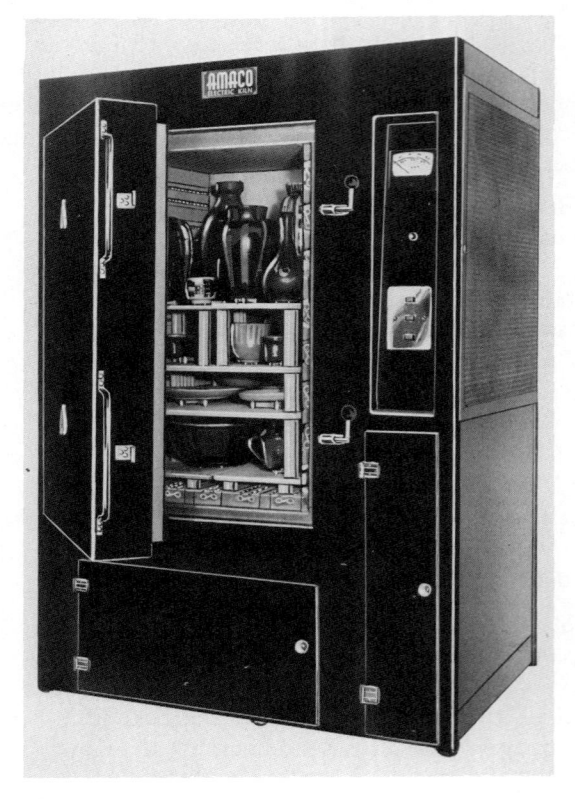

Fig. 62-9. A front-loading electric periodic kiln stocked for glaze firing. Note how the pieces with similar heights are placed together on the same kiln shelf. Glazed ware is usually stilted, as shown. The top surfaces of each kiln shelf has been coated with kiln wash prior to stocking the kiln. (Courtesy American Art Clay Co., Inc.)

that stick to the fired ware. Do not use your fingers to pull off these stilts, since the stilts break easily or your hand may slip and be cut on the sharp stilt mark.

Finishing Fired Glazes

Finishing the fired glaze ware is necessary, even though the piece was dry footed. Any protrusions should be removed to avoid damaging furniture or injuring someone. Grinding will remove these protrusions. Hand grinding with a wet slipstone is recommended.

Water is used with a slipstone to prevent dangerous dust. Remember you are grinding glass. Mechanical grinding wheels can also be used with water and light pressure. Machine grinding is necessary only when a large number of pieces must have the stilt marks removed.

Most of the whiteware manufacturers use machines to remove the stilt marks. An automatic machine is shown in Fig. 62-10.

Glaze Defects

Glaze defects are seldom seen until after firing. Then, it is usually too late to correct the defect. Some of the glaze defects can be corrected. Glaze defects can be caused by the:

1. Clay or clay body
2. Application of glaze

Fig. 62-10. Automatic machine removal of stilt and plate-pin marks from fired glaze ware. (Courtesy Shenango China Products)

3. Glaze firing
4. Composition of glaze

Some of the common glaze defects are listed in Table 62-2. Most of the glazes have a firing range of two or three cones. This flexibility helps reduce the possibility for some defects to occur, such as underfiring and overfiring.

Table 62-2. Common Glaze Defects

Name of Defect	Description	Usual Cause	Correction
Bare spots	Little or no glaze in small areas	Improper handling of unfired glaze	Reglaze spots and refire
Blistering	Bubbles that have burst	Glaze applied too heavy	None
Cracking	Piece develops a crack during glaze firing	Glaze applied too heavy resulting in uneven stress between interior and exterior	None
Crawling	Glaze gathers into globs with some bare areas	Oil or grease on your hands when handling the ware prior to glazing	Reglazing and refiring may correct
Crazing	Network of fine cracks over the entire piece	Glaze shrinks more than the clay. The glaze cracks because it is under too much stress and is only a thin coating.	None
Overfiring	Pooling of glaze in the bottom of hollow ware and/or flowing off the bottom of tall pieces	Kiln too hot	None
Pinholes	Very small holes in the glaze	Dust on the ware when glazed. Gas bubbles released during firing due to moisture or carbonate, i.e. single firing of glazed greenware.	Refiring at one cone higher may smooth out the glaze.
Rough surface	Rough scratchy surface resembling sandpaper	Glaze too thin or underfiring	Reglaze if too thin and refire.
Shivering	Glaze falls off the piece after firing	Clay or clay body shrinks more than the glaze	None
Underfiring	Orange peel texture or rough scratchy surface	Kiln fired at a lower cone	Refire at a higher cone

Unit 63: Utilization

Glazes may be a manufactured product, or they may be used to make other products. The packaged glazes are products that are sold wholesale and/or retail. When dinnerware manufacturers make and/or use a glaze, it becomes part of their final product.

Manufacture

Glazes are used in the manufacture of many clay products. Some of the structural clay products are glazed, and most of the whiteware products are glazed.

Structural clay products that may be glazed are:

1. Face brick
2. Structural tile
3. Wall tile
4. Mosaic tile
5. Sewer pipe (salt glaze)

Face brick, such as a stretcher, usually have only one side glazed. Some of them such as a header, will also have one end glazed for use at corners. Structural, wall, and mosaic tile usually have only one surface glazed. Sewer pipe is completely covered by the salt glaze.

Whiteware products that are glazed include:

1. Art pottery—plates, pots, mugs
2. Hobby ceramics—decorative pieces
3. Dinnerware
4. Sparkplug insulators
5. Porcelain crucibles
6. Ball mill jars
7. High-voltage electrical insulators
8. Sanitary ware

Hobby ceramics, dinnerware, and sanitary ware frequently have a glaze covering the entire piece, both interior and exterior. The other glazed product examples listed above may be dry footed. Dry footing is often used for the decorative contrast on art pottery. Insulators are dry footed to simplify firing and to improve bonding with metal parts during final assembly. Porcelain products, such as crucibles and ball-mill jars, are usually dry footed. The dry foot allows the piece to be placed on a kiln shelf for total surface support.

Terms

There are three coatings that are used in a manner similar to the glazes. These are engobes, underglazes, and overglazes. They are used on art pottery, hobby ceramics, and dinnerware. All are usually used with glazes.

Engobe (pronounced on-gobe) is a layer of white or colored slip or clay applied as a liquid to the surface of a piece. It is used to change the color or as a decoration. Engobes do not flow when they are fired, and they may be applied to greenware and bisque ware. Their composition is different for each type of ware. They can either be white or colored by adding the same colorants used for the glazes. A clear transparent gloss glaze is usually applied over the engobe.

Underglaze means that the glaze is used under a clear transparent gloss glaze. The underglazes are colorants used for a decorative effect. They do not flow when they are fired. The underglazes are primarily composed of metallic oxides that have been calcined and ground. Commercially prepared underglazes are recommended.

The liquid underglazes can be used on greenware or bisque ware. Read the labels carefully to determine whether it is to be used on greenware or bisque ware and whether one or three coats are to be used.

Underglazes in a semimoist state, crayon, and pencil form are available. The crayon and pencil are used only on bisque ware.

Overglaze means that it is used over a fired glaze. The overglazes are fused by an additional firing at a lower temperature, such as *C/020-017*, than the glaze firing. The precious metals, such as gold and platinum, are typical overglazes. These are primarily used on bone and fine china dinnerware.

Commercially prepared liquid and semimoist overglaze colors are also available. These are primarily used for china painting, as by a hobbyist.

Decoration

Decoration and glazes are closely related subjects. The two major types of decoration are:

1. Art pottery or hobby ceramics methods
2. Industrial methods

The art pottery or hobby ceramic methods are essentially handwork. They are creative and often unique. There are so many of these methods that several books have been written on the subject. Most of the pottery books also include an ample coverage of decorative techniques. These methods can be grouped according to the type of ware:

1. Green—includes plastic clay, leatherhard clay, and greenware
2. Bisque
3. Glaze fired

Industrial methods are primarily done by machines. Hand decoration work is used for small orders, special orders, and some overglazes, such as applying a gold band. Many of the industrial methods are listed in Table 63-1. Some of these are shown in Figs. 63-1 through 63-7.

Construction

The glazes are not used in construction. The glazed structural clay products are used in construction. A list of structural clay products that may be glazed was included at the beginning of this unit.

Table 63-1. Typical Industrial Decorative Methods

\multicolumn{3}{c}{Type of Ware to Which Decoration is Applied}
Green[1]	**Bisque**	**Glaze Fired**

Sprigging	1. Underglaze
	   a. Air brush
	   b. Decal
	   c. Stamping[2]
	2. Template stenciling
	3. Silk screen
	4. Murray process
	5. Banding machine
	6. Line decorator[2]
		1. Overglaze
		   a. Decal
		   b. Metallic banding[2]
		   c. Stamping[2]

[1] Includes plastic clay, leatherhard clay, and greenware
[2] May be done by hand

Fig. 63-1. A hand decorator applying a line decoration to bisque ware. The brush is held in one position while the piece is revolved by the banding wheel. (Courtesy Shenango China Products)

Fig. 63-2. An automatic line decorating machine. Only bisque ware has the strength to be handled by this type of machine. (Courtesy Shenango China Products)

Fig. 63-3. Closeup of an automatic line decorating machine. Underglaze is applied by an enclosed roller on the right-hand side of the plate. (Courtesy Buffalo China, Inc.)

Fig. 63-4. A silk-screen process decorating machine for bisque ware. These machines are usually semiautomatic. (Courtesy Shenango China Products)

Fig. 63-5. Closeup of a semiautomatic silk-screen process machine. (Courtesy Buffalo China, Inc.)

Fig. 63-7. A semiautomatic banding machine using a spray gun to apply underglaze to bisque ware. (Courtesy Buffalo China, Inc.)

Fig. 63-6. A semiautomatic rubber-stamp machine. This printing machine is a letter press that has been adapted to accommodate the thickness of flatware. (Courtesy Buffalo China, Inc.)

Unit 64: Disposition

The unfired glazes can be disposed of by either reclaiming or discarding. The fired glazes are part of a structural clay or a whiteware product. Thus, the fired glazes are the disposition problem for the respective clay industry. Disposing of the clay products has been discussed in previous sections.

Reclaiming

Some methods for reclaiming the glazes are discussed here. The liquid glazes that have thickened or dried out during storage can be reclaimed. The thickened glazes can be reclaimed by slowly adding water while hand mixing. A glaze that has dried out will require time for the additional water to soften the materials. Then it can be remixed.

An overspray occurs when a glaze is applied with a spray gun. There are two ways that can be used to reclaim the overspray.

The overspray is allowed to dry on the interior of the spray booth. Periodically, the dried material is scraped off and remixed with sufficient water. This is a relatively safe procedure, if only one base glaze and one color have been used. Usually, several base glazes and/or colors have been used in the spray booth. When this unknown mixture is mixed with water and used, the results are totally unpredictable. Thus, it is called a "surprise" glaze. The art potters tend to use this method of reclaiming.

An industrial firm usually uses only one base glaze, for example, a clear transparent gloss glaze. They reclaim the overspray by scraping it off when it is dry, or it may be washed off with water. Another industrial reclaiming method uses a waterfall to collect the overspray.

Discarding

Occasionally, a glaze must be discarded. For example, students frequently discard the glazes that they have made when their course ends. Discarding the liquid glazes can be done without creating problems.

The liquid glazes should be poured into a trash can containing wastepaper. The paper absorbs the water. Then the glaze is discarded with the trash.

A glaze can be discarded outdoors. Dig a shallow hole, pour the glaze into it, and refill with dirt. This is the safest and best method of discarding reasonable amounts of liquid glazes. The glaze materials will slowly remix with the earth. Remember that glazes are composed of earthy, inorganic, and nonmetallic materials.

Liquid glazes should never be poured into a sink or floor drain! The glaze materials will clog the plumbing. Liquid glazes are composed of about 50% solids. The solids are primarily very fine particles, such as 200-mesh, that will settle out of solution into a dense mass. The dense mass clogs the sink traps and drain lines that are not easy to clean. Therefore, correct discarding is very important!

SECTION XVI

Porcelain Enamel Industry

Unit 65: Material Science

Four glassy materials enter into ceramic products (Table 65-1). Glassy bond and glazes were discussed in Sections XII through XV. The glazes and porcelain enamels are both glassy coatings. The major difference between the glazes and porcelain enamels is the material being coated. The glazes are fired onto clay products.

Table 65-1. Glassy materials

Glassy Material	Product Examples
1. Glassy bond	Clay flowerpots
2. Glaze	Dinnerware, such as plates, cups, and saucers
3. Porcelain enamel	Appliance housings, such as a kitchen range
4. Glass	Bottles and jars

Definition

Porcelain enamels are glassy coatings fused to a base material by the use of heat. The base material may be either glass or metal.

Porcelain enamels serve several functions:

1. Protect
2. Decorate
3. Sanitary
4. Smooth surface

Technically, all the porcelain enamels serve these functions, except decoration. Decoration is a relative thing that depends upon your point of view. Therefore, you could say that all the porcelain enamels are also decorative.

Materials

The materials used to make porcelain enamel frits include:

1. Glass former
2. Flux
3. Opacifier
4. Colorant

All the porcelain enamel frits contain a glass former and one or more fluxes. Opacifier and colorant are optional materials.

Glass Former

The glass former is silica or silicon dioxide (SiO_2). Silica has one major disadvantage. It melts at about 3110°F (1710°C).

Flux

One or more fluxes are used to lower the melting temperature of silica. Fluxes are also used to lower the melting temperature of the frits, so that the base material is not harmed by the heat. Typical fluxes are listed in Table 65-2.

Note: The glass former and fluxes usually make a clear transparent frit. Any additional materials, such as opacifier and colorant, simply act as modifiers.

Opacifier

An opacifier reduces the transparency of the frit, until

Table 65-2. Typical Fluxes Used in Porcelain Enamel Frits

Material	Raw Formula
1. Borax	$Na_2O \cdot 2B_2O_3 \cdot 10H_2O$
2. Cryolite	Na_3AlF_6
3. Fluorspar	CaF_2
4. Lead oxide	PbO
5. Soda ash	Na_2CO_3
6. Zinc oxide	ZnO

it is translucent or opaque. The more opacifier used, the greater the opacity. Antimony oxide, tin oxide, titanium dioxide, and zirconium oxide are typical opacifiers.

Colorant

The colorants give color to the porcelain enamel frit. The metallic oxides and/or carbonates are used for this purpose. Basically, the same colorants are also used in the glazes. For example, cobalt carbonate gives a blue color to the frit. Refer to Table 60-2 for the common colorants. *Note:* The glass former, fluxes, and opacifier and/or colorant, if used, are fritted. The frit process is discussed later in Unit 67.

Floating Agent

A floating agent helps to keep the porcelain enamel frit in suspension when it is mixed with water for liquid application. The floating agents are optional. Clay and organic gums are used for this purpose.

Base Material

A porcelain enamel frit is fused by heat to the *base material*. The two types of base materials are glass and metals. The metals that are porcelain enameled include cast iron, sheet steel, aluminum, copper, gold, and silver.

Unit 66: Procurement

Procurement of raw materials for the porcelain enamel industry is diagrammed in Fig. 66-1. The porcelain enamel frits are made of two or more materials. In fact, six materials may be used in a frit.

The number and variety of materials used in the frit discussed in Unit 65 usually makes purchasing, rather than procurement, more feasible. Therefore, few, if any, porcelain enamel frit manufacturers are involved in procurement. Procurement, for these companies, becomes a purchasing job rather than an extraction process. Purchasing is an important role.

Ceramic Raw Material Industry

The flow sheet (see Fig. 66-1) also represents the ceramic raw material industry. This industry specializes in the extraction type of procurement, beneficiation processes, and selling materials to other manufacturers. The glaze, porcelain enamel, and glass firms are among those ceramic industries that frequently purchase raw materials.

Prospecting, developing, and winning were discussed earlier in Section III—*Ceramic Raw Material Industry*. All three types of winning may be used, due to the variety of raw materials needed. Typical winning processes are shown in Figs. 66-2 through 66-4.

Stockpile

The porcelain enamel frit manufacturers seldom stockpile the won raw materials. This type of stockpiling is done by the ceramic raw material industry. The beneficiated raw materials are stockpiled by the frit manufacturers. Beneficiation is the first transformation process.

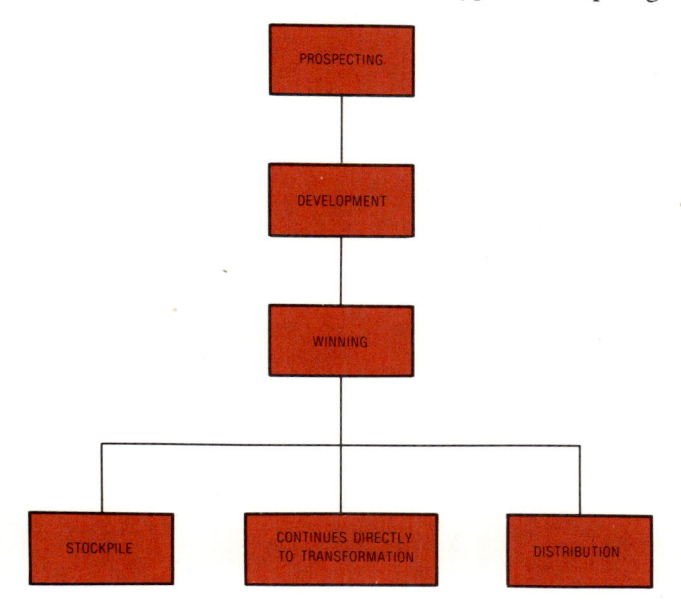

Fig. 66-1. A typical vertical flow sheet for procurement of raw materials used to make porcelain enamel frit. These processes may be done by the porcelain enamel industry, but some companies do their own procurement. The ceramic raw material industry does most of the procurement, some transformation (for example, beneficiation), and sells powdered raw materials to the porcelain enamel industry.

Fig. 66-2. Typical superficial winning equipment—a crawler-type tractor with a blade and ripper. (Courtesy Cedar Heights Clay Co.)

Fig. 66-3. The equipment used for subterranean winning is frequently specially designed. A typical example is the low-profile lead-acid battery powered truck above. (Courtesy Lead Industries Association, Inc.)

238 / *Modern Industrial Ceramics*

Fig. 66-4. Subaqueous winning is frequently done with barge-mounted cranes. This crane is equipped with a dragline bucket. (Courtesy ASARCO Incorporated)

Unit 67: Transformation

Transformation of raw materials is diagrammed in Fig. 67-1. These processes are usually done by the ceramic raw material industry. Most of the porcelain enamel manufacturers purchase their powdered materials from this industry. These powdered materials are then used to make porcelain enamel frit.

Frit Manufacture

Frit is ground-up glass. It is used to make the glassy coating on porcelain enamel products. There are several reasons for using frit:

1. Fritting process makes the soluble compounds insoluble.
2. Fritting process makes the toxic compounds nontoxic.
3. Frit allows faster firing, so base material is not harmed.
4. Frit is easy to handle.

The transformation of powdered raw materials into frit is diagrammed in Fig. 67-2. The powdered raw materials are dry batched. Then, the materials are melted in a large crucible or frit furnace. The molten glass is poured into water (Figs. 67-3 and 67-4). This quenching or rapid cooling causes the glass to shatter into small pieces. Next the frit is dried. Then it is ground into a fine powder. Magnetic separation removes any iron contamination. Classifying by particle size completes the manufacture of frit.

Porcelain Enamel Industry

The porcelain enamel industry has three divisions:

1. Sheet steel
2. Cast iron
3. Specialty

The base material is different for each division. There are also two other differences:

1. Preparation of base material
2. Method of frit application

All the base metals have one limitation. The smallest radius for bends is 3/16-in. (4.8 mm). During firing, the heat concentrates on the sharp edges, causing the porcelain enamel to flow away.

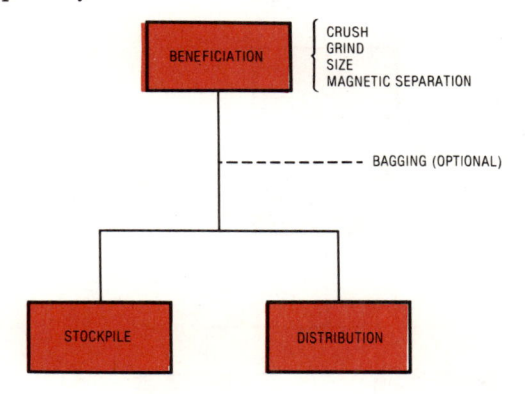

Fig. 67-1. A typical vertical flow sheet for transformation of raw materials used to make porcelain enamel frit. These processes are usually done by the ceramic raw material industry. Powdered raw materials are sold to frit manufacturers.

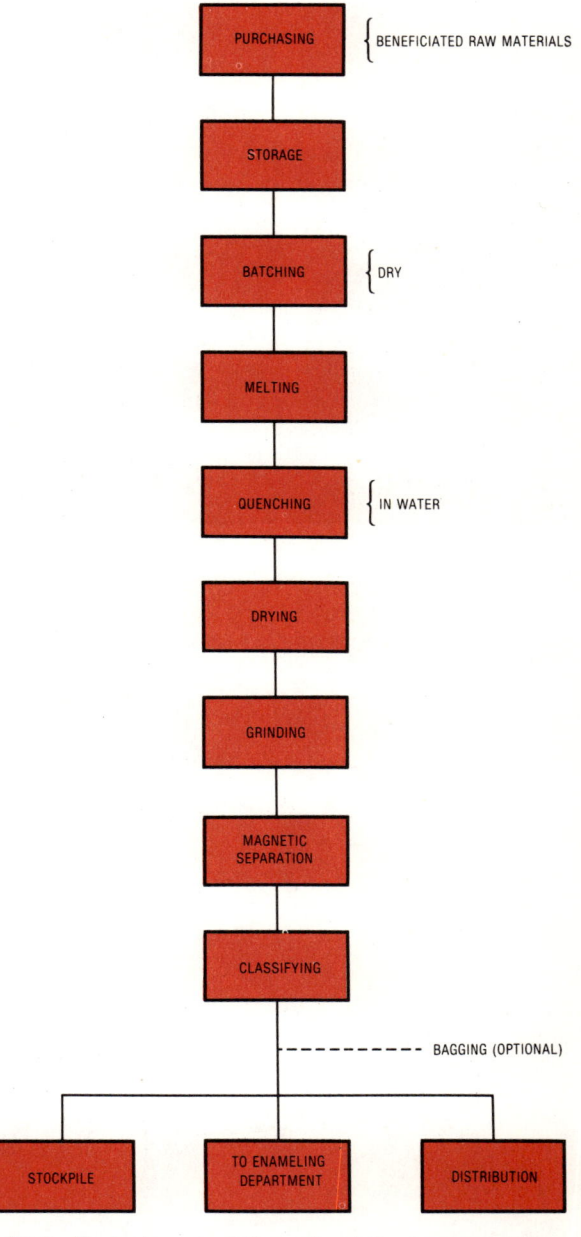

Fig. 67-2. Typical vertical flow sheet for transformation of powdered raw materials into frit. Some porcelain enamel product manufacturers make their own frit, while others purchase it. There are firms that specialize in making only frit.

240 / *Modern Industrial Ceramics*

Fig. 67-3. Molten frit pouring from a furnace. Note the use of external burners to keep the frit molten. (Courtesy Ferro Corporation)

Fig. 67-4. Molten frit is poured from the furnace into water. This quenching or rapid cooling causes the glass to shatter into small pieces. (Courtesy Ferro Corporation)

Sheet Steel

The manufacture of sheet steel products is diagrammed in Fig. 67-5. The forming department is responsible for shaping the base metal by bending, pressing, and similar processes.

The formed metal pieces are frequently moved through the enameling process on a conveyor. For example, the pieces are hung on an overhead conveyor. This makes mass production possible.

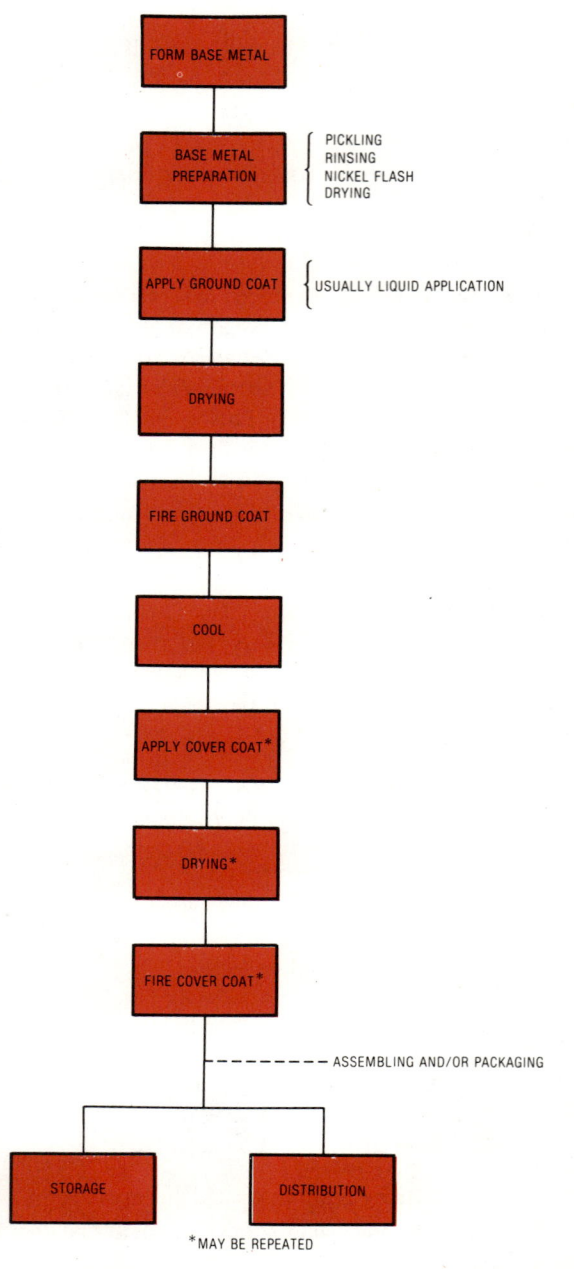

Fig. 67-5. Typical vertical flow sheet for the manufacture of sheet steel porcelain enameled products, such as appliance housings and cookware.

First, the base metal is cleaned to prepare it for enameling. Preparation usually involves four steps:

1. Pickling in a dilute acid to clean the metal
2. Rinsing in water to neutralize the acid
3. Nickel flash by dipping into nickel salts solution. The thin plating of nickel prevents further metal oxidation.
4. Drying

The ground coat of frit is usually applied as a liquid. The method of application is either spraying or dipping. The ground coat bonds to the base metal when fired. It usually contains cobalt. Cobalt gives the ground coat a blue color and helps fusion.

After drying, the ground coat is fired. Continuous tunnel kilns are frequently used. The pieces are moved through the tunnel kiln by conveyor. The firing time is only a few minutes. Fast firing prevents harm to the base metal. The temperature varies from 1400–2000°F (760–1093°C), depending on the size and thickness of the base metal.

The pieces are cooled before the cover coat is applied by spraying or dipping. After drying, the cover coat is fast fired. The cover coat is the exterior porcelain enameled surface. It may be either white or colored.

A very low-carbon sheet steel that requires only a cover coat is now available. Eliminating the ground coat reduces cost.

Cast Iron

The manufacture of cast iron products is diagrammed in Fig. 67-6. A foundry shapes the base metal (iron) by making castings. The cast iron piece is prepared for enameling by snagging and sandblasting. Snagging is a rough-grinding process.

Then the cast iron piece is preheated in a kiln to a red heat. When the hot piece is removed from the kiln, the ground coat of dry frit is immediately applied. The frit is applied by sifting through long-handled sieves. The frit fuses to the hot cast iron. It is smoothed out when the piece is returned to the kiln.

This process is repeated to apply the cover coat while the piece is hot. Then it is refired. The cover coat may be repeated.

Periodic kilns are primarily used for enameling cast iron. The firing time varies from a few minutes to 30 or more minutes, depending on the size and thickness of the product. Firing temperatures vary from 1500–2000°F (816–1093°C).

Product assembly and/or packaging, if necessary, completes the manufacturing.

Specialty

Specialty porcelain enameled products have two subdivisions:

1. Nonferrous metals
2. Glass

The manufacture of specialty nonferrous metal products is diagrammed in Fig. 67-7. The nonferrous base metals include aluminum, copper, gold, and silver. These metals have a chemical affinity for glass. Thus, a cobalt ground coat is unnecessary.

The base metal is shaped by cutting and forming. Precut blanks can be purchased. The typical base metal is 18-gauge copper. Most of the nonferrous base metals can be cleaned with steel wool.

A liquid protective coating is applied to the back surface and dried. The coating prevents oxidation of the metal when fired. Frit is applied with a small sieve, if dry, or with a brush, if liquid, to the top surface. Then, it is

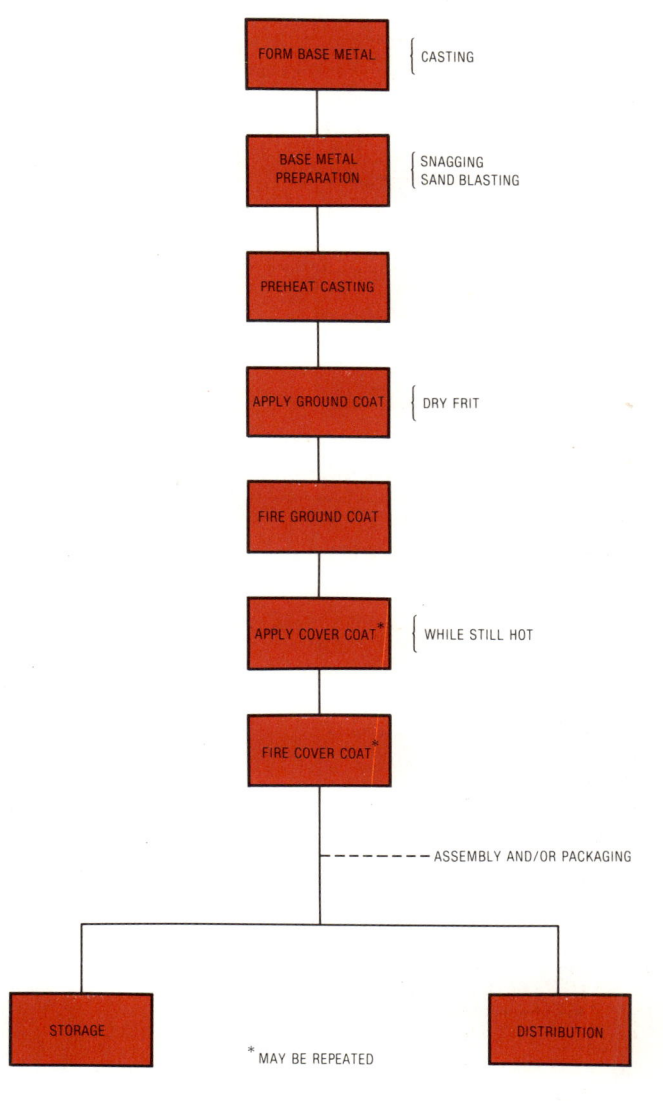

Fig. 67-6. Typical vertical flow sheet for the manufacture of cast iron porcelain enameled products, such as a bathtub.

242 / Modern Industrial Ceramics

dried if a liquid is used. The piece is fired in a hot periodic electric kiln for 2 to 3 minutes. Typical firing temperatures are:

1. Aluminum—1000°F (538°C)
2. Copper—1550°F (843°C)
3. Silver—1550°F (843°C)

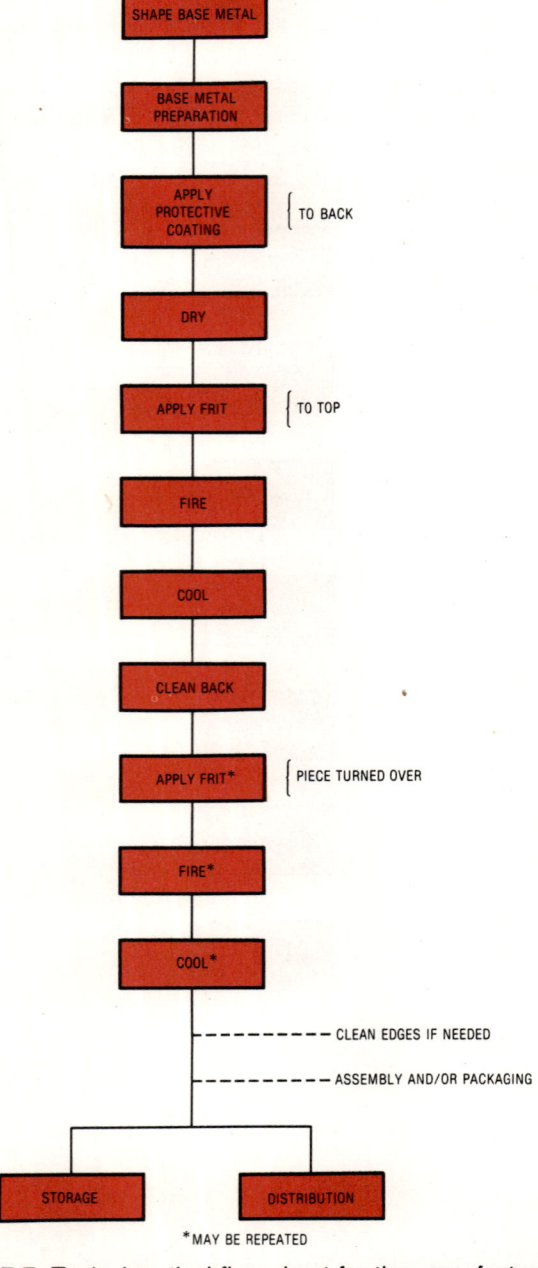

Fig. 67-7. Typical vertical flow sheet for the manufacture of specialty, nonferrous metal, porcelain enameled products, such as jewelry. Both surfaces, front and back, are enameled, as indicated in the flow sheet.

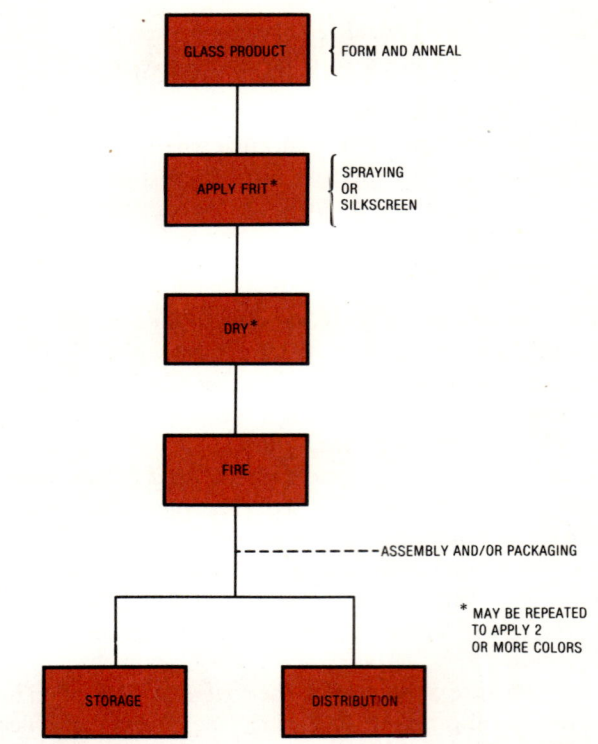

Fig. 67-8. Typical vertical flow sheet for the manufacture of specialty, glass, porcelain enameled products, such as colored glass blocks and bottle labels.

When the piece is cool, the surface that had the protective coating is cleaned with a dilute acid or steel wool. Frit is applied to the clean base metal. Then, the piece is refired for 2 to 3 minutes and cooled. The last two steps may be repeated. This allows a variety of designs.

Oxide forms on the edges of thin pieces during firing. This can be removed with a rubber abrasive. Assembling and/or packaging, if needed, complete the manufacture.

The manufacture of specialty glass products is diagrammed in Fig. 67-8. Forming and annealing of the glass products are discussed in Section XVII—*Glass Industry*. A newly made glass product is clean. Thus, base material preparation is not necessary.

Glass products are usually moved through the entire enameling process by a conveyor. The frit is mixed with a binder and water. This liquid mixture is applied by spraying. Glass blocks and flat sheets of glass are typical products enameled by spraying one coat.

The frit is mixed with an organic binder and drier for silk-screen application. Bottle labels and signs are typical products enameled by the silk-screen process. The drier quickly sets the color, so that additional colors may be immediately applied. Thus, the drying time is usually brief.

Continuous tunnel kilns are usually used to fire the pieces. The pieces may be moved through the kiln on a woven stainless steel wire conveyor. The firing temperature is 1000–1100°F (538–593°C). The firing time is longer than the 2-3 minutes used for the nonferrous metals. The firing time depends on the size of the glass product. Usually, it will take several hours.

The firing process is also different. The base metals are placed in a hot kiln and quickly fired. Glass must be heated and cooled gradually (Fig. 67-9). The firing process for glass starts at room temperature, is raised to maximum temperature, and is gradually cooled. Cooling re-anneals the glass. Assembly and/or packaging, if needed, complete the manufacture.

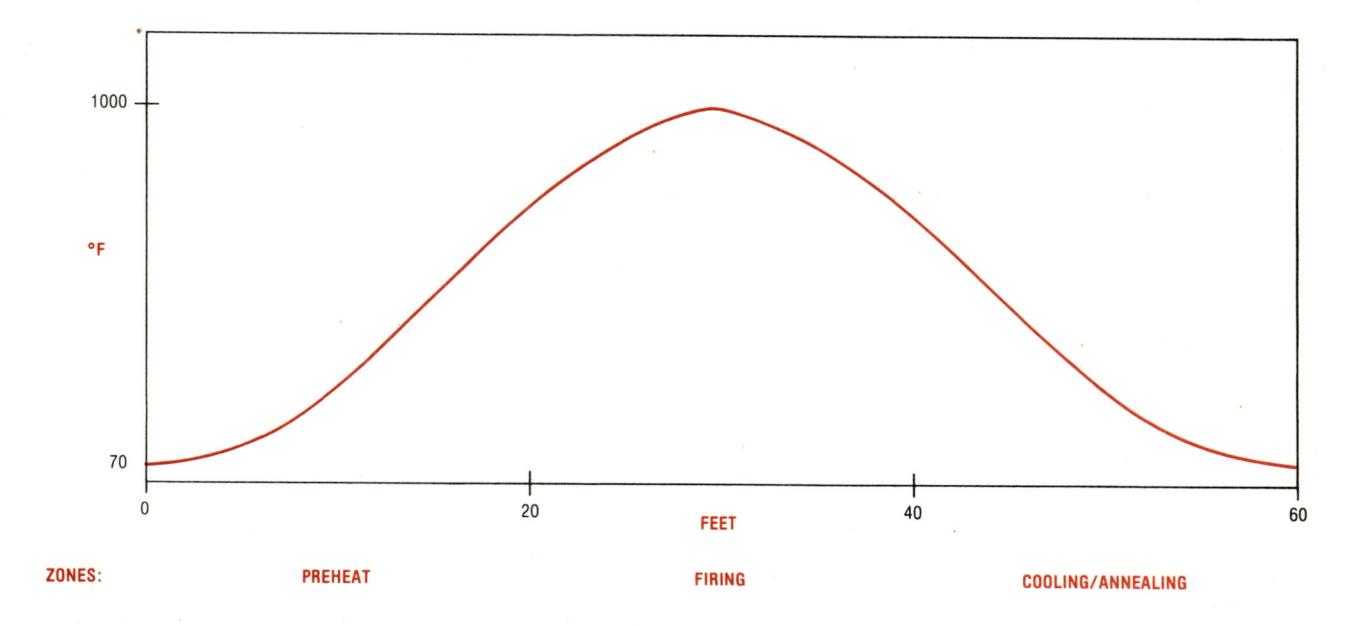

Fig. 67-9. A typical firing curve for porcelain enameled glass products. Note the three zones, if a tunnel kiln is used. These could be considered periods when the periodic kiln is used.

Unit 68: Utilization

Porcelain enamel frit is a manufactured product. Frit is used to make the porcelain enameled products.

Several firms specialize in making the frit. They sell the frit to the companies that produce the finished products. Some of the firms package small quantities of frit for sale to educational institutions, artists, and hobbyists.

Manufacture

The three divisions of the porcelain enamel industry were discussed in Unit 67; they are:

1. Sheet steel (Fig. 68-1)
2. Cast iron (Fig. 68-2)
3. Specialty (Fig. 68-3)

Each division manufactures porcelain enamel products.

Sheet Steel

Porcelain enameled products manufactured by the sheet steel division include:

1. Appliance housings and panels
2. Glass-lined tanks
3. Building panels
4. Sinks
5. Cookware

Appliance panels include those used for the interior and exterior, such as ranges, refrigerators, dishwashers, washers, and dryers (see Fig. 68-1). Hot water tanks are glass-lined tanks with a heating mechanism. The glass-lined tanks are used in the milk, beverage, and food industries. Cookware includes pots, pans, skillets, and trays.

Cast Iron

Porcelain enameled products manufactured by the cast iron division include:

1. Bathtubs
2. Sinks
3. Stoves

Porcelain enameled bathtubs are available in a variety of shapes and sizes. Sinks include those made for kitchens and bathrooms (see Fig. 68-2). Both wood and coal burning stoves use porcelain enamel, usually in dark colors.

Fig. 68-1. Appliance panels for both the interior and exterior of washers, dryers, dishwashers, ranges, and refrigerators are porcelain enamel. The wall and base cabinets in this laundry room are also porcelain enamel. Another ceramic product, quarry tile, was used for the floor. (Courtesy American Olean Tile Company)

Fig. 68-2. Cast iron sinks and bathtubs have porcelain enamel coatings. Note the rich color and glassy porcelain enamel surface. Another ceramic product, wall and floor tile, was used to construct this attractive bathroom. (Courtesy American Olean Tile Company)

Fig. 68-3. Two porcelain enamel products—a coffee mug and glass bottle label. The porcelain enamel label is fused by heat to the glass bottle. An attractive and durable label results.

Specialty

The porcelain enameled products manufactured by the specialty division include:

1. Nonferrous metals
    a. Jewelry
    b. Cookware
    c. Signs
2. Glass
    a. Blocks
    b. Chalkboards
    c. Bottle labels (see Fig. 68-3)
    d. Tumblers
    e. Kitchenware

Porcelain enameled jewelry is usually made with copper, silver, and gold. Aluminum is occasionally used for jewelry. Aluminum is frequently used to make porcelain enameled cookware, such as pots, pans, skillets, fondue dishes, and woks. A *wok* is a bowl-shaped cooking utensil used especially in the preparation of Chinese food. Signs may be made with aluminum.

Most of the porcelain enameled glass products listed are self-explanatory, except for the kitchenware. Kitchenware includes mixing bowls, casserole dishes, and baking dishes.

Construction

Manufactured porcelain enameled products may be used in construction. These can also be classified as:

1. Sheet steel
2. Cast iron
3. Specialty

Sheet steel includes building panels and silos. The building panels are flat. They are used for both interiors and exteriors, such as those of service stations. Silos are used for the storage of hay, corn, and grain. Feed mills and many farms, such as dairy farms, use silos.

Cast iron bathtubs and sinks are installed when a home or apartment building is being constructed (see Fig. 68-2).

Glass blocks are a specialty product. They are assembled and mortared together, as in a brick wall. Actually, they may be a part of a masonry wall. Glass blocks are usually used in place of a window, to provide direct or diffused light.

Unit 69: Disposition

Porcelain enamel products can be disposed of either by reclaiming, or by discarding. Ideally, everything would be reclaimed, rather than discarded. Unfortunately, discarding does occur.

Reclaiming

Products can be reclaimed either by reuse or by recycling. Both these methods are recommended.

Reuse

Bathtubs, sinks, and appliances can be sold as used products. Frequently, they are reused for their originally intended purposes in camps, summer homes, and renovation projects. They may be used in new ways; for example, bathtubs and sinks can be used as feed or watering troughs for livestock. Used refrigerators that can be repaired or still operate properly are reused as a second refrigerator in many homes and camps (Fig. 69-1). There are other uses for refrigerators that are not in operation; for example, they can be used as damp and drying cabinets for clay or for smoking meat and fish.

Glass-lined hot water tanks can also be reused in new locations, as in camps. Sometimes the tanks can be repaired and reused.

Used porcelain enameled silos can be disassembled, moved to a new location, and reassembled. Then, they could be used as originally intended.

Used or old porcelain enamel kitchenware can be reused as originally intended in a camp; or it may be collected as antiques, (Fig. 69-2). Even though this type of kitchenware is not a true antique, it is considered a collectible item.

Recycle

All the porcelain enameled metal products can be recycled for the metal. This includes sheet steel, cast iron, and the nonferrous metals. Individuals can initiate the recycling by selling these items to a scrap and/or junk dealer. The dealer stockpiles and eventually sells the various metals to the companies that use them. Usually, the metal is melted and molded into new shapes. Therefore, you would never recognize the dishwasher that was recycled into cookware.

Glass products can also be recycled. Many communities have collection bins for this purpose. Returnable bottles frequently have porcelain enamel labels. These bottles should be returned for the deposit money, rather than be recycled.

Discarding

Discarding is the easy way to dispose of a porcelain enamel product. Usually, it is also the expensive method for you. Since you are throwing away a product that can be recycled, you are throwing away money.

The discarding of old and broken products continues even today. If you must throw away an item, discard it properly! Improper discarding creates more problems.

Fig. 69-1. Old or used refrigerators can be reused as a second refrigerator or as a refrigerator for beverages.

Fig. 69-2. Old or used porcelain enamel kitchenware can be reused in camps, or it can be collected. Several large collections of this type of kitchenware have been assembled.

SECTION XVII

Glass Industry

Unit 70: Material Science

Four glassy materials enter into the ceramic products (Fig. 70-1):

1. Glassy bond
2. Glaze
3. Porcelain enamel
4. Glass

Glassy bond, glaze, and the porcelain enamels were discussed in Sections XII through XVI.

Glassy Material Industries

The major glassy material industries are glaze, porcelain enamel, and glass. Glaze and porcelain enamel are both glassy coatings. They are both relatively thin coatings. Glass is thicker than the glassy coatings. Therefore, the glass products are made of only one material, and they are homogeneous.

Definition

Glass is a product of fusion caused by high temperatures. It is cooled rapidly from the molten state to a rigid condition without the formation of crystals. Glass is usually transparent; it is an inorganic nonmetallic amorphous material. This is a shorter but less descriptive definition. Amorphous means without crystals (Fig. 70-2).

Fig. 70-1. Four glassy materials in ceramic products are glassy bond, glaze, porcelain enamel, and glass (left to right).

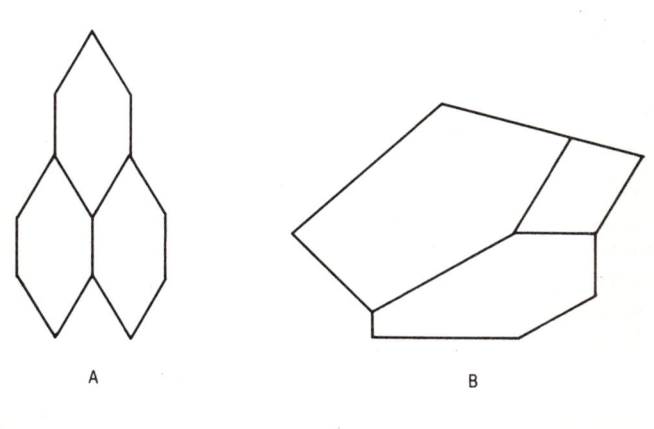

Fig. 70-2. A simplified representation of crystal and amorphous materials. A material containing crystals has order to its structure (A). Note how the structure repeats itself. A material without crystals does not have order to its structure (B). Note how the structure does not repeat.

247

Properties

Glass is chemically inert to most materials. The major exception is hydrofluoric acid (HF). This acid is used to chemically etch or frost glass, as in incandescent light bulbs, for example.

Some of the major physical properties of glass are listed in Table 70-1.

Table 70-1. Major Physical Properties of Glass

Physical Property	Explanation
1. State of matter	Supercooled liquid
2. Amorphous	Without crystals
3. Isotropic	Optical properties are the same in all directions
4. Flexible	Can be made flexible—for example, glass fibers
5. Brittle	Can be brittle or break into jagged pieces—for example, bottles
6. Compressive and tensile strength	One of the strongest materials

Materials

Both the natural raw materials and synthetic materials are used to manufacture glass. The natural materials were primarily used in the past. Today, more of the synthetic material is being used, due to recycling.

Raw Materials

The natural raw materials used in glass include:

1. Glass former
2. Flux
3. Decolorizer
4. Colorant
5. Firing agent
6. Strengthen
7. Brighten
8. Opacifier

The *glass former* is silica or silicon dioxide (SiO_2). The ideal raw material is very pure (99%). Silica has one major disadvantage. It melts at about 3110°F (1710°C). One type of glass, by its composition—fused silica, is made with only silica. All the other types of glass, by its composition, are made with the glass former and flux.

One or more *fluxes* are used to lower the melting temperature of silica. A common flux is soda ash or sodium carbonate (Na_2CO_3). Unfortunately, the glass made from silica and soda ash is water soluble. This is known as water glass or sodium silicate. Therefore, a second flux, limestone, is added to make the glass insoluble in water. The result is soda lime glass. The lead compounds, boron compounds, and dolomite are typical of the fluxes used to make the other types of glass.

The raw materials used to make glass may contain iron oxide as impurities. A small amount of iron oxide causes a green tint in glass. A *decolorizer* is added to complement the green tint. This makes a seemingly colorless glass. Manganese dioxide is frequently used.

Glass can be *colored* by metallic oxides and/or carbonates. Basically, the same colorants are also used in glazes and porcelain enamels. For example, cobalt carbonate makes a blue glass; a small amount of iron oxide makes a green glass; and manganese dioxide makes a purple glass.

Fining agent(s) help gather and expel the gas bubbles when a glass is molten. Potassium nitrate and arsenic compounds are typically used.

Glass is *strengthened* by using alumina or aluminum oxide (Al_2O_3). Alumina also acts as a stiffener when the glass is molten.

Lead compounds and barium carbonate are used to *brighten* glass. These are used in lead glass or crystal.

An *opacifier* reduces transparency until it is translucent or opaque. Zirconium, tin oxide, phosphorus, and antimony compounds are used.

Synthetic Materials

Cullet is the synthetic material used to make glass. It is the waste glass, rejects, and recycled glass products. Cullet speeds melting, since it has been previously melted.

Types of Glass

There are two types of glass—natural and synthetic. Natural glass includes obsidian, fulgurites, and teletites. There are six major types, by their composition, of synthetic glass (Table 70-2); they are:

1. Soda lime
2. Lead
3. Borosilicate
4. 96% silica
5. Fused silica
6. Aluminosilicate

Table 70-2. Types of Glass by its Composition

Type Of Glass	Typical Materials	Notes	Typical Uses
1. Soda lime	Silica sand Soda ash Limestone Cullet	Made with inexpensive materials. Mass production. Used for about 90% of all glass products.	Containers, bottles, window glass.
2. Lead	Silica sand Litharge Soda ash Pearl ash Soda niter Feldspar	Clarity, brilliance, and resistance to electricity.	Crystal glassware, eye glasses.
3. Borosilicate	Silica sand Borax Boric acid Aluminum oxide	Resistance to temperature change, acid, and electricity.	Oven ware, laboratory glassware.
4. 96% silica	Silica sand Borax	High resistance to temperature change, acid, and electricity. High ultraviolet transmission.	Chemical ware, sun lamps.
5. Fused silica	Pure silica sand	Electrical insulation and resistant to severe temperature changes.	Aircraft camera high-frequency electrical insulation
6. Aluminosilicate	Silica sand Aluminum oxide Borax Boric acid Limestone Magnesium oxide	Extreme hardness and heat resistance.	Top-of-stove ware, high-temperature thermometers.

Unit 71: Procurement

Procurement of raw materials for the glass industry is diagrammed in Fig. 71-1. Glass can be made from one or more materials. In fact, six or more materials may be used to make glass.

The number and variety of materials used in glass, as discussed in Unit 70, usually makes purchasing, rather than procurement, more feasible. Thus, few if any of the glass manufacturers are involved in procurement. Procurement, for these companies, becomes a purchasing job rather than an extraction process.

Ceramic Raw Material Industry

The flow sheet (see Fig. 71-1) also represents the ceramic raw material industry. This industry specializes in the extraction type of procurement, beneficiation processes, and selling materials to other manufacturers. The glass manufacturers are among those ceramic industries that frequently purchase raw materials.

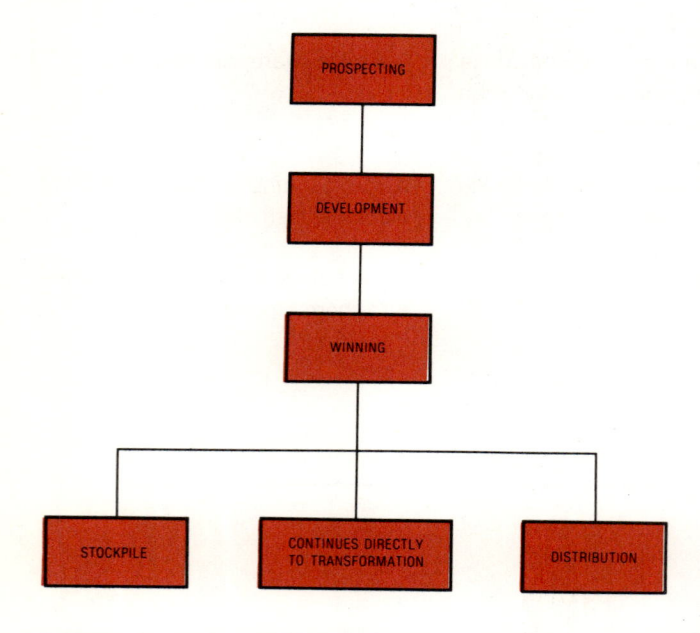

Fig. 71-1. A typical vertical flow sheet for procurement of raw materials used to make glass. These processes may be done by the glass industry, but few companies do their own procurement. The ceramic raw material industry does most of the procurement, some transformation (for example, beneficiation), and sells powdered raw materials to the glass industry.

Prospecting, developing, and winning were discussed in Section III—*Ceramic Raw Material Industry*. All three types of winning may be used, due to the variety of raw materials needed. Typical procurement processes are shown in Figs. 71-2 and 71-3.

Stockpiles

Glass manufacturers seldom stockpile the won raw materials. This type of stockpiling is done by the ceramic raw material industry (Fig. 71-4). The beneficiated raw materials are stockpiled by the glass manufacturers. Beneficiation is the first transformation process. Transformation is discussed in Unit 72.

Fig. 71-2. A small-core drilling trailer-mounted unit being used for a limestone survey. This is typical of equipment used for prospecting. (Courtesy New York College of Ceramics at Alfred University)

Fig. 71-3. A "shot" of explosives blasting rock loose from the face of a quarry. Blasting is frequently used in superficial winning. (Courtesy National Limestone Institute, Inc.)

Procurement / 251

Fig. 71-4. Typical open-air stockpiles of raw materials. (Courtesy National Limestone Institute, Inc.)

Unit 72: Transformation

Transformation of raw materials is diagrammed in Fig. 72-1. These processes are usually done by the ceramic raw material industry. Most of the glass manufacturers purchase their powdered materials from this industry. These powdered raw materials and cullet are used to make glass.

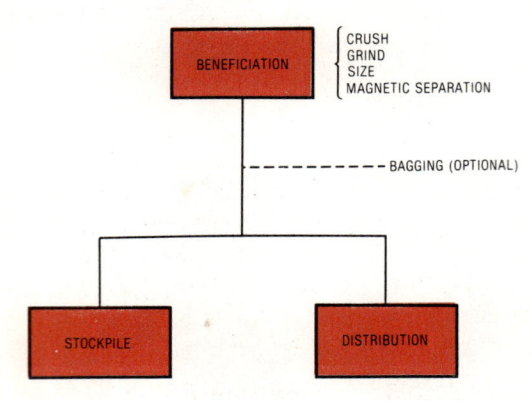

Fig. 72-1. A typical flow sheet for transformation of raw materials used to make glass. These processes are usually done by the ceramic raw material industry. Powdered raw materials are sold to glass manufacturers.

Cullet

Cullet is composed of waste glass, rejects, and recycled glass products. It is usually subjected to beneficiation similar to that outlined in the flow sheet in Fig. 72-2. Removal of metal caps, lids, and rings is necessary when using recycled glass products.

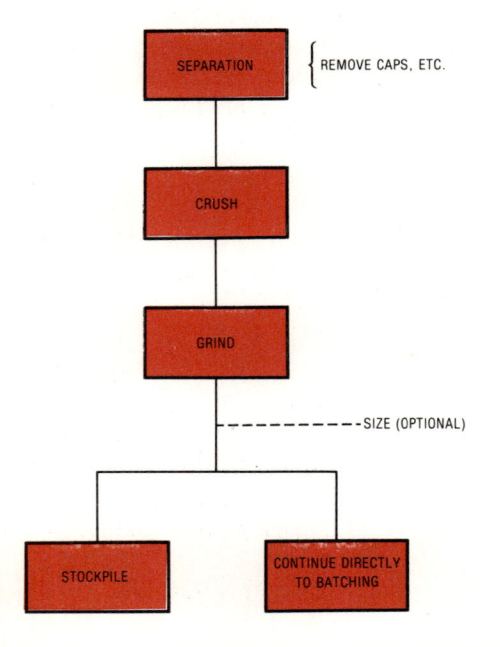

Fig. 72-2. A typical vertical flow sheet for the beneficiation of *cullet*. Recycled glass products are frequently used for cullet. These processes may be done at the glass plant.

Flow Sheets

Transformation of materials into glass products is diagrammed in Fig. 72-3. This is further illustrated by the pictorial flow sheet for the manufacture of glass bottles (Fig. 72-4).

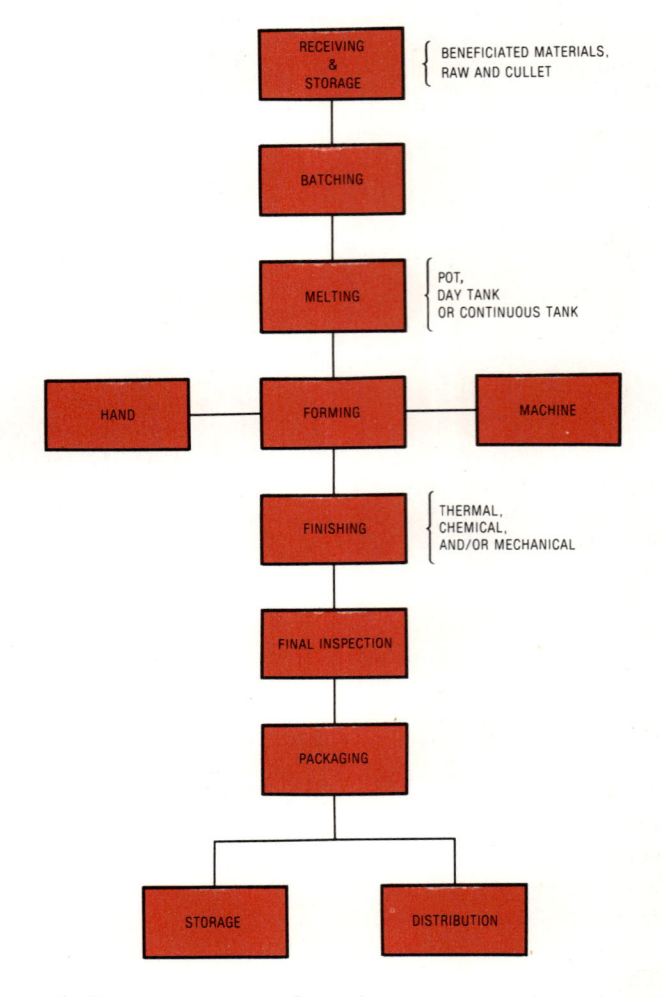

Fig. 72-3. Typical vertical flow sheet for the transformation of materials into glass products.

Beneficiated materials, both the raw materials and cullet, are received at the glass plant. Each material is stored in a separate bin or silo. The cullet may be beneficiated at the glass plant by a separate department. Then it is transferred to storage.

Batching involves weighing and mixing the materials. Usually, the glass batch is dry mixed.

Melting

The glass batch is melted in a refractory container. The container may be a glass pot, day tank, or continuous tank. The type of container is determined by the amount of molten glass that is needed.

252

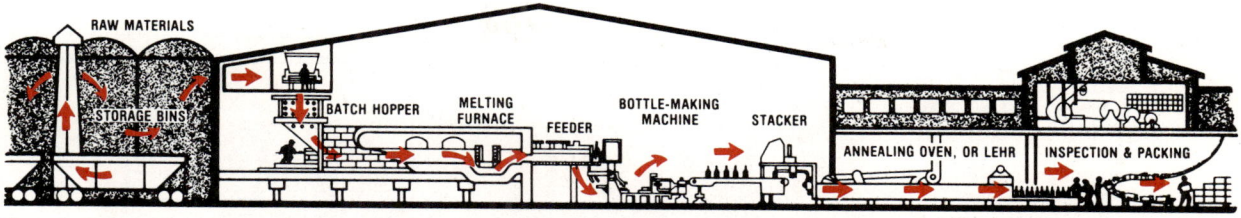

Fig. 72-4. A pictorial flow sheet for the manufacture of glass bottles. This basic flow sheet would be nearly the same for all the glass products. Only the glass forming machine, labeled bottle-making machine above, would change for other glass products. (Courtesy Glass Packaging Institute)

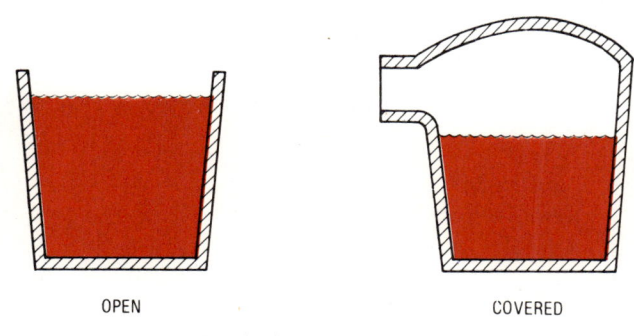

Fig. 72-5. Full-section drawings of glass pots. They are open (left) and covered (right). A refractory crucible can be used as an open glass pot.

Glass Pot

Glass pots are not widely used today. They are not capable of producing large quantities of molten glass.

The two types of glass pots are open and covered (Fig. 72-5). Crucibles are used as small, open glass pots. These are used for test melts and small batches. A crucible containing the powdered glass batch is placed inside a kiln and heated until molten. Then, the crucible is removed from the kiln for pouring or other processing.

Several closed pots may be set inside a circular furnace. They remain in place. The opening (see Fig. 72-5) is used for charging and removing the molten glass. Charging means to fill with the powdered-glass batch.

Glass pots are periodic. They have to be charged, melted, the glass removed, and recharged.

The pots are used for tests, small batches of colored glass, optical glass or special batches, and hand forming.

Day Tank

A day tank is stationary and periodic. The glass is melted in batches like the pot. The major differences between a pot and a day tank are:

1. Pots are movable. Day tanks are stationary.
2. Heating—pots are heated from all directions. Day tanks are heated from the top. Thus, the glass is heated from the top or exposed surface downward.

The name *day tank* is derived from the time required to remove the molten glass—a normal working day. Then, the tank is recharged, heated, and the molten glass is ready for the next working day. Today, the day tanks are frequently used for hand forming glass products.

Continuous Tank

This tank operates continuously. It is used to make large quantities of molten glass for machine forming.

Basically, it is a large refractory tank resembling a covered Olympic-size swimming pool. The tanks are built of refractory bricks that have a chemical nature similar to glass. This is necessary, since the glass is in contact with many of the refractories.

Heat is supplied by side burners with an exhaust on the opposite side. The flames flow across the top surface of the glass. The direction of the heat flow is reversed about every 30 minutes. Natural gas or oil is used for the fuel.

The batched materials are fed into one end, called a doghouse. The batch becomes molten in the melting section of the tank. This is the largest part of the tank. A bridge and throat prevent the floating material from passing into the fining section. The molten glass is then fed through the forehearth to the forming machines. (Fig. 72-6).

Forming

Today, the molten glass is formed into products either by hand or by machines. The first synthetic glass products were hand formed.

Hand Forming

The phrase *hand forming* is misleading. The molten glass is not actually formed with the hands. The forming is done with hand tools.

Hand glass forming is also known as "offhand" or "freehand" glass blowing. The four major types of hand glass forming are: (1) blowing, (2) pressing, (3) drawing, and (4) special methods.

The major steps for hand forming most of the glass products include:

1. Gather molten glass blowpipe
2. Roll on a marver to give it a preliminary shape

253

254 / *Modern Industrial Ceramics*

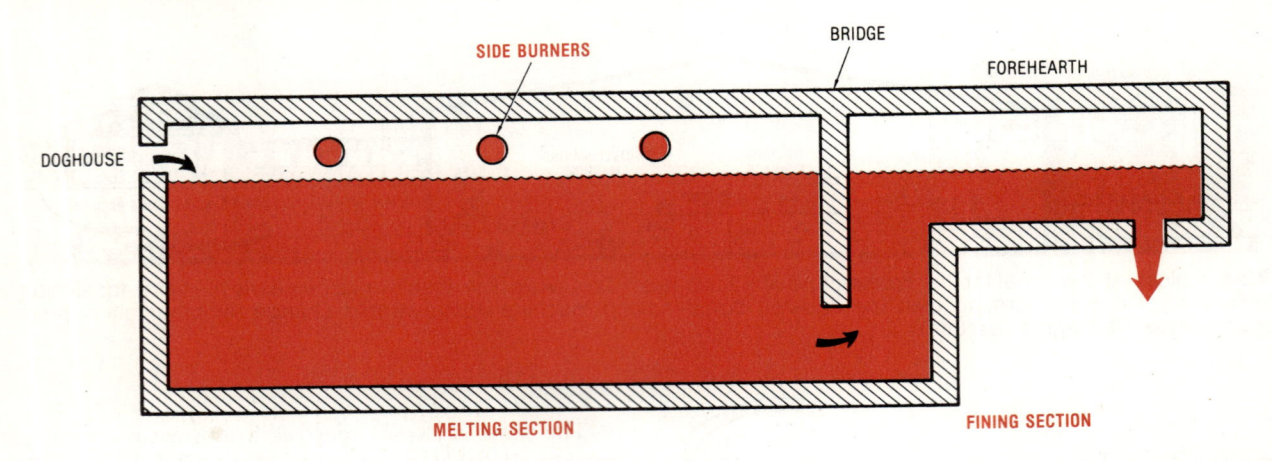

Fig. 72-6. A simplified elevation drawing of a continuous glass tank. It is a large refractory tank resembling a covered Olympic-size swimming pool.

3. Reheat
4. Shape—blowing
5. Repeat reheating and shaping if necessary
6. Transfer to a pontil
7. Reheat
8. Finish edge
9. Remove from the pontil
10. Anneal

These steps are diagrammed in Fig. 72-7.

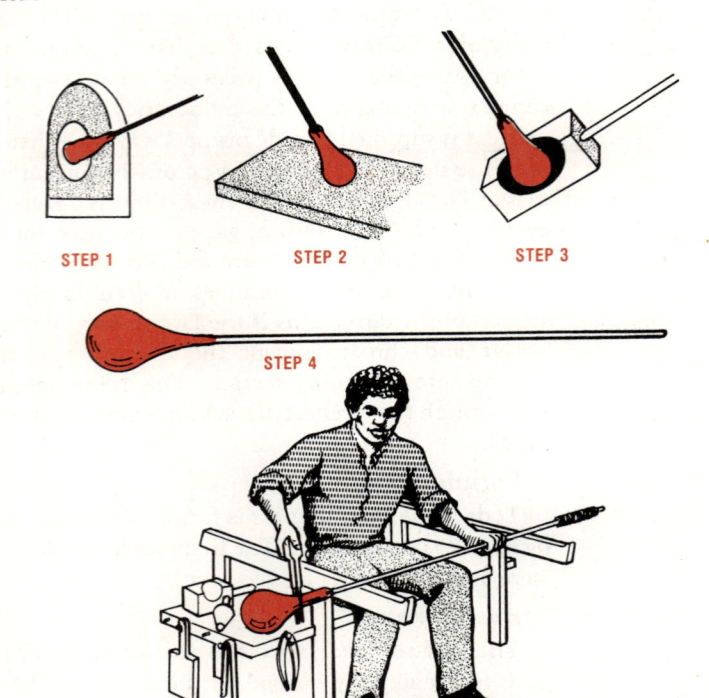

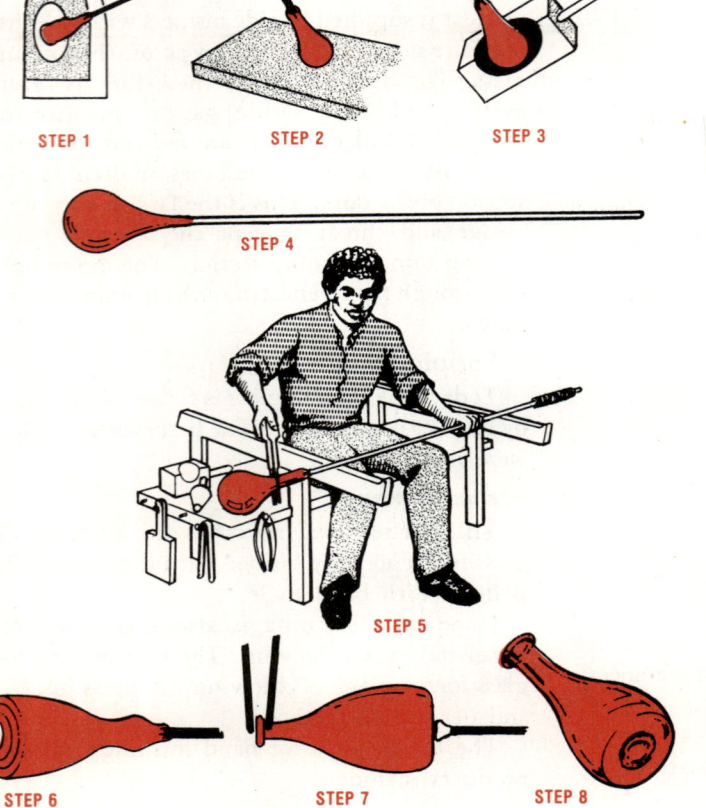

Fig. 72-7. Glass blowing includes many steps typical of hand forming:
- Step 1. Gathering and reheating
- Step 2. Rolling on marver to give it preliminary shape
- Step 3. Shaping in hollowed out wooden block
- Step 4. Air is blown into the blowpipe to make a bottle. The blowpipe is a hollow iron pipe that is bell-shaped at the gathering end.
- Step 5. Glass blower's bench. The arms allow the blowpipe to be rotated. This keeps the molten glass in shape. A "jack" is being used to narrow the bottle's neck.
- Step 6. Transferring to pontil that has a small amount of molten glass on the end. The pontil is a solid-iron rod.
- Step 7. Finishing the top edge after removing the blowpipe.
- Step 8. Removing the pontil leaves a scar that may be ground off after annealing.

(Courtesy Glass Packaging Institute)

Transformation / 255

Machine Forming

A continuous tank is used to provide molten glass for machine forming. Feeding molten glass from the forehearth of the tank is done by several methods. One method is the *gob feeder*. The molten glass flows through an opening in the bottom of the forehearth. This is cut off in "gobs" by shears (Fig. 72-8). A metal chute guides the gob to the forming machine.

The major types of machine glass forming are:

1. Pressing
2. Blowing
3. Drawing
4. Rolling
5. Float
6. Special

Pressing—Glass products formed by pressing include pie plates and baking dishes. The special iron or steel mold is heated and coated with oil. A heated mold aids forming. The oil acts as a separator. The mold has three parts—male, female, and ejector. The male part presses a gob of glass into the female or cavity part. The bottom of the cavity is a separate part that ejects the object after pressing.

Blowing—Glass containers are formed by blowing. These include bottles and jars. The major types of glass blowing machines are:

1. Press-and-blow (Figs. 72-9 through 72-11)
2. Blow-and-blow
3. Vacuum-and-blow
4. Ribbon (Fig. 72-12)

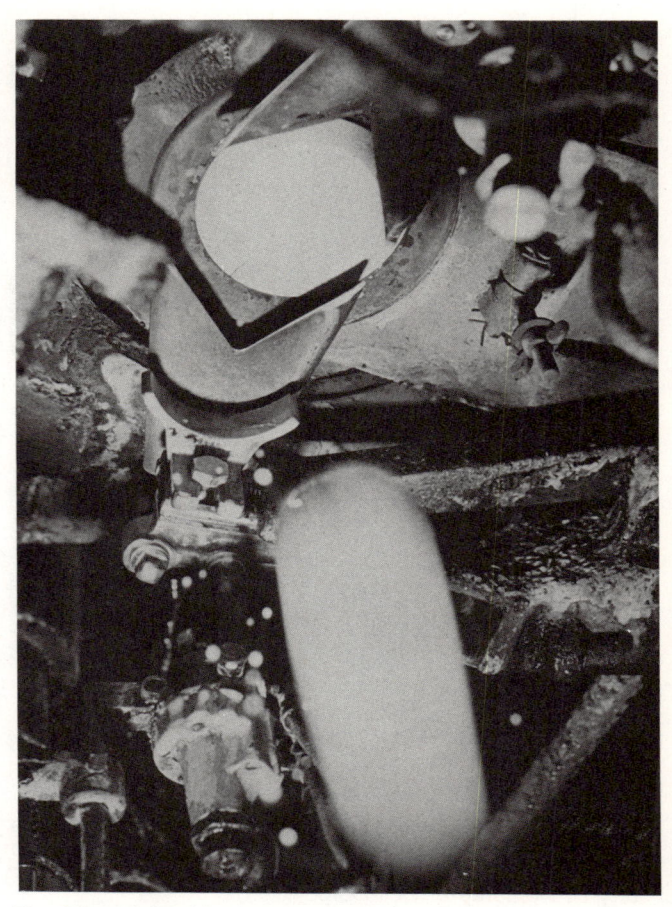

Fig. 72-8. A "gob" of molten glass has just been cut off by shears. Note the opening in the bottom of the forehearth. A metal chute will guide the bog to the forming machine. (Courtesy Glass Packaging Institute)

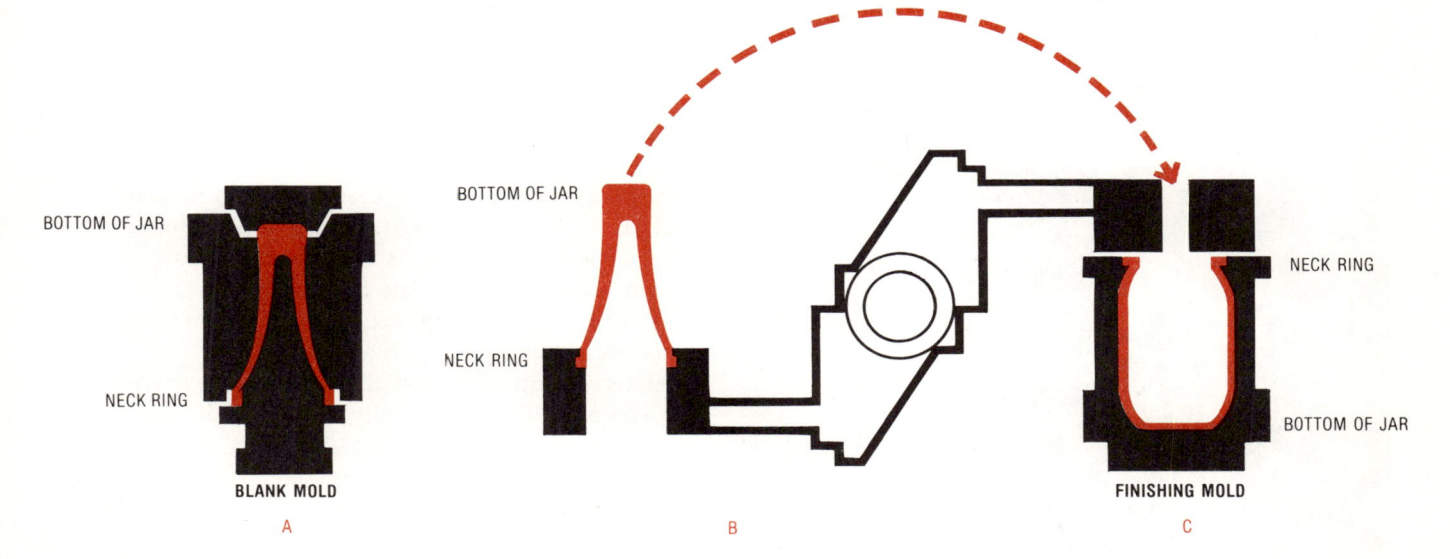

Fig. 72-9. The press-and-blow machine process. The gob is pressed to form the parison in the blank mold in an upside-down position (A). The blank mold opens and flips the parison over by its neck ring (B). The parison is reheated during transfer into the finishing mold where compressed air blows the glass to the final shape (C). (Courtesy Glass Packaging Institute)

Fig. 72-10. Transferring parisons into finishing molds in a press-and-blow machine. Compressed air then blows the glass into the final shape complete with trade name. (Courtesy Gary D. Demaree of Ball Corporation)

Fig. 72-11. Transferring newly formed jars from the finishing mold to the conveyor that moves them to the lehr for annealing. (Courtesy Gary D. Demaree of Ball Corporation)

These machines also use special iron or steel molds that are heated and coated with oil. The first two listed are gob fed. The first three machines listed use two types of molds—blank and finishing. The ribbon machine uses only the finishing mold.

All four of the machines make containers in two stages—*parison* and *blowing*. A parison is the initial shape. Air is blown into the molten glass in the second stage.

The basic glass blowing machine process involves (see Fig. 72-9):

1. Parison formed, usually in a blank mold.
2. Parison reheated during transfer to the finishing mold.
3. Container blown to final shape in finishing molds.

The *press-and-blow* machine is gob fed. The parison is formed by pressing. Wide-mouth and regular jars are made by this machine (see Figs. 72-10 and 72-11).

The *blow-and-blow* machine is also gob fed. The parison is formed by blowing. Pint, quart, and larger bottles are made by this machine.

The *vacuum-and-blow* machine pulls the molten glass from the forehearth into the blank mold by vacuum. The parison is formed by blowing. A variety of bottle sizes are made by this machine.

The *ribbon* machine is fed a glass ribbon. Molten glass flows from the bottom of the forehearth through two rollers. The rollers produce the ribbon that is conveyed to a blow head. Then, the parison is formed by blowing without a blank mold (see Fig. 72-12). Incandescent lamp envelopes and the spheres used for Christmas ornaments are made by this machine.

Drawing—Window or sheet glass, rod, and tubing are formed by drawing. Sheet glass is drawn either horizontally or vertically. The horizontal method draws a continuous sheet upward and out of the forehearth over rollers that feed it into the lehr for annealing. After annealing, the continuous sheet is cut into standard sizes and packed 50 square feet (4.6 square meters) to each box. The vertical process means that the continuous sheet is drawn vertically out of the forehearth and through the lehr. This requires a height equal to the height of a two- or three-story building.

Glass rod and tubing are also drawn either horizontally or vertically. They are usually drawn continuously from the forehearth and through the lehr. Then, they are cut to standard length (4 feet) (1.2 meters) and packaged. Tubing is made by drawing the molten glass over a bell that introduces compressed air to maintain the inside diameter.

Rolling—Plate glass is formed by rolling. The molten glass continuously flows horizontally from the forehearth through two large rollers and through the lehr. Then,

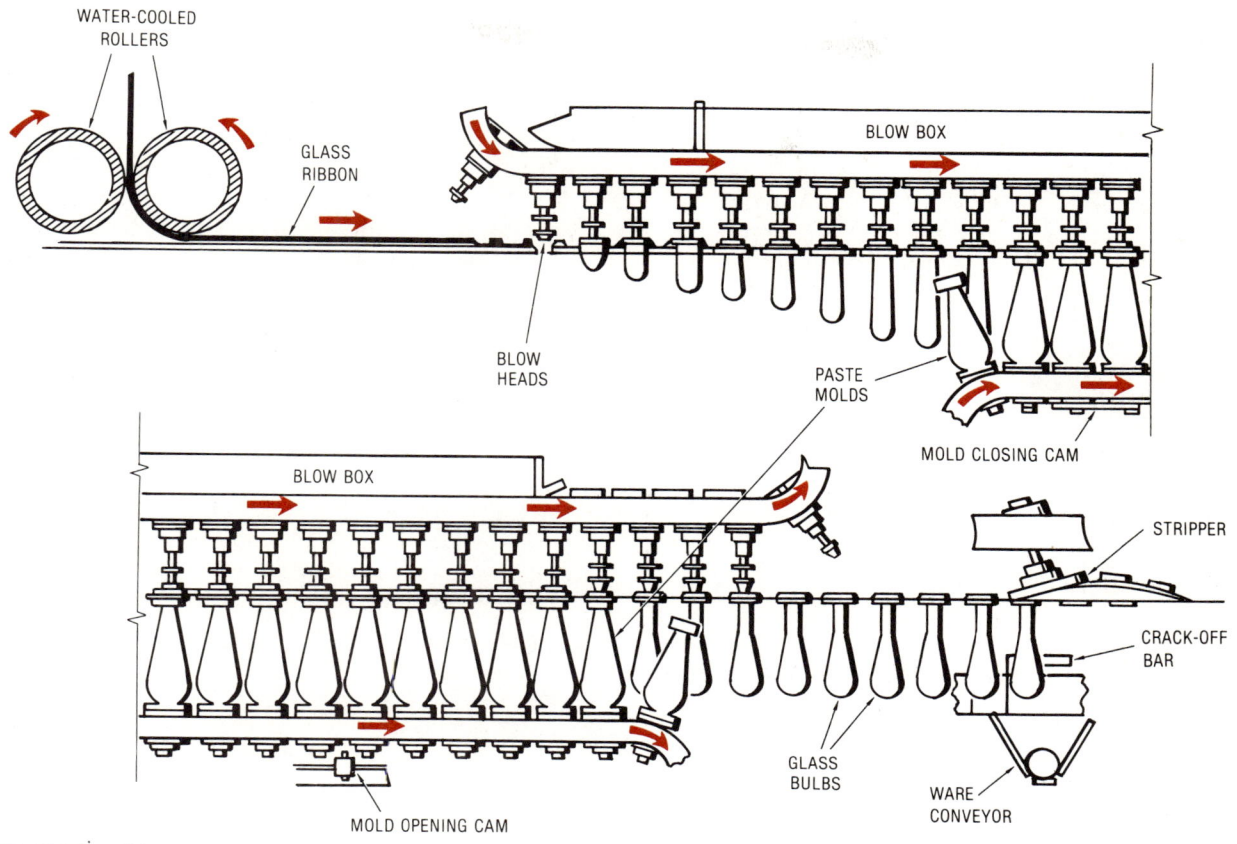

Fig. 72-12. The ribbon machine forms parisons without a mold. The final shape is blown in a finishing mold. The product has a thin wall, such as incandescent lamp envelopes and the spheres for Christmas ornaments. (Courtesy Corning Glass Works)

very large pieces are cut. These are ground and polished on both surfaces to make them parallel or optically perfect.

Float—Plate glass is also formed by the *floating* process. First, the glass is rolled to thickness. Then, it passed through the float unit and lehr as a continuous piece. As plate glass, it is then cut and stored.

The glass floats on molten *tin* inside the float unit. This fire polishes one glass surface. The top surface is fire polished by the controlled atmosphere. Then, the glass is cooled and conveyed directly into the lehr. Thus, the float process eliminates grinding and polishing.

Special—There are two types of glass fibers—continuous and discontinuous. The *continuous* glass fibers are used for textiles. The *discontinuous* fibers are used for glass wool insulation. Continuous glass fibers are formed by drawing (Fig. 72-13). Discontinuous glass fibers are formed by blowing (Fig. 72-14). These glass fibers are not annealed.

Funnel-shape glassware is formed by centrifugal casting—for example, a television picture tube funnel. This machine is gob fed. The gob is dropped into a spinning mold that creates centrifugal forces which make the glass flow upward (Fig. 72-15). This forms a wall of uniform thickness.

Finishing

Almost all the glass products require one or more finishing processes. The types of processes are:

1. Thermal
2. Chemical
3. Mechanical

The term *thermal* means that heat is used; the term *chemical* means that a liquid chemical is used; and the term *mechanical* means that tools and/or machines are used to remove glass during finishing.

The most frequently used finishing is the thermal process of annealing. In fact, after forming, nearly all the glass products go immediately to the lehr for annealing (Fig. 72-16).

During annealing, the glass is reheated slightly and slowly cooled to remove stress (Fig. 72-17). Forming often creates the stress. Thus, annealing is a controlled cooling process.

A lehr operates continuously. It resembles a straight-line tunnel kiln. There are three ways in which lehrs differ from tunnel kilns:

258 / *Modern Industrial Ceramics*

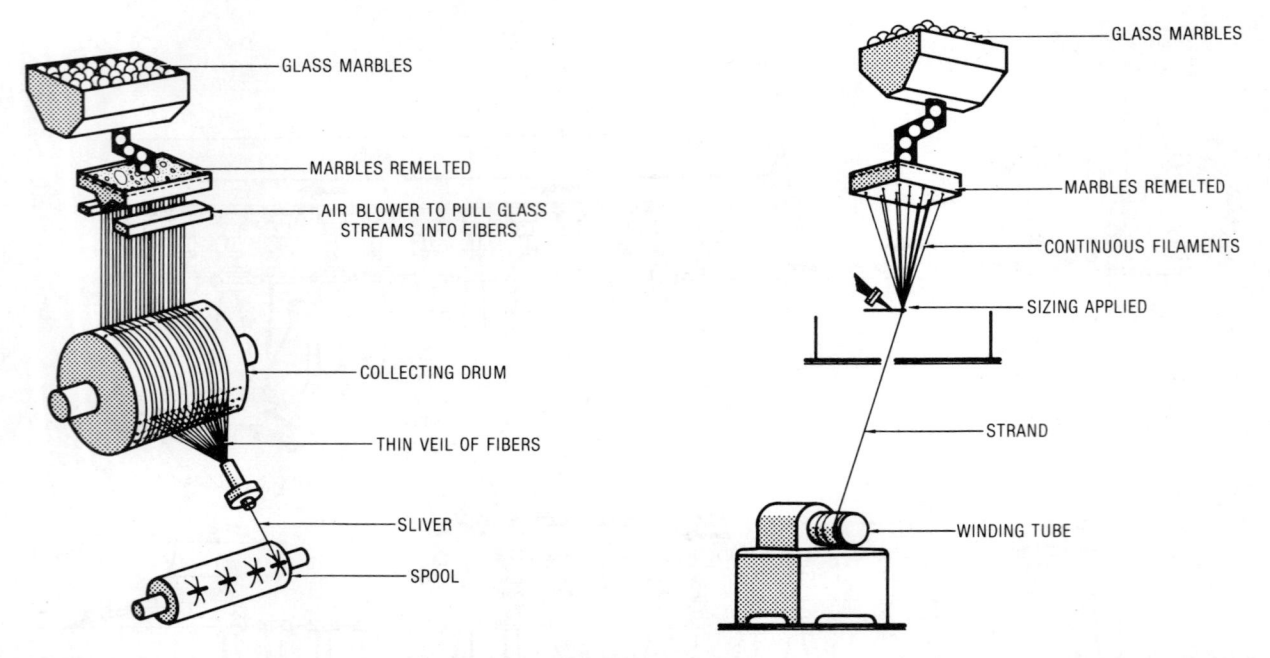

Fig. 72-13. Two methods of making continuous glass fibers. Both use glass marbles that can be easily fed to the remelting unit. Continuous fibers are used for textiles. (Courtesy Owens-Corning Fiberglas Corporation)

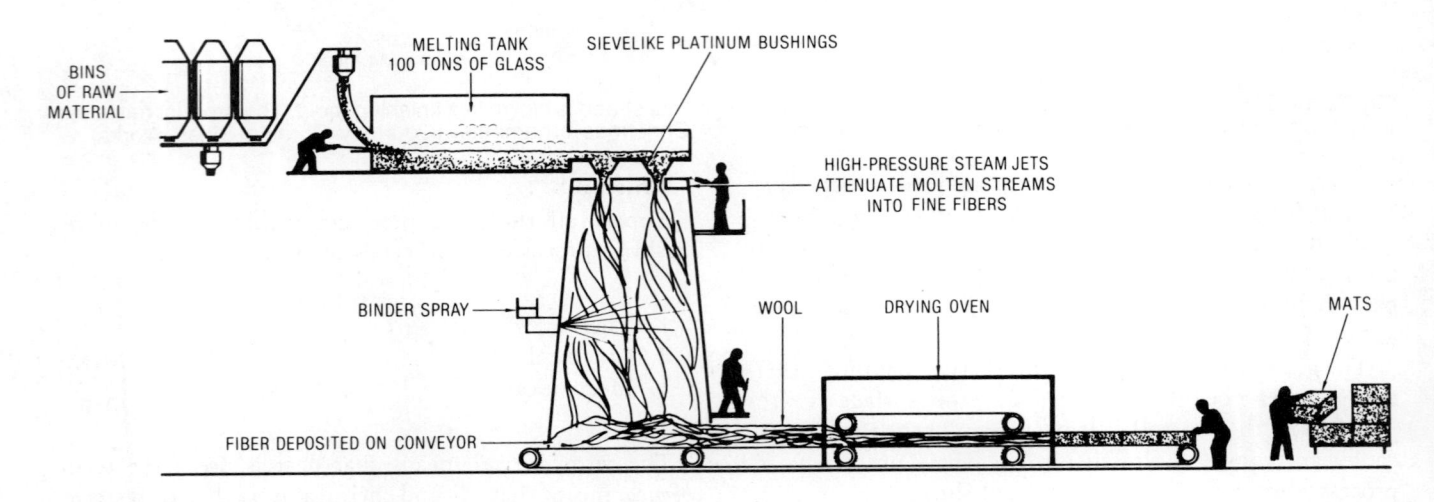

Fig. 72-14. Making discontinuous glass fibers by blowing. These are used as glass wool insulation. (Courtesy Owens-Corning Fiberglas Corporation)

1. Conveyance—Usually, a continuous stainless steel belt moves the glass products (Fig. 72-16).
2. Temperature—Usual maximum temperature is about 1000°F (538°C).
3. Entrance temperature—A lehr is hot at its entrance and gradually cools through its length (Fig. 72-17).

Other finishing processes that glass products require are listed in Table 72-1.

Completion

Final inspection, packaging, storage, and distribution complete the manufacture of glass products. Actually, inspection occurs at several intervals during the transformation processes (Figs. 72-18 and 72-19).

Glass–Ceramic

Transformation of materials into glass–ceramic products is diagrammed in Fig. 72-20. The name *glass–ceramic* means that the materials have been melted, formed as glass, and converted by special treatment to become crystalline. These products begin as a special glass batch that includes a nucleating agent such as titanium dioxide. A nucleating agent converts into tiny crystals when heated. The materials are batched, melted, formed, and finished, like the other glass products. They are a transparent amber color after finishing or annealing. Then, a second thermal finishing process called *ceraming*, a reheating process, changes the glass to an opaque crystalline ceramic.

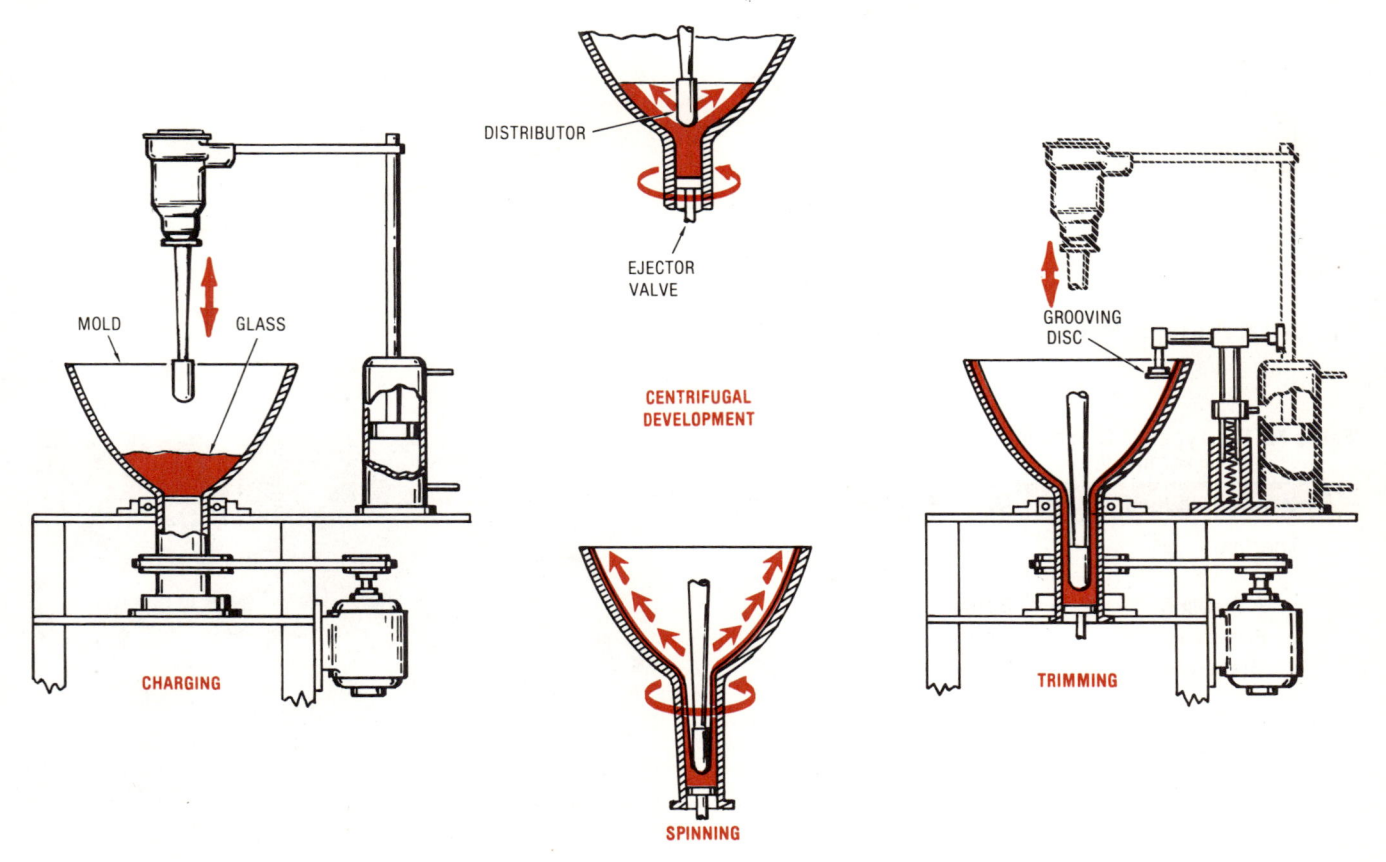

Fig. 72-15. Centrifugal casting television picture tube funnel. The steps are charging (left); spinning, which causes centrifugal development (center); and trimming (right). (Courtesy Corning Glass Works)

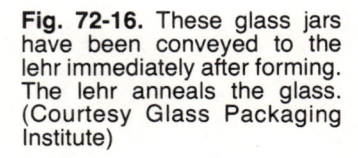

Fig. 72-16. These glass jars have been conveyed to the lehr immediately after forming. The lehr anneals the glass. (Courtesy Glass Packaging Institute)

260 / *Modern Industrial Ceramics*

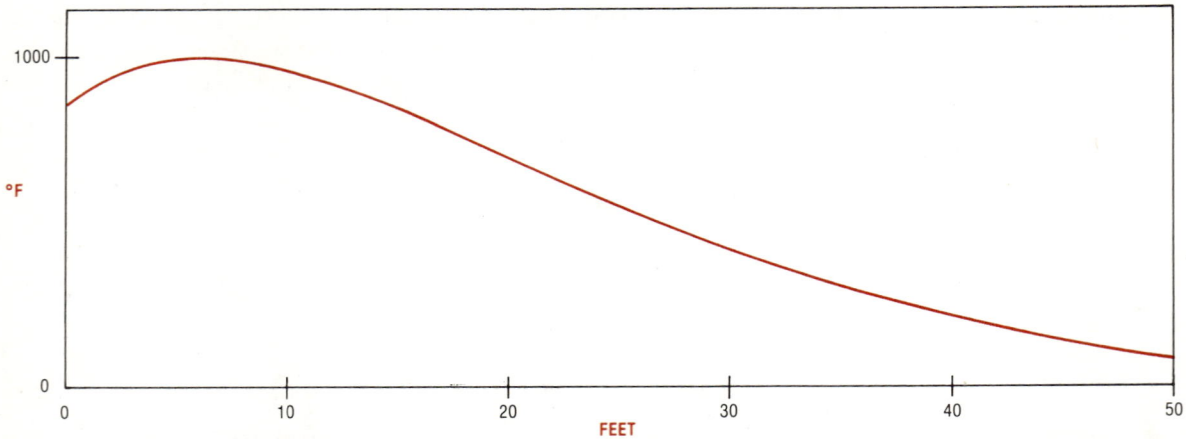

Fig. 72-17. A typical annealing curve for a lehr. The horizontal axis (in feet) represents the length of the lehr.

Table 72-1. Types of Glass Finishing Processes With Major Examples

Finishing Process	Glass Product Example
1. Thermal	
a. Annealing	Bottles
b. Fire polishing	Drinking glasses
c. Glass-to-metal seal	Incandescent lamp filament
d. Lampworking or scientific glass blowing	Novelty items, such as animals Condensers
e. Metalizing	TV picture tube screen
f. Porcelain enamel	Bottle labels
g. Sealing	Car windshield (laminated safety glass)
h. Slumping	Car windows
i. Tempering	Doors (tempered safety glass)
j. Trimming	Laboratory beakers
2. Chemical	
a. Acid etching	Incandescent lamp envelope
b. Staining	Heat lamp bulbs
c. Tempering	Train windows
3. Mechanical	
a. Cutting	Window or sheet glass
b. Engraving	Crystal
c. Grinding/polishing	Lenses
d. Sandblasting	Trademarks

Fig. 72-18. Electronic inspection for defects. This is usually performed after the annealing process in the lehr. (Courtesy Gary D. Demaree of Ball Corporation)

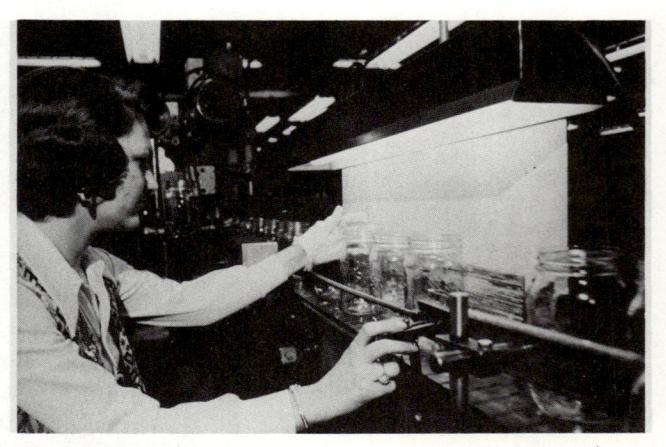

Fig. 72-19. Final visual inspection of the product prior to packaging. (Courtesy Gary D. Demaree of Ball Corporation)

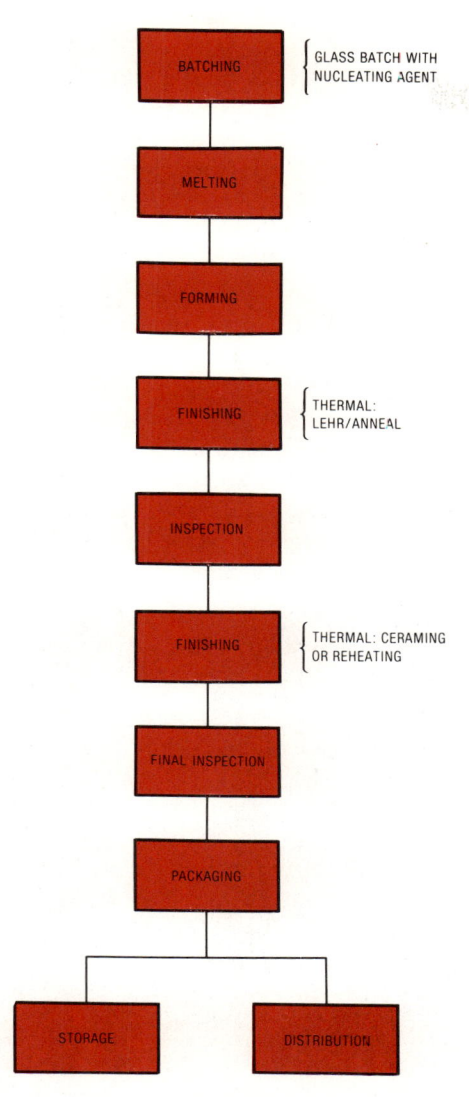

Fig. 72-20. Typical vertical flow sheet for the transformation of materials into glass-ceramic products. These products are used in the home, industry, and military.

Unit 73: Utilization

This unit is concerned with synthetic glass. It is a manufactured product.

Manufacture

Manufactured glass products function as containers, flat, or specialty items. Most of these products are containers.

The containers include bottles and jars in a large variety of colors, shapes, and sizes (Fig. 73-1). Glass dinnerware, cooking dishes (Fig. 73-2), drinking glasses, mugs, ash trays, incandescent bulbs, Christmas ornaments (Fig. 73-3), and laboratory ware are also containers. These are only some of the many glass containers in use today.

Fig. 73-1. Glass containers, such as these bottles and jars, are made in a large variety of colors, shapes, and sizes. Food, beverages, cosmetics, and drugs are some of the products packaged in glass containers. These containers are soda lime glass. (Courtesy Glass Packaging Institute)

Flat glass includes window glass or sheet, plate glass, lenses (Fig. 73-4), and mirrors. Specialty glass includes the textiles, insulation, fiber optics, and foam.

The preceding discussion helps to develop our awareness of the many glass products in use today. Another way of identifying these products is by the *type* of glass (Table 73-1). Two types of glass, by composition, are shown in Figs. 73-1 and 73-2.

Construction

Some of the manufactured glass products are used in construction. These can be classified as:

1. Container
2. Flat
3. Specialty

The containers include the glass pipe drain lines for laboratory sinks. Glass pipes are also used in food and chemical systems, and in the automatic milking systems found on dairy farms.

Window or sheet glass, plate glass, and glass blocks are flat products frequently used in construction. Plate glass is used for picture windows and store display windows.

Glass wool insulation and foam glass are both specialty products. Both these products are used to insulate walls, ceilings, heat ducts, and pipes.

Fig. 73-2. Glass measuring cups with metric calibrations. Pyrex ® is a Corning trade name for borosilicate glass. (Courtesy Corning Glass Works)

Fig. 73-3. Christmas ornaments made of glass. The glass spheres are made by the ribbon machine. (Courtesy Corning Glass Works)

Fig. 73-4. Photochromic glass is sensitive to the ultraviolet rays in sunlight. These sun glass lenses automatically adjust their tint to the brightness of the day. They are darkest when the sun is brightest, but they fade to the lighter tints when less protection is needed. Thus, indoors, at dawn or dusk (left) these sun glasses have only a light fashion tint; but outdoors (right) they automatically become darker. The darkening and lightening action never wears out. (Courtesy Corning Glass Works)

Table 73-1. Uses of Glass by Types of Glass, by Composition

Type of Glass	Typical Uses	Type of Glass	Typical Uses
1. Soda lime	Containers, window glass, mirrors, tumblers, pitchers, ashtrays, glass blocks, incandescent bulbs, photo flash bulbs.		high-tension insulators, heat exchangers.
		4. Silica (96%)	Chemical ware, home appliances, germicidal lamps, sun lamps.
2. Lead	Crystal glassware, eye glasses, optical lenses, neon tubes, capacitors, radiation shielding windows.	5. Fused silica	Delay line, aircraft camera high-frequency electrical insulation.
3. Borosilicate	Oven ware, laboratory glassware, industrial glass piping, gauge glasses, incandescent lamp enclosures, pharmaceutical containers,	6. Aluminosilicate	Top-of-stove ware, chemical ware, high-temperature thermometers, combination tubes, water-level gauge glasses.

Product Information

Some useful facts concerning glass products are included for your information. This information is useful to all glass consumers.

Window or sheet glass is cut to standard sizes. It is packed 50 square feet (4.6 square meters) per box.

Plate glass is usually thicker than sheet glass. The usual thickness is ¼ in. (6.4 mm). This is cut to order. It is sold by the square foot.

Some of the glass products are produced with a trade name on them. These are registered by a particular company. Some of them are listed in Table 73-2.

Glass-ceramic products are used in the home, industry, and military. Home uses include dinnerware, cooking/serving dishes, counter savers, and smooth-top electric ranges (Fig. 73-5). Industrial uses include gas heating units. Missile radomes are a military use. A radome is a housing that is transparent to radar waves used to guide missiles.

Table 73-2. Examples of Glass Trade Names

Type of glass	Trade Name	Company
1. Borosilicate	*Fire King*	Anchor Hocking
	Glasbake	Jeannette Glass Co.
	Kimax	Owens-Illinois
	Pyrex	Corning Glass Works
2. Silica (96%)	*Vycor*	Corning Glass Works

Fig. 73-5. Two domestic uses of glass-ceramic are the cooking/serving dish and smoothtop electric range shown above. (Courtesy Corning Glass Works)

Unit 74: Disposition

Glass products can be disposed of either by reclaiming or by discarding. Technically, all glass can be reclaimed.

Reclaiming

The glass products are reclaimed either by reuse or by recycling. Both of these methods are recommended.

Reuse

Returning the money-back deposit on glass bottles is an easy way to reuse glass. Frequently, the empty jars are reused for storage containers in the kitchen, garage, and shop. No-deposit glass bottles and jars are also cut off to make craft products. These include planters, bowls, coasters, and candle holders.

Fig. 74-1. Old bottles may be considered antiques and/or collectibles.

Fig. 74-2. Old jars may be considered antiques and/or collectibles.

When we discontinue the use of glass dinnerware and cookware in our homes, it can be reused at camps or it can be sold. There is a market for this type of ware for replacement pieces.

Old glass bottles, jars, and dishes are in great demand. They are often considered antiques and collectibles (Figs. 74-1 and 74-2).

Recycle

Containers, bottles, and jars are the glass product most frequently recycled. They may be used for cullet that is remelted to make new glass products. The cullet is separated by color when it is used to make new bottles and jars. Color separation is unnecessary when the cullet is used to make glass wool insulation.

Glass containers may be crushed and used to make the following (Fig. 74-3):

1. Glasphalt
2. Building blocks and bricks
3. Glass terrazzo floors
4. Chicken grit
5. Mulch

Actually, any color and type of glass can be used for the above products.

Glasphalt is a road paving material. It is asphalt that is made with crushed glass, instead of crushed stone.

Fig. 74-3. This crushed glass was made from no-deposit containers. It will be mixed with other materials to make glasphalt, blocks, bricks, and/or terrazzo. Crushed glass is also used as chicken grit and mulch. (Courtesy Glass Packaging Institute)

The recycling collection methods include:

1. Collection centers (Fig. 74-4)
2. Bottle drives, like paper drives
3. Removing glass from incinerator ashes
4. Mechanically separation from solid waste

Unfortunately, the recycling collection methods are not available to everyone. Therefore, discarding does occur.

Discarding

Discarding is the easy way to dispose of the glass products. It is the least desirable of the two methods. We should make every effort possible to reclaim, rather than discard, glass. In fact, the potential uses for reclaimed glass far exceed the supply now or in the near future.

If you must throw away any glass products, discard them properly. Improper discarding disfigures our country.

Fig. 74-4. Glass containers at a recycling collection center. They may be recycled as cullet that will be remelted or crushed. (Courtesy Glass Packaging Institute)

SECTION XVIII

Abrasives Industry

Unit 75: Material Science

There are three major product divisions in the abrasives industry. They are *loose*, *coated*, and *bonded*. Some of the abrasives firms produce only one product. Other firms produce two or three of the products.

Definition
An abrasive is a ceramic material. An abrasive can *abrade* another material when the proper motion and pressure are applied. Abrade means to rub off, wear away, or cut.

Properties
The abrasives are chemically inert substances. This means that they do not chemically react with the material being abraded.

There are several physical properties that the abrasives possess. These include:

1. Hardness
2. Tough
3. Fracture or break
4. Friable

Hardness is a measure of the resistance of a material to a mechanical force. Hardness, in terms of Moh's scale of mineral hardness, is resistance to being scratched (Table 75-1).

Tough means that the abrasive is not easily cut. Abrasives should *fracture* or break with sharp edges. *Friable* is the ability to fracture.

Terms
The *grit* indicates the size of the abrasive particle or the mesh number. This is the number of holes per linear inch in a sieve.

The *grain* indicates the type of abrasive material. These are both natural and manufactured materials.

Table 75-1. Moh's Scale* and Hardness of Common Abrasives

No.*	Mineral	Common Abrasives
1.0	Talc	
2.0	Gypsum	
3.0	Calcite	
4.0	Fluorite	
5.0	Apatite	
6.0	Feldspar (orthoclose)	
7.0	Quartz	Flint—7.0
		Garnet—7.0–7.5
8.0	Topaz	
		Emery—8.5–9.0
9.0	Corundum	
		Aluminum oxide—9.4
		Silicon carbide—9.6
10.0	Diamond	

* Moh's Scale of Mineral Hardness—(A scratch test or a measure of the mineral's resistance to a mechanical force.)

Materials

The materials used to make the abrasive products are:

1. Grain
2. Bond
3. Backing
4. Reinforcement

Grain

The abrasive grains are both natural and manufactured. The natural abrasive grains are listed in Table 75-2. Flint, garnet, and emery are the commonly used natural abrasives.

The synthetic abrasive grains are listed in Table 75-3. Silicon carbide and aluminum oxide are the commonly used synthetic abrasives.

Silica sand, coke, sawdust, and salt are the materials used to make silicon carbide. Bauxite, coke, and iron filings are the materials used to make aluminum oxide.

Bond

A bond adheres the abrasive particles together or to a backing material. Bonds are used in the coated abrasive products and in the bonded abrasive products. Glues and synthetic resins are used to bond the abrasive particles to the backing material.

Vitreous and nonvitreous bonds are used in the bonded abrasive products. The vitreous materials form a glassy bond when fired. Flint, feldspars, clay, and frits are used for this purpose. Shellac, sodium silicate, synthetic resins, and rubber compounds are used for nonvitreous bonds.

Backing

The backing provides stability for the coated abrasive products. It is the material that is coated. Only the coated abrasive products require a backing material. The typical backing materials are:

1. Paper
2. Cloth
3. Fiber
4. Combination of paper, cloth, and/or fiber
5. Wood
6. Metal
7. Nylon
8. Plastic

Paper, cloth, fiber, and a combination of these materials are the commonly used backing materials.

Reinforcement

Reinforcement adds strength to the nonvitreous bonded products. It is used only in these products. Nylon, fiberglass, and burlap fabric are typical reinforcement materials.

Table 75-2. Natural Abrasive Grains

Grain	Composition	Mineral
1. Flint	SiO_2	x
2. Garnet	$3FeO \cdot Al_2O_3 \cdot SiO_2$*	x
3. Emery	Al_2O_3	x
4. Corundum	Al_2O_3	x
5. Diamond	C	x
6. Pumice	Lava	
7. Rottenstone	$CaCO_3$	
8. Tripoli	$CaCO_3$	
9. Tin oxide	SnO_2	
10. Cerium oxide	CeO_2	

* Approximate composition

Table 75-3. Synthetic Abrasive Grains

Grains	Composition
1. Silicon carbide	SiC
2. Aluminum oxide	Al_2O_3
3. Crocus	FeO
4. Rouge	FeO
5. Diamond	C

Unit 76: Procurement

Procurement of the natural raw materials for the abrasives industry is diagrammed in Fig. 76-1. These raw materials are the natural abrasives, silica sand, bauxite, and some materials used for the vitreous bond.

Some of the abrasives companies are involved in procurement. These are the firms that procure natural abrasives. A firm may specialize in procuring and beneficiating a single abrasive material, such as garnet.

The majority of the firms that use raw materials purchase them from specializing firms. The manufacturers of silicon carbide and aluminum oxide purchase their silica and bauxite materials from the ceramic raw material industry.

Prospecting, developing, and winning were discussed earlier in Section III—*Ceramic Raw Material Industry*. All three types of winning may be used, due to the variety of raw materials (Figs. 76-2 through 76-4). Superficial winning is used the most. Garnet, silica sand, and bauxite are won by this method. Large quantities of these materials are stockpiled until they are needed for transformation (Fig. 76-5).

Purchasing

Purchasing of their materials is preferred by many of the coated and bonded abrasives manufacturers. The variety of materials used in these products makes purchasing necessary. Often, the materials are purchased in large quantities. These are usually shipped in bulk (Fig. 76-6).

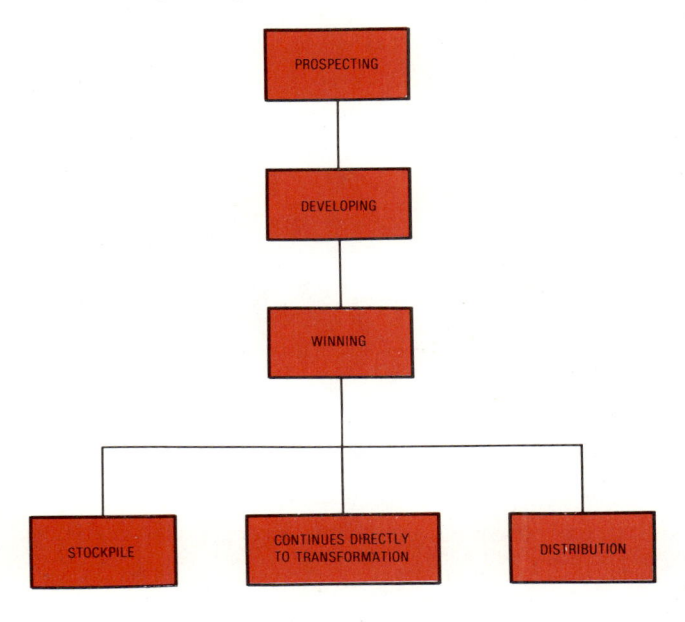

Fig. 76-1. A typical vertical flow sheet for procurement of raw materials used by the abrasives industry.

Fig. 76-2. A closeup of a dipper-stick power shovel bucket. This machine is frequently used for excavating abrasive raw materials such as bauxite.

Fig. 76-3. This low-profile lead-acid battery powered truck was especially designed for subterranean winning. Its height is typical of the machines used for underground mining. (Courtesy Lead Industries Association, Inc.)

Fig. 76-4. A cutterhead suction dredge used for subaqueous winning. The cutterhead is raised out of the water at the left-hand side. This dredge is 12½ ft (3.8 meters) in depth, 46 ft (14 meters) in width, and 208 ft (63.4 meters) in length. (Courtesy Great Lakes Dredge & Dock Co.)

Fig. 76-5. Large quantities of abrasive raw materials are stockpiled until needed for transformation. Stockpiles are usually protected from the weather by a building similar to the one above. (Courtesy American Olean Tile Company)

Fig. 76-6. Large quantities of abrasive raw materials are purchased and shipped in bulk. Hopper cars are frequently used for bulk transport. (Courtesy R. T. Vanderbilt Company, Inc.)

Unit 77: Transformation

Materials are transformed by the abrasives industry. They are transformed into synthetic abrasives, loose abrasives, coated abrasives, and bonded abrasives.

Synthetic Abrasives

Silicon carbide and aluminum oxide are the commonly used synthetic or manufactured abrasives. Silicon carbide was developed in 1891 by Dr. Edward G. Acheson. C. Jacobs developed aluminum oxide in 1899.

Silicon Carbide

Silica sand, coke, sawdust, and salt are the materials used to make silicon carbide. The silica (SiO_2) combines with the coke (C) to form silicon carbide (SiC):

$$SiO_2 + 3C \xrightarrow{\Delta} SiC + 2CO \uparrow$$

The sawdust burns out during heat treatment. This leaves a porous structure that allows the carbon monoxide (CO) gas to escape. The salt (NaCl) helps to remove the iron impurities in the silica as volatile ferric chloride.

A resistance-type electric furnace is used to make the silicon carbide crystals. The furnace has a long low rectangular or trough-like shape (Fig. 77-1). Each end has a large graphite electrode for heating by electrical resistance. Refractory brick are used for the bottom, sides, and ends. The top is open to allow release of the fumes.

The four materials are mixed together. A charge of these materials is placed inside the furnace with coke through the center to connect the electrodes (Fig. 77-1). The silicon carbide crystals form through the center during the one or more days of heating. Then the furnace is cooled and emptied before repeating the cycle. This is a periodic process (Fig. 77-2).

The silicon carbide crystals are black or green. Then, the crystals are transformed into loose abrasives.

Fig. 77-1. Full-section elevation drawing of a typical electric-resistance furnace used to make silicon carbide. Typical dimensions are: height, 8 ft (2.4 meters); width, 8 ft (2.4 meters); and length, 20 ft (6.1 meters).

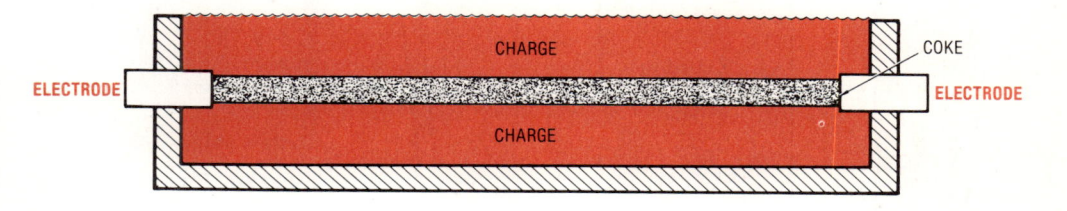

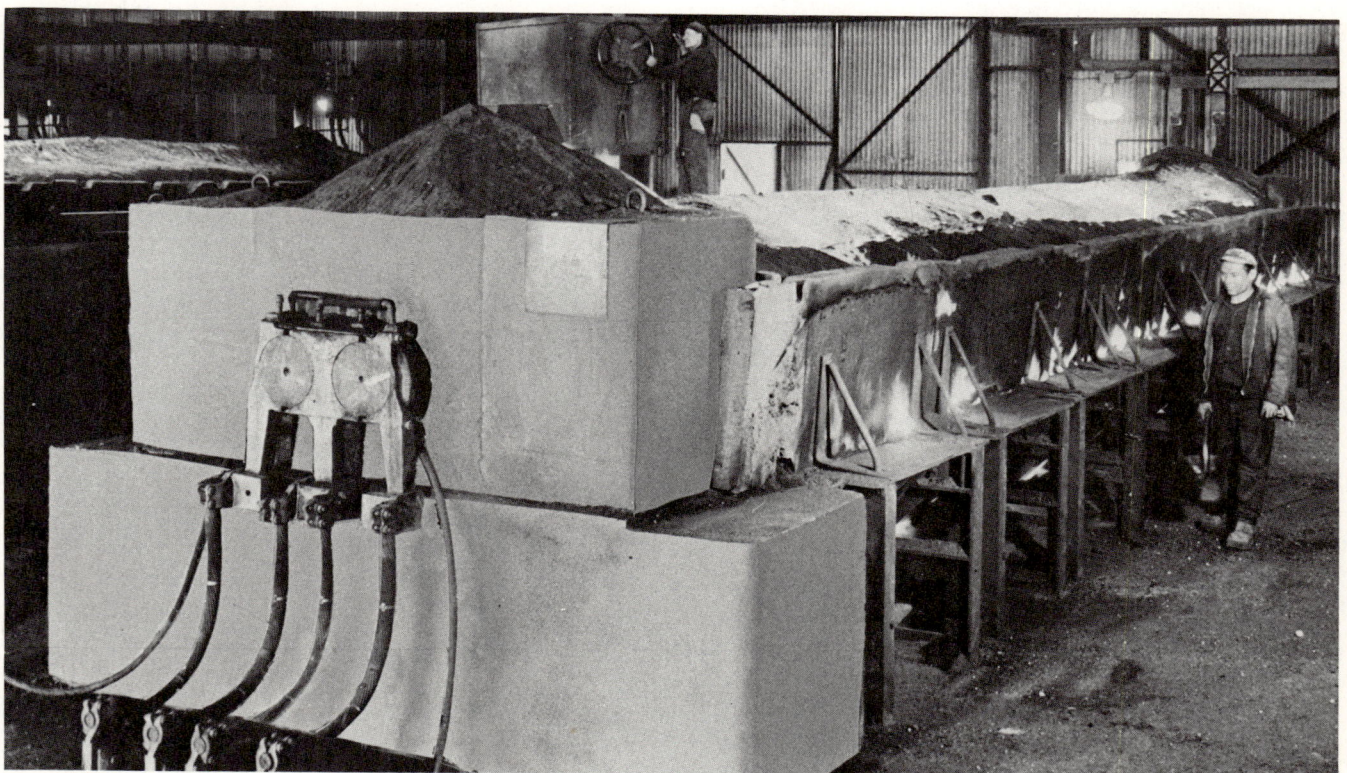

Fig. 77-2. Silicon carbide crystals are formed in the intense heat of this electric-resistance furnace. Note the trough shape, open top, and one of the two electrodes. (Courtesy The Exolon Company)

Aluminum Oxide

Bauxite, coke, and iron filings are the materials used to make aluminum oxide. First the bauxite is calcined to remove the surface water. Then, it is mixed with the coke and iron filings. When heated, the coke reduces the impurities in the bauxite. Then, the impurities combine with the iron and settle to the bottom of the furnace.

An electric-arc furnace is used to make aluminum oxide. It is shaped like a large crucible. Th crucible wall is water-cooled. Graphite electrodes enter the furnace through the open top (Fig. 77-3).

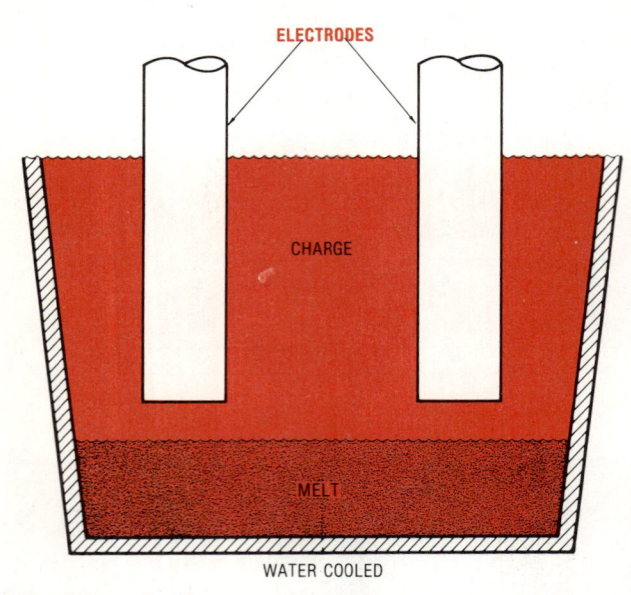

Fig. 77-3. Full-section elevation drawing of a typical electric-arc furnace used to make aluminum oxide. Typical dimensions are: height, 6 ft (1.8 meters); and diameter, 10 ft (3.1 meters).

An electric arc provides heat to melt the charge. The electrodes are slowly raised, until the melt fills the furnace (Fig. 77-4). Then, it is cooled to form crystals, the crystals are dumped, and the cycle is repeated (Fig. 77-5). This is a periodic process.

The aluminum oxide crystals are brown, white, or ruby. Next, the crystals are transformed into loose abrasives.

Loose Abrasives

Transformation of the natural abrasives and the synthetic abrasive crystals into loose abrasives is diagrammed in Fig. 77-6. This transformation involves only the beneficiation processes.

The abrasive material is first crushed to produce sharp irregular particles. Jaw and gyratory crushers are used for primary crushing. Double-roll crushers are used for secondary crushing.

Fig. 77-4. Pouring molten aluminum oxide from an electric-arc furnace. (Courtesy The Exolon Company)

Fig. 77-5. Removing the furnace shell from cooled aluminum oxide. Next, this will be transformed into loose abrasive grain. (Courtesy The Exolon Company)

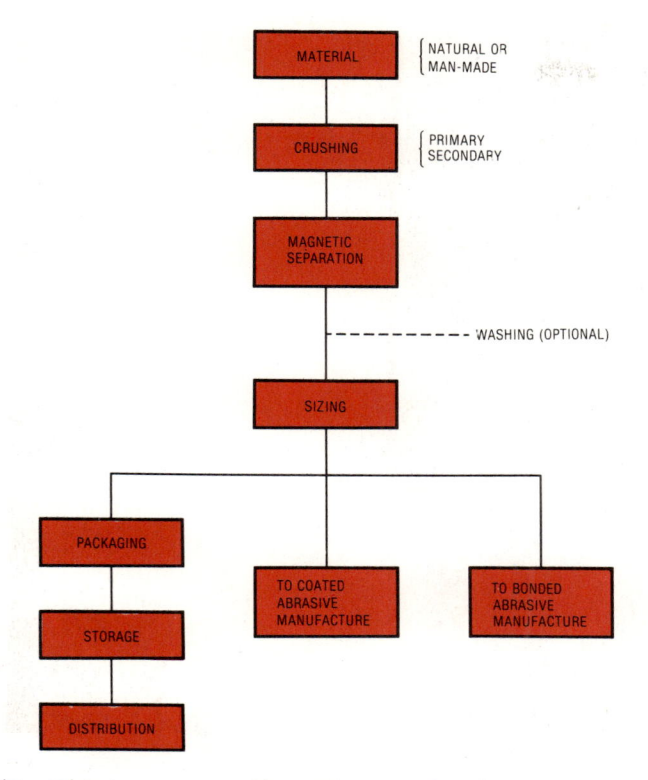

Fig. 77-6. A typical vertical flow sheet for the transformation of natural abrasives and manufactured abrasive crystals into loose abrasives.

Magnetic separation removes the iron impurities. Washing is an optional process that removes any dust. The particles are then separated according to size. This process uses sieves of different mesh sizes.

The abrasive grit is used in packaging or in the manufacture of the coated and bonded products. Storage and distribution complete the manufacture of the loose abrasives.

Coated Abrasives

The manufacture of coated abrasive products is diagrammed in Fig. 77-7. This manufacturing process is illustrated pictorially in Fig. 77-8.

Loose abrasive grain, bond in the form of an adhesive, and large rolls of backing material are used to make these products. The cloth backing requires preparation before coating. Preparation involves stretching, dying, and filling the pores.

In Fig. 77-8, the pictorial diagram shows how the large rolls of backing material are used to maintain a continuous process. Printing the trade name, size of grit, and type of grain on one side of the backing material is done with a fast-drying ink. Then, the first or base coat of adhesive is applied to the side opposite the printing.

Loose abrasive grain is then applied by either gravity or electrostatically. The grain is spread from a hopper by gravity (see Fig. 77-8). This is used for open-coat products. *Open-coat* means that the backing is not completely covered by the grain. Electromagnets pull the grain into the adhesive when electrostatic coating is used. This is used for closed-coat products. *Closed coat* means that the backing is completely covered by grain.

Predrying causes the base coat of adhesive to dry. A heating unit is used. A second or anchor coat of adhesive is applied next. This is also dried by a heating unit. Rerolling completes the continuous process. The large rolls can be stored until needed or taken to cutting.

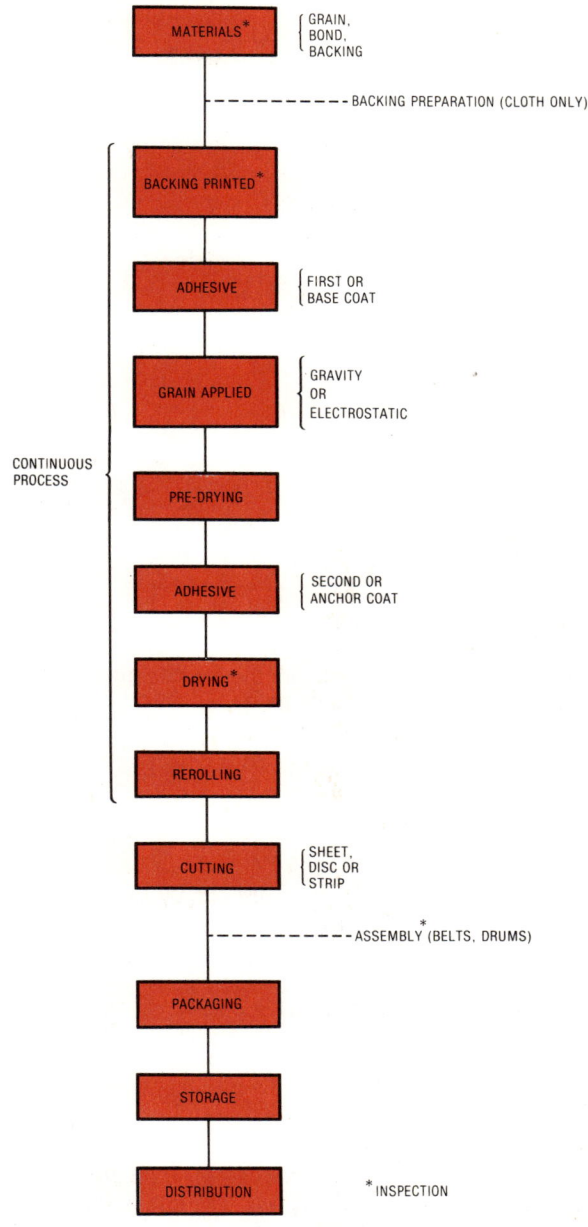

Fig. 77-7. A typical vertical flow sheet for the manufacture of coated abrasive products.

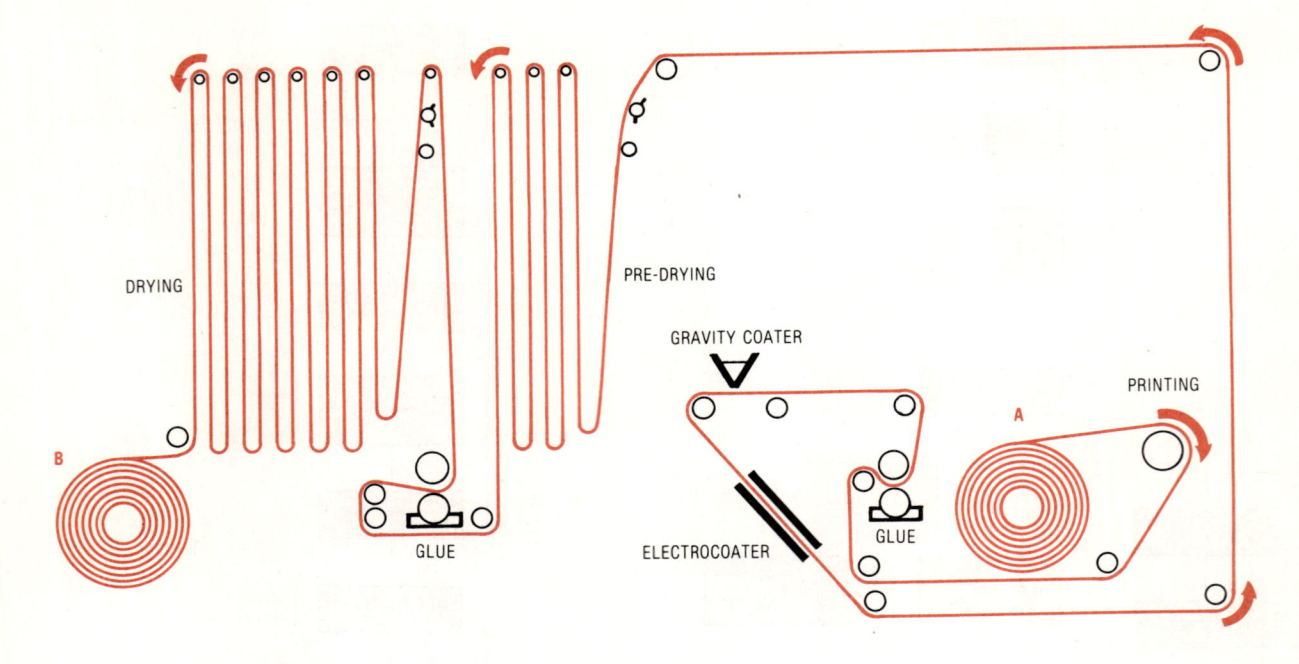

Fig. 77-8. A pictorial flow sheet for the manufacture of large rolls of coated abrasives. The rolls are then cut into sheets, discs, and strips. The process starts with a large roll of backing material (right) and ends at rerolling (left). Gravity coating is used for open-coat products and electrostatic coating for closed-coat products.

The sheets, discs, and strips are cut by machines. The sheets and discs are cut by punch presses, and the strips are cut by knives. Belts and drums are assembled from some of the sheets and strips. The belts are assembled on presses, and the drums are wound on lathes. Packaging, storage, and distribution complete the manufacture of the coated abrasive products.

Bonded Abrasives

The manufacture of bonded abrasive products is diagrammed in Fig. 77-9. Abrasive grain and bond are used to make all the bonded products. These materials are weighed and dry batched for thorough blending. Wet batching is done only if water or some other liquid is used. The amount of water is very small (10% or less by weight). This produces a mixture of small pea-size lumps or pellets.

Vitreous Products

Vitreous bonded products are formed by dry pressing. The same process is used to make refractory shapes and wall tiles. Usually, 10% or less water by weight is mixed with the grain and bond for this forming process. The water serves as a binder when the materials are pressed. Drying removes most of the water. At this point the product is fragile. Firing completely removes the water.

The glassy bond is also formed during firing. This bond makes the product durable.

Both periodic and continuous kilns are used. Electric, updraft, and shuttle periodic kilns are used. The tunnel kiln is the only continuous kiln used.

Nonvitreous Products

Nonvitreous bonded products are also formed by pressing. Reinforcement fabric is used for some of these products. A hydraulic press forms the product. Moderate heat is usually required to cure the bonding material. Curing may be done concurrently with pressing or after pressing.

Both types of bonded products may require assembly of bushings or shafts. Some products, especially the vitreous bonded grinding wheels, require truing. This is a cutting process that makes sure the wheel will be balanced and revolve true.

Inspection and testing insure quality and safe products. For example, the wheel shapes are tested for balance and rpm.

Identification information may be stenciled onto the product or added as a paper blotter, as seen on many grinding wheels. Packaging, storage, and distribution complete the manufacture of the bonded abrasive products.

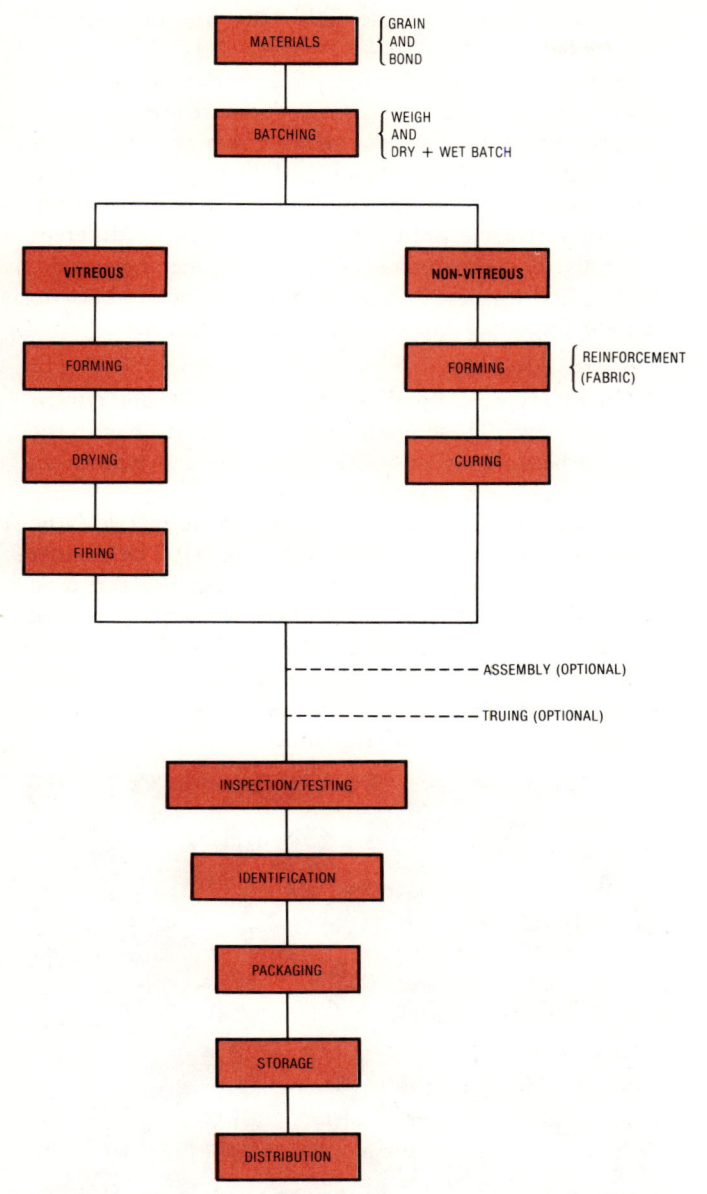

Fig. 77-9. A typical vertical flow sheet for the manufacture of bonded abrasive products.

Unit 78: Utilization

The abrasives industry has three major product divisions. They are loose, coated, and bonded abrasives products. These are manufactured or synthetic products.

Manufacture

Most of the loose abrasives manufactured each year are used to make coated and bonded products (Figs. 78-1 and 78-2).

Loose Abrasives

The loose abrasives are also used for sandblasting, cleaning, grinding/polishing, tumbling, and barrel finishing. Glass products, dimension-stone products (Fig. 78-3), brick or stone buildings, cast iron, and automobile body repair are some of the items that are sandblasted. Cleaning spark plugs is actually another use of sandblasting. Natural gems, dimension stone, glass products, and metals are some of the materials ground and/or polished. Natural gems are tumbled. Metal parts are barrel finished. Most of the tumblers and barrel finishers operate like a ball mill.

Coated Abrasives

Today, it is technically incorrect to call the coated abrasive products "sandpaper." This means that sand-size particles are glued to a paper backing. Flint paper is the only coated product made today that might resemble "sandpaper."

The types of coated abrasive products are sheets, strips, discs, belts, and drums. These products are used by hand, portable machines, and stationary machines. The sheet and strips are used by hand. All five types of coated abrasives are used on portable and stationary machines.

Coated abrasive products are used on many different materials. These materials include wood, metal, ceramic, plaster, and leather. Special uses of the coated abrasives include fingernail boards and pencil pointers.

Bonded Abrasives

The two types of bonded abrasive products are: *vitreous* and *nonvitreous*. These products are usually made with the two synthetic abrasives—silicon carbide and aluminum oxide.

The vitreous-bonded products are made in two basic shapes—wheels and slipstones (Fig. 78-4). The thinner wheels are used for cutting and the thicker wheels are used for grinding. These wheels are made in a large variety of shapes, sizes, and hardness (Fig. 78-4). The slipstones are used for hand grinding and shaping. They are made in several shapes. Among these shapes are cylindrical, disc, rectangular, and tapered slipstones.

Fig. 78-1. Enlarged photograph of loose silicon carbide grain. Note its irregular shape and sharp edges. Silicon carbide is used loose and to make coated and bonded products. (Courtesy The Exolon Company)

Fig. 78-2. Enlarged photograph of loose aluminum oxide grain. This is used loose and to make coated and bonded products. (Courtesy The Exolon Company)

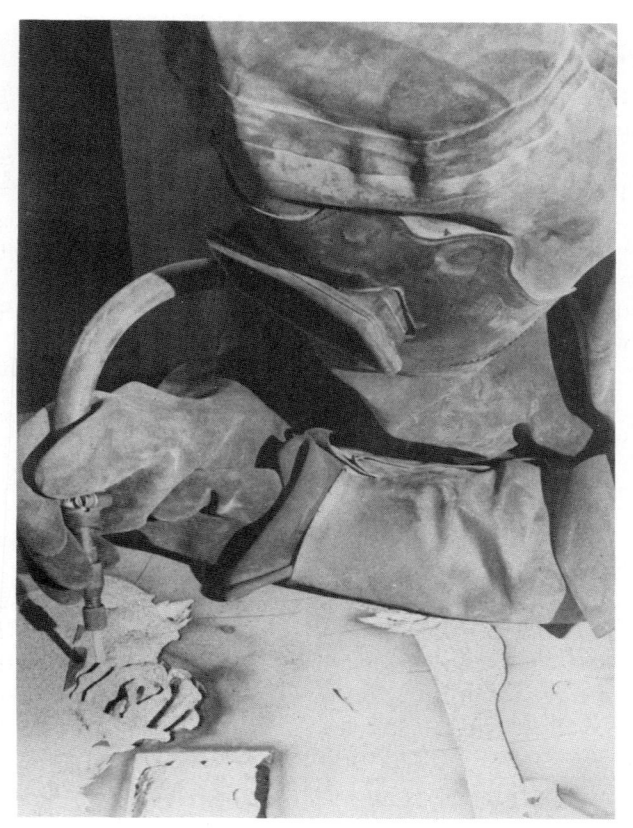

Fig. 78-3. Sandblasting a dimension-stone monument. Note the protective clothing. (Courtesy The Barre Granite Association)

Fig. 78-4. Vitreous-bonded abrasive products are made in the two basic shapes shown above—wheels and slip stones. The two rectangular products are slip stones. These products are usually made with silicon carbide (SiC) and aluminum oxide (Al_2O_3). Three grits for each shape and their manufactured crystals are shown in the second and third rows above. Silicon carbide is at the left, and aluminum oxide is at the right. (Courtesy The Exolon Company)

The nonvitreous-bonded products are made in two basic shapes—wheels and discs. The edge of the thin wheels is used for cutting. The discs are usually thicker than the wheels. The flat or cupped side of the discs is used for grinding. They are used on portable machines for grinding off the concrete-form marks or excess metal.

The vitreous-bonded products are used for cutting, grinding, and sharpening. The major uses are for grinding and sharpening. The nonvitreous-bonded products are used for cutting and grinding.

The hard materials, such as ceramics and metals, are frequently cut by bonded products. Stone, concrete, brick, tile, refractories, and glass are the typical ceramic materials cut by these products. Grinding is also done on these hard materials. Typical examples are dimension stone, natural gems, terrazzo, glaze stilt marks, glass, and metals. Tools with metal cutting edges require sharpening.

Construction

Abrasives are not considered to be construction materials. The three types of abrasive products are the tools used to do construction work. They are used as cutting tools to remove part of the original material. A chip is produced by the cutting action.

Unit 79: Disposition

Abrasive products can be disposed of by reclaiming and discarding. Ideally, all these products would be reclaimed rather than discarded. However, the product ultimately wears out to the point that hardly any material remains to be reclaimed.

Reclaiming

Abrasive products are reclaimed either by reuse or by recycling. Both methods are healthy for our economy and environment.

Reuse

The loose abrasives recovered from sandblasting, tumbling, and similar processes are usually worn down into very fine particles. These particles can be reused as very fine abrasives for various types of hand grinding work.

The worn coated products can be reused as a finer abrasive. This is normal practice when used on machines. When the coated product becomes too worn or rips, it should be removed from the machine. Then, it can be cut into smaller pieces and used for hand sanding.

The worn bonded products, such as grinding wheels, can be reused for hand grinding. The flat side can be used as a slipstone.

Recycle

The worn-out and/or contaminated loose abrasives can be recycled as soil conditioners, to sand icy walks and pavements, and as a masonry sand for mortar.

Some of the worn-out coated products, such as belts and discs that have good backing material, are recycled. These products are recoated with adhesive and grain.

The vitreous-bonded products can be recycled as riprap or hard fill. The quantity available for these purposes is usually insufficient, unless it can be combined with other worn-out materials.

The nonvitreous-bonded rejects are recycled at the manufacturing plant. The bond and reinforcement, if used, are burned out in a kiln at high temperatures. The metal bushings and the grain are recovered when cool. This method of recovery could be used with all the worn-out nonvitreous-bonded products.

Discarding

Discarding is the traditional method used to dispose of the abrasive products. This tradition is a difficult habit to break. However, we must make every effort to reclaim our used, old, and/or worn-out products.

If you must throw away any product, discard it properly. Improper discarding is a blight on our country.

Index

A

Abrade, 267
Abrasives
  bonded, 3, 267, 274, 276
    nonvitreous, 274, 276
    vitreous, 274, 276
  coated, 3, 267, 273, 276
  definition, 267
  industry, 3, 267
    disposition, 278
    material science, 267
    procurement, 269
    transformation, 271
    utilization, 276
  loose, 3, 267, 273, 276
  properties, 267
  synthetic, 271
  terms, 267
Admixtures, 154
Agglomeration, 37
Aggregate
  coarse, 3, 154
  fine, 3, 154
  tests
    colorimetric, 162
    sieve analysis, 162
    silt, 162
Alabaster, 93
Aluminum oxide, 272
Aragonite, 122
Art pottery, 1
Artist, ceramic, 3
American Society for Testing Materials (ASTM), 66, 146, 191
Anhydrite, 93
Animal power
  removal of overburden, 31
Annealed, 83
Assembled precast concrete products, 162
Assembling, 37

B

Autoclave, 96
Awareness of products, 42

Background, 137
Backing, 268
Balance, ecological, 28
Bank run, 162
Baroques, 61
Basalt, 16, 48
Basaltware, 214
Base material, 236
Batching, 69, 175
  amounts of materials, 106
  clay, 196
  completion, 107
  concrete, 155
  dry, 225
    press, 176
  glaze, 225
  manual, 196
  mechanical, 197
  plaster, 106, 114, 118
  preparation, 106
  soft mud, 175
  stiff mud, 175
  wet, 225
Batch tests
  air content, 163
  compression, 163
  slump, 162
Batteries, 33
Beneficiation, 38, 69, 96, 125, 141, 175, 196
Binder, 220
Biosphere, 27
Bisque
  firing, 86, 170
  stacking for firing, 86
Block and case mold, 116
Blocks
  cinder, 158
  concrete, 158
  decorative, 158
  precast, 158

  regular, 158
Blow-and-blow machine, 254
Body, clay
  china, 192
  earthenware, 193
  porcelain, 193
  semivitreous ware, 193
  stoneware, 193
  whiteware, 193
Bond, 184, 268
Breadth, 3
Bricks, 165, 185
  terms, 184
  veneer, 185
Brighten, 248
Broken stone, 45, 47
Brushing glaze, 227
Bulldozer, 32
Burning, 142
  process, 143

C

Cabochons, 61
  cuts, 63
  typical procedure, 61
Calcareous marl, 122
Calcine, 96, 125
  purpose, 97
Calcite, 122
Calcium
  carbonate, 129
  hydroxide, 129
  oxide, 129
Careers
  ceramics, 11
  winning or mining, 30
Cast iron, 241, 244
Castable refractories, 71
  premixed, 71
Casting box, 114
Cement
  Keene, 97, 100
  Portland, 3, 4, 153
Ceramic
  artist, 13

280 / Index

basic terms, 1
chemistry, 21
designer, 13
engineer, 13
major industries, 3
major products, 4
materials, 15
product applications, 4
products, 1
professional engineer, 14
raw material industry, 3, 24,
    27, 38, 42, 46, 194, 196,
    222, 237, 250, 269
scientist, 14
Ceramics
  careers, 11
  definition, 1
  hobby, 1
  industrial, 1
  studying, 4
Ceramists, 11
Chalk, 122
Changes
  chemical, 144, 167
  physical, 143, 167
Chart
  circular, 87
  strip, 87
Chemical, 19
Chemistry, ceramic, 21
China
  bone, 214
  everyday, 214
  fine, 214
  hotel, 215
  institutional, 215
  sanitary, 215
Chum, 110
Classifying, 37, 39, 175, 196
Clastic, 19
Clay
  body, 192
  colluvial, 167
  formation, 167
  formulas, 169
  industries, 3, 165, 182, 217
    differences, 165
    products, 165
    similarities, 165
    type of products, 165
  material science, 167
  minerals, 167
  primary, 167
  secondary, 168

singular purchased, 192
  states, 170
Clinker, 146
Closed coat, 273
Coarse aggregates, 3
Coat
  closed, 273
  open, 273
Colluvial clay, 167
Colorant
  cement, 154
  clay body, 192
  glaze, 220
  porcelain enamel, 235
  glass, 248
Colorimetric test, 162
Comminution, 37, 175, 196
Completion, 208
Compound, 21
Concrete, 3
  batching, 155
  blocks, 158
  combination, 162
  definition, 133, 153
  flow sheet, 158
  industry, 133, 147, 153
    material science, 153
  job site tests, 162
  manufacturing, 158
  monostructures, 161
  precast, 158
  prestressed
    post tensioning, 160
    pretensioning, 160
  regular, 158
  reinforced, 159
  special, 161
  terms, 157
Cone
  pyrometric, 88
  number, 88
Conditioning, 37
Construction, 38, 73, 99
    130, 149, 184, 232, 245,
    262, 277
  concrete, 161
Consumer information, Portland
    cement, 147
Continuous kilns, 71, 77
  definition, 82
  glass tank, 84, 253
  lehr, 84
  major types, 82
  rotary, 82

  tank, 80, 253
  tunnel, 82
  vertical, 83
Coral, 122
Core, 15, 20
  drills, 29
Cottle, 115
Course, 184
Creative glass blower, 13
Crushed, 38
  and broken stone, 45
    disposition, 52
    industry, 47, 59
    material science, 47
    procurement, 49
    raw materials, 48
    transformation, 50
    utilization, 51
  stone, 45, 47
Crushers
  continuous pressure, 38
  gyratory, 38
  hammermill, 38
  impact, 38
  jaw, 38
  reciprocating pressure, 38
  roll, 38
  single-roll, 38
Crushing, 38
  primary, 39
  secondary, 39
Crust, 15, 20
Cullet, 252
Curing
  air, 157
  water, 157
Cutter, 178
Cutting, 37, 201
Cycle, rock, 20

D

Day tank, 80, 253
De-airing, 175
Debris, unconsolidated, 27
Decolorizer, 248
Decoration, 231
Defects
  scumming, 183
  structural clay products, 183
Delta, 23
Depth, 30
Designer, ceramic, 13
Development, 123

definition, 29
people, 30
work, 30
Dimension stone, 45
disposition, 59
industry, 53
material science, 53
procurement, 25, 53
raw materials, 53
transformation, 53
utilization, 56
Dipping glaze, 226
Direct carving, 109
Discarding, 43, 75, 101, 150, 190, 216, 234, 246, 266, 278
Dispose of materials, 24
Disposition, 24, 43, 52, 61, 75, 101, 131, 150, 190, 216, 234, 246, 265, 278
problems, 43
Distribution, 183
Dolomite, 121
Downdraft kiln, 79
Drain casting, 204
Drape mold, 108, 116
Drawing glass, 256
Dredging, 35
Drilling, 29
Drills
core, 29
rotary auger, 29
well, 29
Dry press
batching, 175
forming, 178, 201
wall, 99
Drying, 179, 205
air, 69
forced, 69

E

Earth, 15
Earthy, 1, 15
Ecological balance, 28
Ecology, helping, 139
Electric kiln, 78
Element, 21
Engineer
ceramic, 13
professional, 14
Engobe, 231

Equations, 23
Exploratory
prospecting, 28
program, 28
Extracting, 27
Extraction
definition, 27
methods, 28
Extruding, 203
Extrusion, 69

F

Faceted, 61
Faceting, 61
Facetor, 61
Fauna, 28
Feeder, gob, 256
Feldspar, 16, 167
Fettle line, 115
Fine aggregates, 3
Fines, 51
Fining agent, 248
Finishing glass
chemical, 257
mechanical, 257
thermal, 257
Firebrick, 66
Fireclay brick, 66
Fired color, 170
Firing, 4, 70, 181, 206
bisque, 86, 170
glaze, 86, 170, 228
oxidizing, 183
procedure, 86
reduction, 183
special, 208
temperatures, 181, 207
types, 182
Flashing, 183
Float (glass), 257
Floating agent, 236
Flora, 27
Flow sheet
glass, 253
glaze, 224
manufacturing concrete products, 158
Flux, 218, 235, 248
Foliate, 19
Foliation, 19
Forming, 37, 69, 176, 197, 253
blowing, 255
coil, 197

dry press, 178
extrusion, 69
hand, 197, 253
machine, 201, 255
other methods, 69
pinch, 197
pressing, 69, 255
slab, 198
soft mud, 178
stiff mud, 176
Forms, concrete, 155
Formula, 23
Fracture, 267
Friable, 267
Frit, 220
definition, 239
manufacture, 239
Furnace
electric
arc, 80
resistance, 80

G

Geological studies, 28
Glasphalt, 265
Glass
aluminosilicate, 4
blower, 13
borosilicate, 4
ceramic, 258
completion, 258
composition, 4
definition, 218, 247
fibers
continuous, 257
discontinuous, 257
finishing, 257
forming, 253
hand, 253
machine, 255
former, 218, 235, 248
fused silica, 4
industry, 3, 131, 242, 247
disposition, 265
material science, 247
procurement, 250
transformation, 252
utilization, 262
lead, 4
materials, 248
pot, 80, 253
properties, 248
scientist, 14

soda lime, 4
tank, 84, 253
  doghouse, 84
types of, 3
  by composition, 248
  silica, 4
Glass-ceramic, 258
Glassy
  bond, 4, 217
  coatings, 4
  glass, 3
  glaze, 3
  material industries, 3, 217
  porcelain enamel, 3
Glaze, 180
  application, 226
  calculation, 219
  defects, 229
  definition, 217
  finishing, 229
  firing, 86, 170, 228
  industry, 3, 206
    disposition, 234
    material science, 217
    procurement, 222
    transformation, 224
    utilization, 231
  leadless, 219
  materials, 219
    soluble, 221
    toxic, 221
  matt, 220
  over, 231
  properties, 219
  salt, 183
  stacking for firing, 86
  under, 231
Gob feeder, 256
Grain
  abrasive, 267
  natural, 268
  synthetic, 268
Granite, 16, 48
Gravel, 47
Greenware, 86
Grinding
  continuous pressure, 39
  definition, 39
  impact, 39
  reciprocating pressure, 39
Grit, 155
  abrasive, 267
Gypsum
  industry, 3, 93, 130, 133
  rock, 93, 99

H

Halite, 15
Hammermill, 38
Hand forming, 197, 253
Handmade ware, 165
Hardness, 38, 170, 267
Harvesting
  definition, 27
  processes, 27
Header, 184
Heat, 3
  treatment, 1, 133
Heavy refractories, 71
High
  periods, 23
  temperature, 1
Hobby ceramics, 1
Hydration, 157
Hydrosets, 3
Hydrosetting, 3, 93, 97, 165
Hydrosphere, 27

I

Igneous rocks, 16, 18
Incandescence, 1
Industrial
  ceramics, 1
  positions, 11, 14
    management, 11, 13
    production, 11, 14
    quality control, 11, 13
    research and development, 11, 14
    sales, 11, 13
Industries
  clay, 3, 165
  glassy material, 3, 217, 247
  hydrosetting material, 3
  portland cement and concrete, 3
  refractory and kiln, 3
  stone, 3
Industry
  abrasives, 267
  ceramic raw material, 3
  concrete, 149, 153
  crushed and broken stone, 47
  dimension stone, 53
  kiln, 77, 125
  glass, 3, 131, 247
  glaze, 3, 217
  gypsum, 3, 93, 130
  lime, 3, 121, 184
natural gem, 60
plaster mold, 103, 198
porcelain, enamel, 3, 235
portland cement, 133
refractory, 65
structural clay products, 166, 171
Inorganic, 1
Insulating refractory materials, 71
Introduction
  clay, 165
  kilns, 77
  plaster molds, 103
Iron, cast, 241, 244
Isostatic pressing, 201, 206

J

Jasperware, 214
Jiggering, 203
  mold, 108, 117
Job titles, 30
Joggles, 115

K

Kaolin, 167
Kaolinite, 167
Keene cement, 97, 100
Kettle, 80, 96
Kiln, 181, 206
  definition, 77
  downdraft, 79
  electric, 78
  industry, 14, 77, 125, 208, 228
  muffle, 79
  operation, 86
  rotary, 82, 96, 125
  shuttle, 79
  tunnel, 82
  types, 75
  updraft, 79
  use, 75
  vertical, 83, 125
  wash, 86
Kirkendale, George A., 30

L

Land reclamation, 43
Lathe, 110
Lehr, 84
Lime

agricultural, 3
industry, 3, 121, 133, 184
    disposition, 131
    material science, 121
    procurement, 123
    transformation, 125
    utilization, 129
quick, 3
slaked, 3
Limestone, 48, 121
  family, 48
Limitation, plaster molds, 103
Lithified, 19
Lithosphere, 27

M

Magma, 16
Major ceramic industries, 3
  products and applications, 4
Management, 11, 13
Mantle, 20
Manual removal of overburden, 31
Manufactured ware, 165
Manufacturing, 38, 51, 56, 73, 99, 129, 148, 184, 244, 262, 276
  concrete products, 158
  glazes, 224, 231
  materials, 192, 220
Marble, 48, 122
Material, 268
  preparation, 175
  science, 53, 60, 65, 93, 121, 133, 153, 171, 191, 217, 235, 247, 267
    clay, 167
Materials, 153, 171, 192, 235
  basic, 236
  dispose, 24
  hydrosetting, 93
  manufactured, 172, 192, 220
  natural raw, 24
  nonplastic, 192
  procure, 24
  raw, 47, 53, 93, 121, 133, 171, 218, 248
  synthetic, 248
  transform, 24
  use, 24
Matting agent, 219
Mechanical power
  removal of overburden, 31
Melting glass, 252

Metamorphic rocks, 19
Methods
  extraction, 28
  prospecting, 28
Minerals
  clay, 167
  definition, 16
Mining, 4, 30
Miscellaneous rock, 51
Mixing, 37
Mixtures, 21
Model
  definition, 108
  designing, 108
  making, 109
  materials, 108
  mount, 108
Moh's scale of mineral hardness, 93
Mold
  block and case, 116
  designing, 114
  drape, 108, 118
  industry, 104
  jiggering, 108, 117
  making, 109, 115
  plaster, 103, 112
  press, 115, 118
    one-piece, 118
    slip-casting, 118
    ram-press, 108, 118
  simple, 112
  spare, 109
  terms, 114
  trimshelf, 109
  types, 112
Monostructures
  concrete, 161
  prestressed, 161
  regular, 161
  reinforced, 161
Mortar, 130, 184
Muffle kiln, 79

N

Natural
  ceramic raw materials, 24
  gems, 45, 60
    disposition, 61
    material science, 60
    procurement, 60
    transformation, 61
    utilization, 61
  raw materials, 24

Nonfoliate, 19
Nonmetallic, 1
Nonplastic materials
  colorants, 192
  fluxes, 192
  glass former, 192
Nonvitreous products, 285

O

Obsidian, 16
Opacifier, 219, 235, 248
Open-coat, 272
Organic, 19
Other forming methods, 69
Overburden
  definition, 28
  methods of removing, 31
  removal, 31
Overglaze, 231
Oxides, 21
Oxidizing, 78, 183

P

Packaging, 225
Parison, 256
Periodic kilns, 71, 77
  day tank, 80
  definition, 78
  downdraft, 79
  electric, 78
    arc furnace, 80
    resistance furnace, 80
  glass pot, 80
  kettle, 80
  major types, 78
  muffle, 79
  shuttle, 79
  updraft, 79
Pin template, 110
Plant location, 139
Plaque, 88
Plaster, 99
  batching, 106
  model making, 109
  mold industry, 103, 198, 200
  of Paris, 97
  types, 104
  water-plaster ratios, 107
  wheel, 110
    chum, 110
Plasterboard, 97, 99

Plasticity, 169
Porcelain enamel
　definition, 218, 235
　industry, 3, 235, 239
　　disposition, 246
　　material science, 235
　　procurement, 237
　　transformation, 239
　　utilization, 244
Porosity, 104, 169
Portland cement, 3, 153
　definition, 133
　gray, 3
　hydration, 147
　industry, 133
　　disposition, 150
　　material science, 133
　　procurement, 138
　　transformation, 141
　　utilization, 147
　particle size, 147
　types, 147
　weight, 147
　white, 3
Portland cement and concrete
　industry, 3
Portland Cement Association, 147
Pot, glass, 80, 253
Potter, 13
Potter's wheel, 199, 202
Pottery, art, 1
Pouring glaze, 227
Powder, 3, 25
Precast concrete products, 158
Preparation, material, 175
Press
　and-blow machine, 256
　molds, 115
Pressing, 69, 200
Primary
　clay, 167
　product, 3, 42, 133
Problems, disposition, 43
Process
　burning, 143
　portland cement
　　dry, 141
　　wet, 141
Processes, optional, 206, 209
Procedures
　batching concrete, 155
　drape molds, 118
　glaze, 225
　moldmaking, 109

one-piece press mold, 118
　slip-casting mold, 118
　simple molds, 112
　using plaster molds, 119
Procure materials, 24
Procurement, 4, 24, 27
　　53, 60, 68, 94, 123, 138,
　　173, 194, 222, 237, 250,
　　269
Product
　information, 213
　primary, 3, 133
　secondary, 4
Production, 11, 14, 38
　continuum, 26
Products
　awareness, 42
　ceramic, 1
　manufactured, 42
　nonvitreous, 274, 276
　primary, 42
　secondary, 42
　structural clay, 3
　vitreous, 274, 276
Professional ceramic engineer, 14
Properties, 121, 247
　after firing, 170
　chemical, 134
　clay particle, 169
　drying, 170
　firing, 170
　glazes, 218
　physical, 104, 134, 168
　plastic, 169
Prospecting, 28, 123, 173
　definition, 28
　exploratory, 28
　　programs, 28
　external signs, 29
　people, 28
　purpose, 28
　what to look for, 28
　where to look, 28
Pug mill, 175
Purchasing, 194, 225, 267
Pyrometer
　indicating, 87
　optical, 87
　recording, 87
　thermocouple, 87
Pyrometric cones, 88
　how plaques are assembled, 88
　how they work, 88
　number, 88

Orton, 91
　plaque, 88
　temperature equivalents, 91

Q

Quality control, 11, 14
Quarry, 53
Quarrying, 53
Quartz, 16
Quartzite, 49

R

Ram-press, 203
　molds, 108, 118
Ratios
　concrete, 155
　volume, 155
　weight, 155
Raw materials, 47, 53, 93, 121,
　　133, 171, 192, 218, 247
Ready mixed concrete, 156
Reclaiming, 43, 75, 101, 150,
　　190, 216, 234, 246, 265,
　　278
Recording pyrometer
　circular chart, 87
　strip chart, 87
Recycle, 75, 101, 190, 216, 246,
　　265, 278
Reduction, 183
Refractories
　castable, 71
　definition, 65
　heavy, 71
　insulating, 71
　materials, 65
　products, 66, 71
　types, 65
Refractory industry
　and kiln industries, 3
　disposition, 73
　material science, 65
　procurement, 68
　transformation, 69
　utilization, 73
Reinforcement, 268
Research and development (R &
　　D), 11, 14
Reuse, 75, 101, 190, 216, 246,
　　265, 278
Ribbon machine, 256

Riprap, 51, 150
Rock
 cycle, 19
 gypsum, 93, 99
 miscellaneous, 48
Rocks
 definition, 16, 47
 igneous, 16, 18
 metamorphic, 18
 sedimentary, 17
 types of, 16
Rolling glass, 256
Rotary
 auger drills, 29
 kiln, 83, 96, 125, 142

S

Saggars, 75, 79
Sales, 11, 13
Sandstone, 15, 49
Satin spar, 93
Scale
 Moh's, 93
 Wentworth, 17, 47
Science, material, 53, 60, 65, 93, 121, 133, 153, 167, 171, 191, 217, 235, 248
Scientist
 ceramic, 14
 glass, 14
Scofield, Charles B., 37
Scumming, 183
Secondary
 clay, 169
 product, 4, 42
Sedimentary rocks, 17-19
Sediments, 19
Selenite, 93
Separation, 37, 41, 196
Set, 157
Shale planer, 173
Sheet steel, 240, 244
Shuttle kiln, 71
Sieve analysis, 162
Silicon carbide, 271
Silt test, 162
Singular-purchased clay, 192
Size, 31
Slake, 125, 197
Slate, 48
Slip, 69, 109
 casting, 200
 casting molds, 115
  one-piece, 118

Slump test, 162
Soft mud
 batching, 175
 forming, 178
Spare, 109, 114
Special glass forming, 257
Specialized equipment, 33
Specialty porcelain products, 241, 245
Spraying glaze, 227
Sprigging, 214
Springs, 28
Stacking
 bisque firing, 86
 glaze firing, 86
 preparing, 86
Stalactites, 122
Stalagmites, 122
States of
 clay, 170
 matter, 15
Steam, 15
Steel, sheet, 240, 244
Stiffener, 219
Stiff mud
 batching, 175
 forming, 176
Stockpile, 51, 139, 173, 195, 222, 237, 250
Stone, 47
 crushed and broken, 45
 dimension, 45, 53
 divisions, 45
 industries, 3, 45
 natural gems, 45, 60
Storage, 183, 225
Strengthened, 248
Stretcher, 184
Structural clay products, 3, 165, 203
 bricks, 165
 definition, 171
 industry, 166, 171, 208, 217
  disposition, 190
  material science, 171
  procurement, 173
  transformation, 175
  utilization, 184
Subaqueous winning, 35, 123, 139
 advantages, 35
 common name, 35
 definition, 35
 disadvantages, 35
Subterranean winning, 33, 123, 139

 advantages, 33
 common names, 33
 definition, 33
 disadvantages, 33
 equipment
  specialized, 33
Superficial winning, 123, 139
 advantages, 30
 definition, 30
 disadvantages, 30
 names, 30
Symbols, 23

T

Tank
 continuous, 83, 253
 day, 80, 252
Taped, 99
Technology, 24
Temperature measurement, 87
Terms, glaze, 231
Test pits, 29
Tests
 aggregate, 162
 air content, 163
 batch, 162
 colorimetric, 162
 compression, 163
 silt, 162
 slump, 162
Thermocouple
 alumel and chromel, 87
 platinum and rhodium platinum, 87
Throwing, 198
Thrown, 1
Tiles, 165, 186
Time, 167
Tough, 277
Transform materials, 24
Transformation, 24, 37, 50, 53, 61, 69, 96, 125, 141, 175, 196, 213, 224, 239, 252, 271
 definition, 24
 systems, 37
Trimshelf, 109, 115
Tunnel kiln
 car, 83
 pusher slab, 83
 roller hearth, 83
Turning, 205
 box, 111
Types of glass, 3

U

Unconsolidated debris, 27
Underglaze, 231
Updraft kiln, 79
Use of
   bricks, 184
   materials, 24
   other, 52, 130
   safety equipment, 30
   structural clay products, 184-189
Using
   ceramic raw materials, 24
   plaster molds, 119
Utilization, 24, 42, 51, 56, 61, 73, 99, 129, 147, 184, 213, 231, 244, 262, 276
   physiological, 42
   psychological, 42
   sociological, 42

V

Vacuum-and-blow machine, 256
Vegetation, 28

Veneer
   brick, 185
   stone, 45
Vermiculite, 167
Vertical kiln, 83, 125
Vitreous, 181
   products, 274
Vitrify, 170

W

Walking, 29
Wall
   dry, 99
   wet, 99
Washing, 50
Water, 3, 30, 154
   absorption, 170
   of crystallization, 97
Weathering, 15, 167
Wedging, 196
Wedgewood
   Basaltware, 214
   Jasperware, 214

Well drills, 29
Wentworth scale, 17, 47
Wet wall, 99
Whitewares, 3, 165
   definition, 166, 191
   differences, 191
   handmade ware, 165
   industry, 119, 166, 191, 217
      disposition, 216
      material science, 191
      procurement, 194
      transformation, 196
      utilization, 213
   manufactured ware, 165
   similarities, 191
Whiting, 129
Winning, 123, 139, 173
   definition, 30
   people, 30
   processes, 30
   subaqueous, 34
   subterranean, 33
   superficial, 30
   type of operation, 30
Won, 30